DISPOSAL OF SEWAGE

AND OTHER WATER-BORNE WASTES

DISPOSAL OF SEWAGE

AND OTHER WATER-BORNE WASTES

KARL IMHOFF, W. J. MÜLLER
and D. K. B. THISTLETHWAYTE

Revised by
Professor Emeritus Dr.-Ing. W. J. Müller M.I.E. Aust., *Lehrstuhl und Institut für Wasserversorgung, Abwasserbeseitigung und Stadtbauwesen, Technische Hochschule, Darmstadt, Germany*

and

D. K. B. Thistlethwayte F.R.A.C.I., M.I.E. Aust., M.I.W.P.C. *Chief Chemist (Research), Metropolitan Water Sewerage and Drainage Board, Sydney, Research Consultant in Public Health Engineering, University of New South Wales*

ANN ARBOR SCIENCE PUBLISHERS INC.
ANN ARBOR · 1971

First published in England by
THE BUTTERWORTH GROUP

First published in the USA by
ANN ARBOR SCIENCE PUBLISHERS, INC.
P.O. BOX NO. 1425,
ANN ARBOR, MICHIGAN 48106. USA.

Library of Congress Catalog Card No. 74-136114

SBN 250 97516 5

Printed in England by Butler & Tanner Ltd
Frome and London

CONTENTS

PART D
DISPOSAL AND TREATMENT OF WASTES FROM RESIDENCES, INSTITUTIONS AND SMALL COMMUNITIES

FOREWORD

THE book *Taschenbuch der Stadtentwaesserung* was first prepared in 1906, now half a century ago. At that time I had undertaken the task of preparing plans for sewerage and sewage disposal of the River Emscher district of which Essen was the centre, including nearly 2 million people and large industrial establishments. I was aided by a small staff of four or five engineers and a chemist.

Taschenbuch der Stadtentwaesserung was written for myself and my fellow workers so that we might have by us, in suitable abbreviated form, all the technology which we needed for our work. In the following year (1907) the first edition was published and its use spread throughout Germany. The Emscher District Board by this time had constructed many kilometres of sewers and some 30 purification plants. The sewers were mostly open ditches sealed with concrete slabs, and the plants (for unstable organic wastes) were two-storey settling and digestion tanks. This type of tank later became popular and such tanks were built all over the world under the names of 'Emscher-Brunnen' or 'Imhoff-Tanks'. For mineral wastes shallow settling beds were developed.

A new edition of the book has appeared about every three years, and each time the progress of sewerage and sewage technology has been written into it up to the known state of knowledge, because for our purposes we wished to have everything that was important in summarized form readily to hand. Close touch was maintained with England and America, and many overseas visitors came to Essen, while after 1909 I made regular visits to England and America in return and I owe much to my colleagues there. Among them I must mention in England Mr. J. D. Watson of Birmingham, Dr. G. F. Fowler of Manchester, Dr. H. T. Calvert of London and Mr. H. C. Whitehead of Birmingham, and in the United States of America Rudolf Hering and George W. Fuller of New York, Leonhard Metcalf and H. P. Eddy of Boston, and Langdon Pearse of Chicago.

The first translation of *Taschenbuch* into English was carried out in the United States of America (Imhoff and Fair, *The Arithmetic of Sewage Treatment Works*, Wiley, New York, 1929), into Polish in 1933, Italian in 1934, French in 1935, Spanish (in Argentina) in 1935 and into Jugoslav in 1950. In 1940 the first edition of the book *Sewage Treatment* was published (Imhoff and Fair, Wiley, New York), and Gordon M. Fair, of Cambridge, Mass., not only translated the eighth edition of the *Taschenbuch* but transposed it and supplemented it to suit American conditions. More particularly the contents included also the methods of estimating the self-purification of natural waters which Fair developed, and these appear in the following edition of the German editions.

In 1922 I ended my work with the Emscher River Board and until 1934 was occupied with the Ruhr-Verband, a similar organization which was set up originally in 1913. The task here was to preserve the quality of the Ruhr River water which afforded water supply for an industrial area with a population of 3·5 million people. Within the 20 years (1913–34) 70 sewage treatment works were constructed. Of special interest are the biological treatment processes using 'activated sludge' or artificial lakes.

During my service with the Ruhr-Verband, Essen, Dr. Ing. Wilhelm Müller worked on my staff (1926–27). Dr. Müller later joined the Sewerage Department of the city of Halle/Saale, Germany, and in 1949 accepted a position with the Metropolitan Water Sewerage and Drainage Board in Sydney, Australia, and later with the Water Supply, Sewerage, and Drainage Department in Perth, Western Australia. Working as an engineer in Australia, Dr. Müller became convinced that the German work should be made available to colleagues in the British Commonwealth, and together with a colleague, Mr. D. K. B. Thistlethwayte, chief chemist of the Sydney Water Board, Dr. Müller undertook a translation into a form suitable for the British Commonwealth, the manuscript of the 16th (1956) edition of the German *Taschenbuch* being made available.

In my judgment Müller and Thistlethwayte have produced an excellent work, and I am convinced this English book will be successful.

KARL IMHOFF

Essen, June 1956

PREFACE TO THE SECOND EDITION

As with the first edition, this book retains its connection with the late Dr. Karl Imhoff's book, *Taschenbuch der Stadtentwaesserung* (the 22nd edition of which was issued recently), but again differs considerably both in arrangement and in the detail of its contents.

Many parts of the book have been rewritten to take account of the greatly expanded data from experiment and practical experience during the intervening ten years or so. Particular attention has been given to the references appended to each chapter. We hope that the book will again prove useful to the specialists of all the disciplines concerned with these important problems.

During the rewriting of this book, our much admired and beloved Master, Dr. Karl Imhoff, Essen, Germany, passed away in his 90th year. We are deeply indebted to him for encouragement and inspiration and for his help and interest in our work with the book.

W. J. Müller, Darmstadt, Germany

1971 D. K. B. Thistlethwayte, Sydney, Australia

PREFACE TO THE FIRST EDITION

This book is based upon a translation from the German book *Taschenbuch der Stadtentwaesserung*, 16th edition, Munich, 1956, by Dr. Karl Imhoff. The arrangement and contents differ, however, from the German work in order to make it more suitable for use in the countries of the British Commonwealth, where natural conditions and tradition call for some modifications from practices common in Europe.

The section on sewerage is kept very brief and confined to details required for understanding the problems related to sewage disposal; on the other hand, the methods of disposal of sewage and water-borne wastes including the methods of sewage treatment are fully described and, in some sections, enlarged substantially. Special care has been taken to outline the techniques regarding disposal of sewage from residences and institutions in unsewered areas and of wastes from different groups of industries.

It is hoped that this book will be useful in a practical way and also as a textbook for engineers, chemists, medical officers and others, and to sewerage and Public Health authorities, river boards and departments of water conservation, as well as to students, universities and institutes in general.

Being based on experience and much information and data carefully selected and evaluated by Dr. Imhoff from many countries in Europe and America during his long course of practice and study in sewage engineering, it can be of great help in solving problems of liquid wastes disposal.

We wish to thank and acknowledge our well-known 'Master of Sewage Engineering', Dr. Karl Imhoff, Essen.

W. Müller
D. K. B. Thistlethwayte

Perth, W.A., and
Sydney, 1956

PART A

INTRODUCTION AND BASIC DATA

1

WASTES ENGINEERING AND ASPECTS OF SANITATION, ECONOMY AND WATER-RESOURCES MANAGEMENT

SEWAGE results from the use of water in dwellings, hospitals, schools and elsewhere for the disposal of waste in kitchens, bathrooms, lavatories, laundries, laboratories, etc.; industrial waste from similar disposal of waste in factories and trade premises for the production, preparation and distribution of food, raw materials, electric power and of a variety of manufactured goods, incidental and necessary to the communal life and industrial activity of today.

Collection and disposal of these wastes may be primarily a matter of environmental health and sanitation. The wastes usually contain substantial proportions of organic and inorganic matter and, if not disposed of properly, will produce serious nuisances such as bad odours, corrosion and killing of fish or beast. Commonly they are also charged with viruses, bacteria and other organisms perhaps specifically dangerous to man, or may contain toxins, carcinogens or other constituents harmful to human, animal or vegetable life or to particular activities such as industry, recreation, cultural life, etc. Therefore, from the point of view of public health and general sanitation in any communal area, proper treatment and disposal of the water-borne wastes is practically essential.

Water-borne wastes are usually disposed of finally into natural surface waters such as rivers, lakes, estuaries and oceans or into groundwater streams or reservoirs. In nearly all cases these waters must serve a variety of purposes, not merely as recipients of waste waters but also as sources of water supply for residence, farm or industry, for bathing, fishing, transport and even for landscaping and scenic purposes.[1] The multiple-purpose usage of natural waters and their importance for widespread human activities make it necessary to take into account the various uses of the waters in determining the wastes engineering requirements for satisfactory waste-water disposal.

Demands for water are increasing throughout the more highly developed and industrialized countries and have in some areas exceeded the supply of water readily available. From these growing demands has developed the realization of the need for better management of water resources within industry, within its associated community and, in turn, within the wider region (water resources area) of which each local community usually forms but a part. Moreover, in densely populated and industrialized areas the unrestricted use of water and the discharge of wastes directly to watercourses produces degrees of pollution of natural waters resulting both in health dangers and intolerable nuisances, such as siltation or scum formation, eventually leading to demands for proper waters pollution control. Finally, the view has developed that if water-pollution control is to be effective and economical, water resources must be managed totally; with respect to water supply, waste disposal and reclamation of water, recovery of waste materials which otherwise would cause pollution, and for water-power production, flood control, navigation and other legitimate water uses.[1, 2]

In general, therefore, water-borne wastes and their disposal must play a very important part in water resources management.[3] The principal idea is that of conservation of water in the best interests of the people of a country. The aim of wastes engineering, then, is to carry the wastes of any activity to a suitable place for disposal, by some engineering means designed so that the principles of conservation of water resources are observed as far as may be feasible and economical, but at the very least so that the wastes are disposed of harmlessly from the viewpoint of public health.

The disposal of water-borne wastes requires substantial engineering works. These must be properly designed, constructed and maintained, and administrative and technical staffs commensurate with their nature are necessary. The costs of construction and the annual expenditure for proper waste-water disposal are usually high, and often there is no apparent financial benefit to the community or organization from which the wastes are discharged, e.g. from a municipal sewerage system or a particular factory. In some cases, the financial burden may be unbearable for a particular industry and in the general interest some sort of assistance may be provided from public funds. The disposal of wastes should thus be related technically and financially to the whole economy of the particular community, industry, region or country, its water-resources management, its urban and rural planning and its industrial and financial resources. Law and administration, existing standards and probable future development are also involved.

Because of the high costs, water-borne waste disposal must be obtained by the most economical means consistent with a proper solution of all the problems involved. This may well be different from place to place, even within one country, because of varying local conditions and circumstances. It must also be realized that the costs of waste-water disposal are not confined to those of engineering design and construction, of maintenance and operation. Secondary costs may arise, perhaps outside the municipality disposing of its wastes, as when expensive water-treatment works become necessary because of the risk of pollution; or nuisance may be caused beyond local boundaries owing to bad drainage or in other ways. On the other hand, new factors in economy may be developed along with waste disposal. Products such as reclaimed water, digester gas, fertilizer and other chemicals, or even food, may add new resources to the economy of industry, community or nation. Usually, however, any utilization of by-products of waste-water treatment reduces the financial burden only little, if at all. Seldom can waste-water treatment actually become profitable for the polluter when judged only by the book-keeper's ledgers.

Taking all together, the proper management of natural water resources, including groundwaters, requires the co-operation of specialists in many specific aspects of hydrology and hydraulics, chemistry and physics, biology, bacteriology, mathematics, geology, etc., in addition to specialized skills required in sound engineering designs for the necessary works, for proper operation and adequate administration of the organizations charged with these responsibilities.[4]

REFERENCES

1. Isaac, P. C. G. 'Water, Waste and Wealth', *Wat. & Wat. Engng* **69** (April 1965) 151
2. Lester, W. F. 'The Water Resources Act 1963 and River Water Quality Management', *J. Proc. Inst. Sew. Purif.* (1966) Pt 2, 103
3. Okun, D. A. 'The future of water quality management', *Wat. Pollut. Control* **67** (1968) Pt 2, 127
4. Stander, G. J. 'Water Pollution Research—A Key to Wastewater Management', *J. Wat. Pollut. Control Fed.* **38** (1966) 774

2

SEWERAGE AND DRAINAGE SYSTEMS

THE PURPOSE of sewerage and drainage systems is to receive the liquid wastes at or near their source and to convey them directly to the disposal works. In particular cases it may be feasible to use open channels, but in general, closed pipelines or covered channels are necessary to proper sanitation.

Water-borne wastes are natural consequences of water supply into homes and factories. In addition to these wastes, natural drainage may require the collection and disposal of rainwaters or groundwaters from residential or industrial areas. The liquid wastes may be classified as follows:

Sullage—this includes any household waste liquids from kitchens, bathrooms and laundries, but excludes faecal matter and urine.

Sewage—this is an inclusive term applied to the mixture of all liquid domestic wastes, including especially human body wastes (faecal matter and urine). The term is used loosely to include the combined liquid wastes discharged from all domestic and industrial sources within a given area. Any sewage containing body wastes may be referred to as foul sewage.

Trade Wastes or Industrial Wastes—this includes any water-borne wastes, or waste aqueous solutions or dispersions, from trade or factory premises other than the 'domestic' wastes (sewage or sullage) from staff rooms, luncheon rooms and lavatories.

Stormwater—this is from surface or subsoil water which may be collected, along with normal sewage, as *combined* sewage or which may enter sewers adventitiously as *infiltration water.*

Groundwater—this is derived from the true groundwater below the groundwater table and may enter sewers by *infiltration.*

Municipal Sewage—this consists usually of a mixture of normal or domestic sewage with some trade wastes, sometimes diluted with storm- and groundwater.

Sewerage systems may be partial or complete, separate or combined:

Partial Sewerage where only part of the water-borne waste is provided for, for example, where only sullage is collected, without faecal matter or urine from water-closets.

Separate Sewerage which is used for the collection of sewage and trade wastes only, surface water being collected entirely by a separate *stormwater drainage system.*

Combined Sewerage where the foul-sewer system is combined with stormwater drainage of all premises, including roofs and yards, also pathways and roadways.

Partially Separate Sewerage where only part of the stormwater is run off to the foul-sewers, e.g. roof water or roof and yard drainage, the remainder, such as roadways, parklands, gardens and unoccupied areas, being drained either naturally or by an additional stormwater drainage system.

Partial sewerage of a community can be regarded as offering only a temporary solution to more urgent problems of sanitation until the construction of a proper sewerage system can be completed. It may be useful also for temporary installations such as hostels or camps, or for individual dwellings or small communities where the cost of complete sewerage installations may not be justified (see Part D).

Combined sewerage systems are especially common in England, Germany and in some other countries. More recently, however, separate sewerage has been favoured, mainly because sewage treatment and river-pollution control are to some extent simplified and have become more reliable and economical. These advantages are more marked, of course, where the ratio of maximum stormwater run-off to maximum sewage flow per acre is relatively large, e.g. in areas where low population density is associated with high average rainfall and high rates of precipitation. These circumstances often apply in tropical and subtropical urban areas.

Partially separate sewerage systems were developed particularly in Great Britain[1] in cases where the separation of all stormwater (especially roof water and the run-off from residential premises) was difficult. This is commonly experienced in urban areas which were fully developed long before modern sanitary engineering practice.[2]

DESIGN OF SEWERAGE SYSTEMS

Sewerage systems consist of a reticulation system of gravity sewers of appropriate sizes and capacities according to the settled areas served and the gradients of the sewer lines, the tributary sewers merging progressively into larger ones in much the same way as natural river systems. The final sewers are described as main sewers or trunk sewers and lead directly to the treatment works or outfall (point of discharge). For proper sanitation, sewers generally have to be constructed as closed pipelines or covered channels; except in some cases, especially for trunk sewers in industrial areas where it may be preferable to construct the main sewers as open channels.[3, 4] For maintenance, cleansing and inspection, the necessary access to the sewers is obtained by manholes properly spaced to facilitate this work; these also may provide permanent or temporary ventilation of the sewerage system as required (see p. 13).

In the designing of sewerage systems, careful attention should be given to the hydraulics of all pipelines, channels, etc., to ensure steady flow lines along all sewers, through junctions and any other structures required. The desired minimum flow velocity must always be sustained (or exceeded) and backwater conditions avoided. Junctions and changes in grade, sewer sizes or direction, as well as drops, require special attention hydraulically; this becomes more and more important as the *age of the sewage* (total time of travel) increases progressively, particularly with increasing septicity (if unavoidable, e.g. in very large systems or, perhaps, in warmer climates).

Where the sewage must be lifted to some higher level, for economy of construction or for other reasons, pumping stations (see p. 10) are used to discharge it through rising mains (p. 11), usually into other sewers at a higher level. Combined or partially separate sewerage systems are usually relieved of excessive flows following storms by means of overflows (p. 13) constructed to divert the excess flow as required. Sometimes overflows are also provided in separate sewerage systems to allow relief in case of emergencies, but they should not be required in separate sewerage systems except in special cases.

A primary requirement of any sewerage system is that the waste should as far as possible be maintained in a fresh condition, that is, containing some dissolved oxygen. Therefore, in addition to the hydraulic requirements, the design of sewerage systems has to take into account the composition, both qualitative and quantitative, of the flowing wastes. Sewage and other unstable or putrescible wastes have the tendency to decompose and become septic even in a relatively short time of flowing through a sewerage system, unless its design and operation ensure that a fresh condition is maintained. It is common experience that *fresh sewage* has many advantages in comparison with *septic sewage*. The latter contains hydrogen sulphide which even in small concentrations may give rise to local odour nuisances and often causes serious damage to sewerage structures. In some cases, especially in hotter climates where higher temperatures favour septic conditions, sewers have actually collapsed owing to bacterial oxidation of the hydrogen sulphide evolved, producing large amounts of sulphuric acid which destroyed metals and concrete.[5–8] Furthermore, a septic condition of sewage makes treatment and disposal more difficult and consequently more costly. The high immediate oxygen demand of septic sewage may lower the efficiency of aerobic biological treatment processes and, therefore, some reconditioning of septic sewage may be desirable before biological treatment is applied. These disadvantages are avoided if the sewage is kept fresh, and this aspect should always be given careful consideration, as it is one of the most important bases for the formulation of a proper sewerage design.[6, 9]

Some of the measures which have to be considered in the design and operation of sewerage systems to keep the flowing wastes always fresh and free from septicity are dealt with below.

GRADES OF GRAVITY SEWERS

Deposits in sewer lines and pumping stations, and the *accumulation of slimes* on sewer or internal pipe-line walls, together with too low a rate of uptake of dissolved oxygen, are the main causes of septicity in sewage. Drains and sewers must, therefore, be laid to grades at least adequate to keep the settleable solids in suspension[10] and, wherever possible, to ensure reaeration of the sewage fast enough to replace the oxygen used up by inevitable fermentation or other biological or chemical reactions.[6] Proper *ventilation* of the sewerage system is another requirement (p. 12), important also for safe working conditions for maintenance staffs.

The grade of gravity sewers should be sufficient to produce *self-cleansing velocities*, namely about 2·5 ft/s (0·8 m/s) at least during periods of peak dry-weather flow.[10] Wherever normal sewer flows are insufficient to prevent deposition of solids, suitable arrangements should be provided for regular flushing or mechanical cleansing (see p. 12). The maximum velocity generally allowed is about 10 ft/s (3 m/s) in order to control the scouring effect of grit and other abrasive materials carried by sewage in suspension.

Where *inverted siphons* are necessary along a sewer line to traverse depressions, pass under watercourses, railway lines, etc., duplicate structures should be provided, one to carry the peak dry-weather flow with a minimum velocity of 4 ft/s (1·2 m/s), the other designed to come into action only when excess flow has to be carried, as in wet weather, by provision of an overflow weir at the inlet end. Cleansing and flushing valves should be installed. The up-leg of the inverted siphon should not be vertical but should rise no more than one vertical to two horizontal to ensure the discharge of sand or silt. For small sewage flows, single inverted siphons may be adequate, but they should be flushed regularly. *Ordinary siphons* (i.e. true siphons) should be avoided but, where unavoidable, the siphon duct should be laid with a slightly rising gradient to allow the travel of air bubbles, with the end sharply down-piped to the gravity sewer so as to prevent accumulation of air or other gases.

SEWAGE PUMPING SYSTEMS

In low-level areas where sewage has to be pumped to a higher level the sewerage system is normally designed as a separate one in order to avoid excessive pumping of surface water. Both high-level and low-level sewers should be designed to maintain freshness and to be self-cleansing.

The sewage collected from low-level areas must be pumped through rising mains (force mains or pressure mains) to high-level systems. In most cases, pumping stations are used, and the sewage often first enters storage wells which serve also as suction wells for the pump or the pumps. Practical experience shows that any dissolved oxygen carried with the inflowing sewage into storage wells may become depleted, so that the subsequent passage through the rising main results in the discharge of septic sewage not only devoid of dissolved oxygen but often containing appreciable or even substantial amounts of sulphides, especially where sewage temperatures are relatively high. Septicization can be avoided by proper design of pumping systems.[6]

Often, especially in smaller pumping stations, sewage pumps are automatically controlled so as to operate intermittently, at intervals determined by the fluctuation of sewage flow. However, the ideal to be aimed at is continuous pumping, that is, a continued adjustment of pumping rates to match instantaneous sewage inflow rates, with storage kept to the minimum. However, where this is not feasible, the compromise arrangements should limit the maximum dry-weather pumping rates to values close to the peak dry-weather inflow rates. Where pumping stations deliver directly to treatment works, pumping arrangements should be designed to ensure that the flow is as continuous and as uniform as feasible.

The influent to the *wet wells* of pumping stations should pass through bar screens having clear openings of about 1 in (25 mm) and should enter the wells with a minimum of turbulence. The arrangements should also ensure that the wells are pumped out completely, preferably during each pumping cycle (but at least once daily) to control deposition, scum accumulation, etc. The floor of the wet wells might be constructed to slope steeply—perhaps approaching 45°—to facilitate this.

Like other sewerage structures, wet wells should be properly ventilated directly into the free atmosphere. If the sewage cannot be kept fresh, continuous forced ventilation may be desirable or necessary to obviate odour nuisance. With septic sewage, the rate of ventilation may have to be relatively high, say up to 10–20 changes of the wet well atmosphere per hour. Deodorization may be required.[6, 11]

Rising mains from pumping stations should be designed similarly to inverted siphons, with proper provision for the release of gases. A minimum flow velocity of 3 ft/s (1 m/s) is recommended. Flushing arrangements are necessary where flows vary widely. Pumping of sewage may favour septicization of the sewage because no oxygen is available to it during passage through rising mains, unless special aeration is provided.[12]

Anaerobic slimes, which normally grow on the walls of rising mains, consist of a densely populated community of commensal bacteria (sometimes also other schizomycetes) which can produce very low redox potentials. Large numbers of sulphate-reducing bacteria develop in these slimes, and sulphate diffusing into them is rapidly decomposed and sulphides formed accordingly, at a rate which increases with increased sulphate content in the sewage. Where sewage is more concentrated in organic matter content, septic conditions and sulphide production tend to develop faster likewise. Hence, when sewage is pumped through rising mains, the rate of sulphide production is increased, other things being equal, if it is high in biochemical oxygen demand (B.O.D.) and in sulphate. Subject to these influences, the rate of sulphide generation depends primarily on the area of slimes over which the sewage flows and the time of its passage over these surfaces.[6, 12] Therefore, rising mains should be as small and as short as feasible, and the average time of passage of sewage, from entry to pumping works to discharge from the rising main, should be minimized.

MAINTENANCE AND OPERATION

An important aspect of sewer design arises from the tendency of some of the suspended matter of sewage to adhere to the slimes which form on the sewer walls below the sewage water level, and for heavier solids to concentrate near the invert level and to form deposits wherever the flow velocity is low enough. As such circumstances favour septicization of sewage, the sewers have been kept as clean as possible by the application of proper structural means (sufficient grades, etc.) and by providing for the regular cleansing and flushing of the sewerage system as required at certain intervals.

In practice, there are many cases where self-cleansing of sewers cannot be obtained structurally because of low gradients in the areas served or, in early stages of development, with sewage flows well below the design flow, or for other reasons. It is then essential to adopt other means to keep the sewers clean. These might be intermittent flushing or direct removal of slimes, sludge deposits or detritus, manually using shovels or with the aid of special apparatus and machines.[6]

Processes used for flushing include the discharge of large volumes of fresh water from natural surface waters or groundwaters or from ordinary supply water, or the intermittent accumulation and discharge of sewage at a rate sufficient to carry all deposits down the sewer to other sections where deposition will not occur. For this purpose it is common practice to install specially designed flushing manholes or flushing chambers upstream of sewer sections where deposits would occur.

The required frequency of cleansing depends on local circumstances. It may be necessary only twice per year, but usually needs to be more frequent, say monthly. In addition, routine inspections and local maintenance should be regular practice, both to prevent actual chokages or other causes of nuisance and to maintain the freshness of the sewage as far as feasible. This control of sewer cleansing and maintaining the fresh state of the sewage may be especially important in the case of separate sewerage systems operated in warmer climates.

VENTILATION

Ventilation of gravity sewers is necessary to permit air movement into and out of the sewer as an unavoidable consequence of the variations of sewage flow levels from time to time, and to provide always a sewer atmosphere containing sufficient oxygen. This movement of air also serves to dilute odorous or harmful gases which may be introduced accidentally into the system, such as town gas containing carbon monoxide, explosive gases from gasoline, etc., or gases produced in and evolved from the sewage (carbon dioxide, methane, hydrogen sulphide, etc.). Proper ventilation reduces potential dangers to a minimum.

Sewers should be freely ventilated through the ventilating pipes of each house drainage system, omitting any boundary or intercepting traps or other water-seals between the drains of premises and the public sewers. The principal vents of the house drains should be carried above roof level, but each individual drainage fitting be separated from the drainpipe and common ventilation system by an efficient water-seal (trap). Additional ventilation of the sewers may be obtained by using ventilating openings through manhole covers, provided no odour nuisance results. In combined sewerage systems, street gullies should not be trapped.

Usually, natural ventilation of the sewers will be sufficient, provided there are sufficient openings for the sewer air to enter and escape freely. Artificial ventilation of sewers is seldom necessary, except for very long lines which cannot be freely ventilated.[13]

STORMWATER DISCHARGES

Treatment of sewage from combined sewerage systems during periods of wet weather is usually limited to a definite capacity relative to the average dry-weather flow, any excess flow being discharged directly to natural drainage systems[14] or bypassed into the treatment works' outfall without treatment. Dilution ratios between one and ten volumes of stormwater for each volume of dry-weather flow are in use or sanctioned by regulation in different countries. The British practice normally requires provision for treatment of all flows up to six times the average dry-weather flow, in accordance with the recommendation of the Royal Commission on Sewage Disposal. In Germany, it is common practice to provide overflow relief in excess of five times peak dry-weather flow and relief ahead of sewage treatment works in excess of two or three times daytime average flow, equivalent to about 2·5–4 times average dry-weather flow. In the U.S.A., where water consumption is much higher and, therefore, the sewage flow greater and more dilute than in Europe, sometimes dilution ratios as low as 2:1 or even 1:1 are permitted at overflow reliefs. Whatever the minimum prescribed by the local authority, considerations with regard to the receiving waters may require provision for higher ratios.[15] It may then be advisable and feasible to provide basins (stormwater tanks) for storage of excess storm flows[16] from which the excess water may gradually be drained back to the sewers after rainfall ceases and sewer flows return to more normal levels (p. 338). Such basins may be designed as sedimentation tanks to provide partial primary treatment even for flows up to maximum sewer capacity.

As the overflow is usually discharged into natural watercourses, it might ease pollution control if the frequency of overflowing or length of overflowing period per year is restricted according to the intensity and duration of rainfalls to be expected in the particular area.[17] For instance, the overflow weir might be laid at a flow level in the sewer according to the run-off of a rainfall of 0·08–0·14 in/h (2–4 mm/h) as suggested for Germany, or 0·21–0·28 in/h (5–7 mm/h) as recommended for Switzerland. The value appropriate for the area in question should be determined by correlating rainfall intensity–frequency data and run-off into the sewers with the local requirements of river pollution prevention.[8]

BIBLIOGRAPHY

Babbitt, H. E. and Baumann, E. R. *Sewerage and Sewage Treatment*, 8th Edn (1958) Wiley, New York

Escritt, L. B. *Sewerage and Sewage Disposal*, 3rd Edn (1965) C.R. Books, London

Escritt, L. B. and Rich, S. F. *The Work of the Public Health Engineer* (1959) McDonald & Evans, London

Fair, G. M., Geyer, J. C. and Okun, D. A. *Water and Wastewater Engineering*, Vol. 1. Water Supply and Wastewater Removal (1966) Wiley, New York

Steel, E. W. *Water Supply and Sewerage*, 4th Edn (1960) McGraw-Hill, New York

Institution of Municipal Engineers, 'Drainage (Sewerage)', *Civil Engineering Code of Practice No. 5* (1950) London

Water Pollution Control Federation, 'Design and Construction of Sanitary and Storm Sewers', *Manual of Practice No. 9* (1969) Washington, D.C.

REFERENCES

1. Watson, D. M. 'Modern Sanitation in Great Britain', *J. Instn civ. Engrs* **13** (1939/40) 125
2. Camp, T. R. 'The problem of separation in Planning Sewer Systems', *J. Wat. Pollut. Control Fed.* **38** (1966) 1959
3. Imhoff, K. 'Offene Abwasserkanäle', *Wass. Abwass.* **1** (1909) 401
4. Carp, H. 'Offene oder geschlossene Abwasserkanäle?' *Gesundheitsingenieur* **72** (1951) 302
5. Parker, C. D. 'Mechanics of Corrosion of concrete sewers by hydrogen sulphide', *Sewage ind. Wastes* **23** (1951) 1477

6. Technological Standing Committee of the Principal Australian Water and Sewerage Authorities, *Control of Sulphides in Sewerage Systems* (1971) Butterworths, Sydney
7. Committee Report, Water Pollution Control Federation, 'Preventing the Corrosion of Portland Cement Concrete by Hydrogen Sulphide', *J. Wat. Pollut. Control Fed.* **39** (1967) 1731
8. Gt Brit. Min. of Technology, Notes on Water Pollution No. 32, *Formation of Sulphide in Sewers* (repr. *J. Proc. Inst. Sew. Purif.* (1966) Pt 6, 594)
9. Müller, W. J. 'Die Frischhaltung des Abwassers in Leitungen', *Gesundheitsingenieur* **73** (1952) 164
10. Final Report of Committee, *Boston Soc. civ. Engrs* (1942) March 4
11. Carlson, D. A. and Leiser, C. P. 'Soil beds for the control of sewage odours', *J. Wat. Pollut. Control Fed.* **38** (1966) 829
12. Pomeroy, R. D. 'Generation and Control of Sulphide in Filled Pipes', *Sewage ind. Wastes* **31** (1959) 1082
13. Gustafsson, B. and Westberg, N. 'Oxygen Consumption and Re-aeration in Sewers', *Proc. 2nd Int. Conf. Wat. Pollut. Res. Tokyo* **1** (1964) 221
14. American Public Works Association, *Report on Problems of Combined Sewer Overflows* (1967) Federation of Water Pollution Control Administration, U.S. Dept. of Int. WP 20–11
15. Akerlindh, G. 'Permissible water pollution at combined sewer overflows', *Sewage Wks J.* **21** (1949) 1059
16. Carter, R. C. 'Improvements in systems of combined sewerage', *Surv munic. Cty Engr* **108** (1949) 743
17. Hörler, A. 'Die Wirkung der Regenauslässe', *Schweiz. Bauztg* **118** (1941), Nr. 20

3

QUANTITIES AND QUALITIES OF SEWAGE AND OTHER WATER-BORNE WASTES

DRY-WEATHER SEWAGE FLOW

THE FLOW of sewage and other water-borne wastes from any community varies from hour to hour and from day to day. Nevertheless, this daily flow tends to follow a typical pattern, the flows from smaller areas (shorter times of travel of sewage) being more variable than those reaching the main or outfall sewers serving larger areas. Depending on the characteristics of the particular area, week-end

Table 3.1

SEWAGE FLOW

Country	imp. gal/hd d		l/hd d	
	*Range**	*Average*	*Range*	*Average*
Great Britain	30–65	40	140–300	180
Germany	25–60	35	110–207	160
U.S.A.	65–125	80	300–600	360
Australia	35–65	45	160–300	200

* In larger cities, the flow may be greater.

or holiday flow patterns may show marked differences from the normal.

The total daily volume of sewage flow likewise varies from area to area, largely according to the water consumption in residential areas including dwellings,[1, 2] local trade and commercial premises, offices, etc., and in some cases also according to industrial water usage for manufacturing goods in excess of local requirements. There is a marked tendency for sewage flows to increase as the living standards of a community are improved. Table 3.1 gives flow figures characteristic of different countries and Fig. 3.1 shows a typical diurnal flow pattern for normal dry-weather flow from a small residential area.

The maximum hourly discharge rate arriving at the treatment works or at outfalls usually lies between $\frac{1}{10}$ and $\frac{1}{20}$ of the daily flow, depending on the time of concentration in the sewerage area served. Consequently, in small areas with short times of sewage travel, the peak flow is relatively greater, and the ratio of 24 : 10 peak/h to average/h flows is typical for small towns, but for very large communities a ratio of 24 : 20 may be applicable; a value of 24 : 14 is often taken as an average for the ratio of total peak/h to average/h flow in medium-sized towns (see Fig. 3.1). In some cases an additional allowance for permanent or seasonal groundwater infiltration is necessary. Special allowances may be required for particular

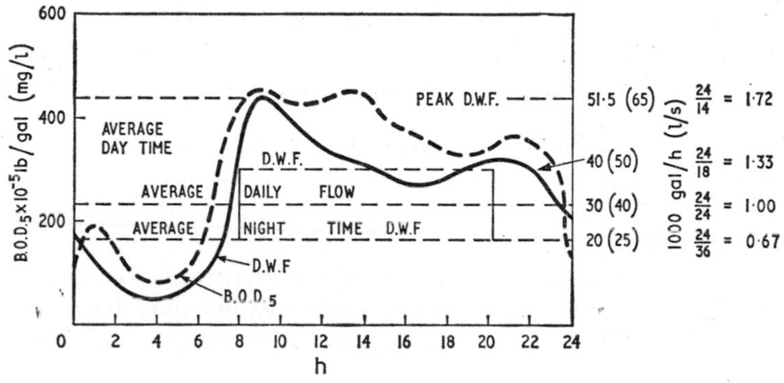

Fig. 3.1. *Hourly variations in sewage flow and B.O.D.*$_5$ *content for residential area (dry weather)*

waste-water discharges from industries into the municipal sewerage system (see below).

WET-WEATHER SEWAGE FLOW

With separate sewerage systems, stormwater is carried by separate drains and channels. Nevertheless, in many areas the wet-weather flow in the foul-sewers is found to be greatly increased by the accidental access of stormwater, either directly or indirectly, and corresponding allowance is required for sewerage and the capacity of sewage treatment works (see p. 337).

In combined sewerage systems, the wet-weather flow is usually many times greater than that in dry weather. Provision for wet-weather flows may include retardation or storage works, overflow arrangements, etc., depending on circumstances or local regulations (see Chapter 15).

GENERAL CHARACTER OF DRY-WEATHER SEWAGE

The matter carried in the sewers serving residential areas includes inorganic and organic solids, in solution or in suspension,[3–6] from the lavatories, kitchens, bathrooms and laundries of family residences with other waste discharges from local trades, schools, hospitals, hotels, etc. Very large numbers of bacteria are usually present in such sewage—hundreds or thousands of millions per millilitre—some of which can cause dangerous illness in man, together with harmful viruses and other disease-producing organisms (fungi, protozoa, helminth ova, etc.).

The character of sewage differs from place to place according to customs and living standards of the population. Table 3.2 shows the usual actual values, rounded off, for an average domestic (residential) sewage.

The content of mineral matter depends largely on the total dissolved solids content of the communal water supply. This may be particularly important in cases where re-use of water for any purpose comes into question.

In combined sewerage systems where stormwater carries some polluting matter into the sewers, higher values may be assumed, e.g. a daily total of 0·16 or 0·17 lb (73 or 77 g) $B.O.D._5$/hd d or even more.

The allowances of Table 3.2 require adjustment when relatively substantial proportions of industrial wastes are discharged into the municipal sewerage system. These additional loadings, both in volume and pollution content, should be taken into account, quantitatively, for example, by using population equivalents (p. 309). Qualitative aspects (including toxic effects) must also be considered (see pp. 304 and 306).

Combining the data from Tables 3.1 and 3.2 provides the elementary analytical data of Table 3.3.

If the sewage flow is greater or less than 40 gal/hd d (180 l/hd d) merely because of different *per capita* rates of water consumption, the composition of the sewage is correspondingly weaker or stronger. The total amount of waste solids contributed per person varies to some extent according to the standard of living, but in communities of comparable standard it is relatively constant per head of population, whereas the rates of water consumption may differ considerably. However, dilution tends to increase the non-settleable suspended solids and the dissolved solids contents relative to the settleable. In the U.S.A. and some other countries, the average water consumption is much greater than in most European countries, and the concentration of the waste solids is usually much lower or, as it may be said, the sewage is weaker.

Of the organic matter in domestic sewage, a little more than one-half is contributed by faeces and urine. The discharge from bath-

Table 3.2

SOLIDS CONTENT OF AVERAGE DOMESTIC (RESIDENTIAL) SEWAGE*

Type of solids	*Mineral*		*Organic matter*		*Total solids*		B.O.D.$_5$†	
	lb/hd d	g/hd d	lb/hd d	g/hd d	lb/hd d	g/hd d	lb/hd d	g/hd d
Settleable	0·03 (0·04)	14	0·09 (0·10)	41	0·12 (0·13)	55	0·04 (0·05)	18 (22)
Non-settleable suspended	0·02 (0·03)	9	0·04 (0·05)	18	0·06 (0·08)	27	0·03 (0·03)	14 (14)
Dissolved	0·11 (0·12)	50	0·11 (0·15)	50	0·22 (0·28)	100	0·05 (0·07)	23 (32)
Total solids	0·16 (0·18)	73	0·24 (0·30)	109	0·40 (0·49)	182	0·12 (0·15)	55 (68)

* Values in brackets are applicable to communities with high living standards.
† B.O.D.$_5$ = 5-day Biochemical Oxygen Demand at 68°F (20°C) (p. 35).

Table 3.3

AVERAGE ANALYSIS, DRY-WEATHER SEWAGE (RESIDENTIAL AREAS WITHOUT SUBSTANTIAL INDUSTRIES)

Basis: 40 gal/hd d (180 l/hd d)

Type of solids	*Solids content**			B.O.D.$_5$ mg/l or lb/gal $\times 10^{-5}$
	Mineral	*Organic matter,*	*Total solid*	
	mg/l or lb/gal $\times 10^{-5}$			
Settleable	80 (80)	230 (250)	310 (330)	100 (130)
Non-settleable suspended	60 (80)	110 (140)	170 (210)	80 (80)
Dissolved	280 (310)	280 (380)	560 (700)	130 (180)
Total solids	420 (470)	620 (770)	1040 (1240)	310 (390)

* For average or moderate standards; communities with high living standards in brackets. The dissolved mineral matter depends largely on the mineral content of the carrying water.

rooms, kitchens and laundries might on average contain about one-third of the total polluting matter in sewage (see pp. 365, 366).

GENERAL CHARACTER OF SURFACE WATER AND WET-WEATHER SEWAGE

With separate sewerage systems, the water carried in surface water (stormwater) sewers during and following rain is contaminated mainly by suspended matter washed off impervious surfaces. Most of this material is of an inert character, but some is organic, and part of this may exert an appreciable oxygen demand. However, organic pollution from such discharges is usually negligible compared with that from storm sewage overflows on combined sewers.[5, 7]

In combined sewerage systems, the first flush from each local area (say during the first quarter of an hour or so of run-off) from rainfall following dry weather may be highly polluted, perhaps similar to average municipal sewage. Afterwards, the pollution carried in by the stormwater normally decreases considerably. The initial discharge of rainwater into the sewers contains most of the pollutants accumulated during dry weather on roofs, roads, etc.; the rising flow also tends to scour deposits from the sewers more than the subsequent falling flow.[8] These effects may be more noticeable after long periods of dry weather or in sewerage systems with relatively flat gradients.

INDUSTRIAL WASTES

From the standpoint of wastes discharge, local service industries and other small-scale trades need to be distinguished from larger manufacturing establishments. The smaller-scale trades and industries supplying the needs of only the local community, or perhaps of a small associated regional area, usually discharge relatively small total volumes of water-borne wastes which, together with the domestic sewage from living areas, represent the residential sewage flow and are normally taken into account in the values for the water-borne waste flow (domestic or residential sewage) and composition (Tables 3.1–3.3).

However, larger industries usually manufacture or process goods in excess of local needs, and these are 'exported' from their own sewerage area into other areas of the country or even abroad. The water-borne wastes from such industries, therefore, are related not to the locality but to the size of other markets, and their quantities and composition will in many cases be much greater and significantly different from the local residential sewage. A separate and thorough investigation of such wastes should always be made with regard to the additional sewerage and drainage to the disposal works. Industrial wastes may be admitted to the public sewerage system for disposal (including treatment) along with residential sewage or may be disposed of separately (usually following some degree of purification, if only screening or other simple process) directly to natural waters. This may be decided upon by the factory management on grounds of economy or convenience, or for other reasons, but it is increasingly determined, in the general interest (see Chapter 1), by regulating authorities such as municipalities or river boards.

The characteristics of water-borne industrial wastes from manufacturing processes cannot be adequately described in general terms, because not only the amounts and rates of flow but, in particular, the nature of the wastes discharged from any one factory are different, even for similar products according to the type of raw materials, the individual factory processes, etc. However, the wastes from any particular type of industry are usually similar enough from factory to factory to be compared directly, and their composition and volume may be shown to be correlated with production rates. Both rate of flow and waste composition tend to vary much more than in municipal sewage, especially where processing is batch-wise.

The data of Table 3.4 are representative of a variety of industries. The volume of sewage arising from canteens, lavatories and other sanitary fixtures is usually relatively so small as to be negligible. It might be noted that wastes from miscellaneous activities such as research institutes, power stations, etc., are commonly regarded as industrial wastes by the regulatory authorities.

Industrial wastes may be classified broadly as:

1. Non-fermentable inorganic and other inert wastes;
2. Fermentable, mainly organic;
3. Toxic, including radioactive, wastes.

Wastes from coal mining (p. 318) are of the first type, comprising generally suspensions of fine coal—non-fermentable mineral matter—in aqueous wastes which commonly contain high concentrations of saline matter. Waste pickling-bath solutions from sheet-metal or galvanizing shops (p. 318) provide other examples, but it should be noted that these may also be toxic, due to acidity, metal content, etc. Many food industries discharge readily fermentable wastes, e.g. those from meat-packing plants, fruit canneries, and this characteristic is frequently most important; alternatively, such wastes may be acid, owing to sulphurous acid,[5] or toxic because of plating-shop wastes from canister manufacture, discharged along with packing-shop wastes. Industries which also commonly discharge toxic wastes include dye-houses and electroplaters, producing, for example, cyanides, sulphides, chromium or copper compounds (p. 319), and chemical factories manufacturing chlorinated hydrocarbons, etc. Again it might be noted that dyehouse wastes (p. 322), for instance, are usually mineralized and may also contain significant amounts of fermentable organic matter. Some industrial wastes are highly coloured, giving rise to aesthetic nuisance, sometimes on this account alone; others are malodorous owing to specific constituents which are otherwise of little or no significance.

Despite these many and various differences in character of industrial wastes relative to domestic or municipal sewage, they may in many cases be characterized by laboratory tests similar to those used for the broad characterization of sewage, such as tests for suspended solids content or oxygen demand—in some cases either chemical or biochemical oxygen demand (C.O.D. or B.O.D.), in others only C.O.D.—or for organic solids content, etc. Where such tests are applicable, the values found provide a basis for calculating the volume of a particular waste as equivalent to so much sewage, and thus to the wastes from so and so many persons (population equivalents, see p. 309). There are, however, types of industrial wastes of special characteristics which preclude any comparison with domestic sewage, e.g. waste phosphoric acid–zinc phosphate

Table 3.4

CHARACTERISTICS OF SOME TYPICAL INDUSTRIAL WASTE DISCHARGES

Industry	*Production unit basis*	*Water-borne waste discharge per unit*		*Typical analytical data*			*Notes*
				Suspended solids, lb/gal × 10^{-5} (mg/l)	*Oxygen demand* lb/gal × 10^{-5} (mg/l)		
		imp. gal	l		B.O.D.$_5$	*Permanganate value*	
1. Steel works	1 ton steel	1 000–2 000	4 500–9 000	100–1 000			suspended solids mainly mill scale
2. Electroplating	1 000 ft² (93 m²), plated	1–20	4·5–90				cyanides, nickel, zinc, chromium; strongly acid to strongly alkaline
3. Paper manufacture	1 ton fibre	10 000–60 000	45 000–270 000	150–1 000	20–100	200–1 000	fibre, clay
4. Laundering	100 lb (45 kg) clothing	100–400	450–1 800	400–1 000	300–1 000	—	detergents
5. Textile dyeing and finishing	1 lb (0·45 kg) cloth	5–100	20–450	200–2 000	100–1 000	100–1 000	mineral salts, dyestuffs, sulphides
6. Wool scouring	100 lb (45 kg) wool	50–200	220–900	2 000–20 000	500–10 000	2 000–6 000	considerable grease up to 10 000 mg/l
7. Tanning	1 lb (0·45 kg) raw hide	2–20	9–90	1 000–5 000	500–5 000		hair, sulphides
8. Abattoirs	1 cattle	50–300	220–1 300	400–1 500	800–5 000	500–2 000	fats
	1 pig, lamb or sheep	20–120	90–540				
9. Milk wastes	1 gal (4·5 l) milk	1–6	4·5–25	100–300	300–2 000	50–300	fats
10. Fruit and vegetable canning	1 ton stock	200–700	910–3200	200–3 000	300–3 000		
11. Brewery	1 gal (4·5 l) beer	6–30	25–130	250–650	400–1 200		
12. Manufacture of pharmaceuticals (vitamins biologicals)	100 lb (45 kg) chemicals	100–300	450–1 300	low	500–10 000	500–5 000	phenols or other organics; total dissolved solids 5 000–50 000 mg/l
13. Coal gas making	1 ton coal	150–350	680–1 500	200–3 000	1 000–6 000		phenols, cyanides, ammonia

solutions from metal treatment plants or radioactive wastes (p. 327).

The waste waters from the use of water for cooling purposes are usually relatively unpolluted. Water is widely used for cooling in many industries, especially in power stations. The discharge of heated effluents into sewerage systems may, owing to the resulting temperature rise, produce undesirable effects such as increasing the rate of decomposition of the organic content and, hence, that of septicization of the sewage. A similar discharge into rivers may result in a reduction of the solubility and thus the availability of oxygen and, furthermore, increase the biological activity in the river where, consequently, critical conditions (p. 79) may develop more readily.

MUNICIPAL SEWAGE

Municipal sewage normally includes some industrial wastes along with the domestic wastes. Even in residential areas there may be laundries, milk depots, butchers and other food-handling shops or factories from which wastes may be discharged which are not purely 'domestic' in character. In towns and cities there are also factories, and accordingly municipal sewage usually contains appreciable quantities of industrial wastes, the proportion amounting to 10–20% at least in the municipal sewage of most cities of Europe and the U.S.A. (Fig. 3.2). In more highly industrialized communities, the

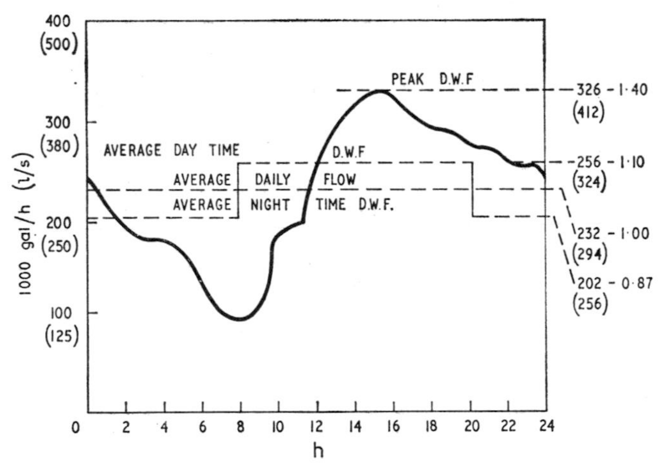

Fig. 3.2. *Typical pattern for daily municipal sewage flow*

proportion may be much greater, specially where relatively large factories are established near smaller cities. In some such cases the relative importance of the local industries and their waste discharges may result in subordination of the city's sewerage and drainage systems to the community's industrial activities.

Tables 3.2 and 3.3, although typical of domestic sewage flow from residential areas including effluents from local trade activities, do not allow for high proportions of industrial wastes discharges. These should be assessed separately on the basis of actual or expected flows and composition, or by applying population-equivalent figures. In areas where future industries of unknown type are to be provided for, a figure of 100–300 population equivalents per acre of industrial area is a common basis for estimates of the future rate of industrial waste discharges.

Even if industrial wastes are pretreated before discharge into municipal sewers according to the regulations (see p. 313) to adjust their composition to the local requirements of proper sewage disposal, the character of municipal sewage can be substantially altered by their admixture. The more important effects of greater industrial activity may include:

1. An overall increase in sewage strength, including both dissolved and suspended matter, organic and inorganic;
2. An alteration in the degree of acidity or alkalinity, such as from metal-industry or milk wastes (acid) or strawboard wastes (alkaline);
3. The addition of foul constituents, such as sulphides from tanneries, or other objectionable matter, as in yeast wastes;
4. The addition of toxic substances, such as in phenolic, electroplating or radioactive wastes.

Before designing sewerage schemes and treatment works for the disposal of mixed domestic sewage and industrial wastes, the figures of sewage flow and composition of the mixed wastes or treated effluents should be considered very carefully, because all calculations for proper treatment and disposal depend on them.

Recently the increasing use of *synthetic detergents*[9, 10] for household purposes and in manufacturing processes has had the result that municipal sewages contain substantial amounts of mixed detergents equal to 20 mg/l total content or more. The presence of surface-active material has caused undesirable effects at sewage treatment works, particularly increased foaming on the surfaces of treatment works and receiving waters, which, *inter alia*, reduces the rate of oxygen transfer from air to water. This has led to the development of 'softer' (more biodegradable) detergents which are decomposed more readily and cause less difficulties with normal sewage treatment processes.[11, 12]

The most important characteristic of municipal sewage and many other water-borne wastes is that they contain *unstable* (fermentable) *organic matter.* Its decomposition may result from the activities of micro-organisms in association with other living organisms (see pp. 125, 197); it is 'biological' or 'biochemical' and, therefore, depends on temperature and other environmental factors and conditions. These natural processes of decomposition may cause nuisances, e.g. odour (p. 135) or corrosion damage to sewerage and other structures (pp. 8, 9), if water-borne wastes are not controlled and disposed of properly, using suitable techniques as described in the following chapters.

REFERENCES

1. *Rep. Wat. Pollut. Res. Bd, 1957* (1958) 40 H.M.S.O., London; PAINTER, H. A. *Wat. Waste Treat. J.* **6** (1956–58) 496
2. HUTCHINSON, G. D. and BAUMANN, E. R. 'Variation of Sewage Flow in a College Town', *Sewage ind. Wastes* **30** (1958) 157
3. FELDMAN, L. A. and McGARVEY, F. X. 'The Moon-base Problem', *Effl. & Wat. Treat. Jnl* **5** (1965) 209
4. HUNTER, J. V. and HEUKELEKIAN, H. 'The Composition of Domestic Sewage Fractions', *J. Wat. Pollut. Control Fed.* **37** (1965) 1142
5. *Rep. Wat. Pollut. Res. Bd, 1960–65* (Investigations of Sewage and Storm Sewage Flow and Composition) H.M.S.O., London
6. 'Domestic Sewage Load per person per day', *Wat. Pollut. Control* **66** (1967) Pt 2, 193
7. LUMB, C. 'The Storm Sewage Pollution Problem', *J. Proc. Inst. Sew. Purif.* (1964) Pt 2, 168
8. GAMESON, A. L. H. and DAVIDSON, R. N. 'Stormwater Investigations at Northampton', *J. Proc. Inst. Sew. Purif.* (1963) Pt 2, 105
9. Min. of Housing and Local Govt Gt Brit., Reports of Standing Committee on Synthetic Detergents (1956–66) H.M.S.O., London; 8th Rep. and Suppl., 1966
10. WALDMEYER, T. 'Analytical Records of Synthetic Detergent Concentrations 1956–1966', *Wat. Pollut. Control* **67** (1968) Pt 1, 66
11. 'Synthetic Detergents and Sludge Digestion', Min. of Technology, Notes on Water Pollution No. 29 (June 1965)(repr. *J. Proc. Inst. Sew. Purif.* (1965) Pt 4, 383)
12. HUSMANN, W. 'Solving the Detergent Problem in Germany', *Wat. Pollut. Control* **67** (1968) 80

4

LABORATORY EXAMINATION OF WATER-BORNE WASTES

PROCEDURES for the investigation of sewage and similar wastes include physical, chemical, bacteriological and other biological methods of examination. Such procedures are used for the following distinct but closely related purposes: (*a*) to determine the character of the wastes; (*b*) to test the purifying capacity of the receiving waters into which the sewage outfall discharges;[1, 2] and (*c*) for the proper supervision of treatment works.[3] The nature, significance and purpose of the more important tests will be briefly described; details of test procedures are to be found in the general references at the end of this chapter.

SAMPLING

The proper investigation of sewages, industrial wastes and other such mixtures requires an appropriate sampling programme as a basis.[1] Each sampling of waste water should be made in accordance with the purpose, or purposes, of the particular investigation, demanding detailed inspection of the whole area involved, followed by careful selection and accurate location of the most suitable sampling sites.[1–4] In sewers and open channels, sampling should be made from sections where practically complete mixing of the flowing wastes is ensured, for example, in turbulent regions below junctions, flumes or weirs. Careful planning is necessary also for sampling influents and effluents to take proper account of varying detention periods within the treatment works tanks, etc.

Because sewage and other water-borne wastes commonly are highly variable in composition, a single ('snap') sample taken out of a flowing stream of the waste may give a false indication of its nature. However, in many cases, especially with suitable choice and planning of sampling sites and testing programmes, a comparatively small number of individual or snap samples can provide the data on which reliable assessments of the characteristics of the waste, its variation and probable average composition can be based.[1, 2] It is always necessary to have at the same time corresponding data for the rates of flow. The number of snap samples required for any particular case depends on the circumstances. Naturally, a large municipal sewerage system discharges sewage which is inherently less variable in flow and composition than a small system of the same general make-up. Likewise, effluents from factories including waste-water discharges from a number of different processes usually vary considerably in flow and composition, often more than municipal sewages, although individual industrial processes may yield fairly constant rates of flow of wastes of comparatively uniform composition. In some cases, a series of samples taken during a single typical week may prove almost as fully descriptive of a waste—in terms of average concentration, range and variability—as many weeks of repeated observations, in other cases it may take months before all the possibilities are realized. This may be of special significance, e.g. for discharge of toxic wastes.

In the particular case of domestic or municipal sewage discharges, there usually occurs a more concentrated, 'stronger', sewage during the period of greatest sewer flow around the middle of the day than in the period of minimum flow (Fig. 3.2), and it would normally require a planned series of about 10–20 samples at least to obtain reliable data for the normal ranges and weighted averages of general characteristics such as solids content and oxygen demand, with the corresponding flow data. Investigations of more specific properties would probably require a further series of samples and tests, the latter preferably correlated with those of the first 10–20 samples.

For a specific short-term investigation to determine the actual mean composition of the total sewage flow reaching a treatment works during a particular period, sufficient data may be obtained by examining a *composite sample*, that is, by making a single 2, 8 or 24 h sample by mixing portions from snap samples taken after short regular intervals of time (say, every 10, 30 or 60 min) in volume ratios proportional to the rates of flow at the time of sampling. The individual samples may be taken by hand or by means of suitable automatic apparatus.[5] Continuous proportional and snap sampling is coming more into use recently as suitable and reliable apparatus is being developed.

Because sewages and many other wastes may alter relatively rapidly during storage at ordinary temperatures, examination in the laboratory should either start at once or should be preceded by some treatment designed to obviate any changes in the characters to be measured, such as adding suitable chemicals or cooling and holding at low temperatures (but above freezing point). Some observations and tests cannot be deferred even by such measures but require testing 'on the spot'. All sampling, transport, storage and testing operations require careful planning and consideration of all details, particularly with regard to the particular properties of the wastes which are to be described or measured.[1]

PHYSICAL TESTS

Physical and chemical investigations of sewages and other water-borne wastes, on the spot or in the laboratory, provide tests both for measuring general properties, such as turbidity, suspended solids contents or oxygen demand, or more specific characteristics, or the content of specific chemical substances or biota, such as temperature, chlorinity or dissolved oxygen contents, or the actual numbers per volumetric unit of defined micro-organisms. The same or similar tests are mostly used in the examination of unpolluted waters; but although the total content of polluting substances in waste waters usually amounts only to a total of the order of 0·1% of solids—dissolved, colloidal and suspended (the latter including both relatively coarse and fine)—the values characteristic of polluted waters differ for most tests considerably from those characteristic of normal unpolluted natural waters.

Temperature is very important because it is a basic determinant of many other aspects, physical, chemical and biological, including rates of decomposition and purification. Temperature differences can cause stratification in streams or lakes and generally must be taken into account in almost every connection, so that it is a sound rule that, wherever any sample is taken, the temperature should be measured on the spot.

The general *condition and colour* of sewage and similar wastes is very important (see Chapter 2). 'Fresh' domestic or municipal sewage has a not unpleasant odour, the colour being usually light- to yellowish-grey. As the dissolved oxygen is used up and the sewage becomes 'stale', the colour darkens, finally to dark grey or black, when the sewage smells of hydrogen sulphide and is called 'septic'. Similar changes may occur with industrial wastes, but all sorts of colours can be observed in the wastes of various industries, and colour and odour may be important as the causes of public nuisance or aesthetic complaint.

Turbidity is caused mainly by highly dispersed (but not soluble) suspended substances of particle size mainly in the colloidal range, and its degree is correlated with the strength of raw wastes such as sewage (and many other water-borne wastes) or with the purification effected by various treatment processes. Measurements are usually referred to standard suspensions of finely powdered silica (Fuller's earth), 1 g/l which after settling leaves about 1 000 units of turbidity; and may be made by direct comparisons of samples and standard suspensions, each under the same conditions, or by measuring the scattering of standardized beams of light in special photometric turbidimeters.

Odour is often very important and may be highly characteristic, e.g. with meat, fish, refinery wastes, etc. Odours due to organic sulphides are highly persistent, even in high dilution, as are some of the acidic types of odours (volatile organic acids, etc.) and some organic amino compounds.

The total solids content comprises the whole amount of pollutional matter (excluding dissolved gases and vapours) and reflects the 'strength' of a particular type of water-borne waste. It may be considered as composed broadly of organic solids (volatile matter) on the one hand, and inorganic (fixed solids or residue, non-volatile on ignition) on the other and, again, as dissolved and suspended matter, the latter being subdivided in settleable and non-settleable (very fine suspended) solids, each in turn including organic and inorganic. The total solids content is measured as the total solid residue (T.S.R.) by evaporation and drying at 220°F (105°C), cooling in a desiccator and weighing. If this residue is then ignited at 1 200–1 290°F (650–700°C), moistened again, redried at 220°F (105°C) and reweighed, the *loss on ignition* represents the organic or *volatile matter* content, and the ignited residue the fixed residue solids (F.R., inorganic solids). The *suspended solids content* is measured by filtration through a crucible with a prepared asbestos mat, a standard ceramic filter crucible or other similar means, determining the dried solids by weighings and differences. Metal-wire sieves may also be used to determine what proportions of the suspended solids are removable by screens of different mesh. Filtration through standard discs of filter paper or other porous material may be used to provide a visual indication of the relative amounts of suspended matter in different samples of wastes.

The *settleable solids* are conveniently determined by means of an Imhoff cone, a specially designed tapered glass vessel of 1 litre capacity, graduated accurately up to about 50 ml (Fig. 4.1). One litre of sewage is transferred to the cone and allowed to settle for a predetermined period. The cone is then rotated gently backwards and forwards to dislodge any flocculent solids which may have adhered to the sides, and the volume of sludge read off after a further 15 min. Recording the values for different settling periods (Fig. 10.8, p. 159) gives the ideal relative sedimentation efficiencies to be expected, and so provides a basis for sedimentation tank design and for the computation of sludge digestion works. The data of such tests are, of course, more reliable if the settleable solids are drawn off from the bottom of the cone, dried and weighed. Normally domestic sewage will produce between about 3 and 12 ml/l from western communities in an Imhoff cone after 2 h settling; this is the standard time for the test. A sedimentation-tank effluent may usually be regarded as satisfactory if no more than 0·5 ml of residual solids settle out after 2 h.

The *non-settleable solids* are estimated by first determining the total suspended matter and subtracting the settleable solids content. (The

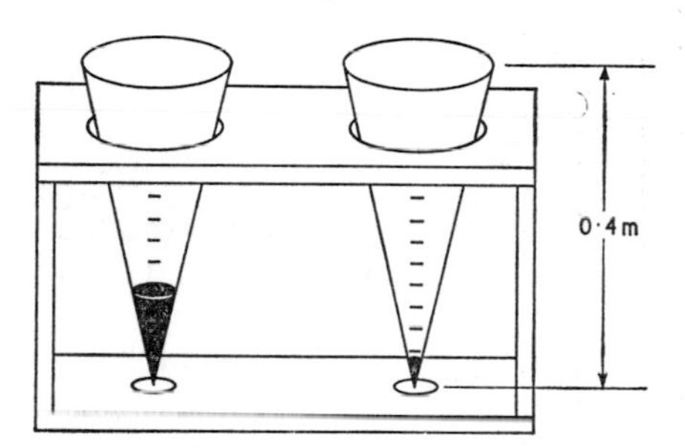

Fig. 4.1. *Imhoff cones*

settleable solids must be weighed, not only measured as sludge volume.)

Dissolved and colloidal solids—The filtrate from filtration through the Gooch crucible contains not only matter in solution but also highly dispersed insoluble matter in colloidal and semi-colloidal form. However, it is customary to regard the solids in the filtrate simply as 'dissolved solids' and their amount may be determined by evaporation of a portion of the filtrate, drying and weighing. The residue is also ignited and again weighed, giving as before loss on ignition, or volatile matter, and fixed residue.

CHEMICAL TESTS

Dissolved oxygen

Tests for *free dissolved oxygen* (D.O.) are not applied commonly to raw wastes or even to primary treatment effluents but are required for the control of activated-sludge works, oxidation ponds, etc., and are of the utmost importance in controlling pollution of natural waters. D.O. testing is also an essential part of the B.O.D. (Dilution Method) test procedure (see below).

Table 4.1a

APPROXIMATE DISSOLVED OXYGEN CONTENTS OF AIR-SATURATED FRESH WATERS (TOTAL SALINITY ~ 300 P.P.M.) based on air of normal composition saturated with vapour at the given water temperature

Elevation above mean sea-level		*Total normal air pressure*		*Oxygen dissolved at equilibrium* (mg/l) *at*						
ft	m	in	mm	32°F (0°C)	41°F (5°C)	50°F (10°C)	59°F (15°C)	68°F (20°C)	77°F (25°C)	86°F (30°C)
0	0	29·9	760	14·6	12·8	11·3	10·2	9·2	8·4	7·6
1 000	305	28·9	734	14·1	12·3	10·9	9·8	8·9	8·1	7·3
2 000	610	27·8	707	13·6	11·9	10·5	9·5	8·6	7·8	7·1
4 000	1 220	25·8	656	12·6	11·1	9·8	8·8	8·0	7·3	6·6
6 000	1 830	23·9	609	11·7	10·3	9·1	8·2	7·4	6·7	6·1
8 000	2 440	22·2	565	10·8	9·5	8·4	7·6	6·8	6·2	5·6
10 000	3 050	20·5	522	10·0	8·8	7·8	7·0	6·3	5·8	5·2
				104°F (40°C)	122°F (50°C)	140°F (60°C)	158°F (70°C)	176°F (80°C)	194°F (90°C)	212°F (100°C)
At sea level		29·9	760	6·5	5·6	4·8	3·9	2·9	1·7	0

Table 4.1b

APPROXIMATE DISSOLVED OXYGEN CONTENTS OF AIR-SATURATED WATERS OF VARYING SALINITY

basis of reference, diluted sea-water, taking saturated normal air at sea level, total pressure 29·9 in (760 mm)

Chlorinity, lb/gal $\times 10^{-5}$ (mg/l) *of chloride chlorine in diluted sea-water*	*Oxygen dissolved at equilibrium*, lb/gal $\times 10^{-5}$ (mg/l) *of diluted sea-water, at*						
	32°F (0°C)	41°F (5°C)	50°F (10°C)	59°F (15°C)	68°F (20°C)	77°F (25°C)	86°F (30°C)
0	14·6	12·8	11·3	10·2	9·2	8·4	7·6
5 000	13·7	12·1	10·7	9·6	8·7	8·0	7·2
10 000	12·8	11·3	10·0	9·0	8·2	7·5	6·8
15 000	11·9	10·5	9·4	8·4	7·7	7·0	6·4
20 000	11·0	9·8	8·8	7·9	7·2	6·5	5·9

Special sampling techniques are necessary to obviate changes in dissolved oxygen content during sampling or testing, both for under- and supersaturated waters; the water temperature also has to be measured and changes of temperature between sampling and testing must be avoided. In most cases it is desirable, if not essential, that determinations be completed in the field immediately after sampling. The method of Winkler, with modifications, is commonly used. That of Miller is simpler for field work, but suitable apparatus must be used for reliable results.[6] Polarographic and direct[7, 8] electrometric methods of determination also have been developed.

The solubility of oxygen in water depends on both the temperature and the partial pressure of the oxygen, also, to a lesser degree, on the mineral content of the water. Table 4.1 shows the dissolved oxygen content of fresh and saline waters saturated with respect to atmospheric oxygen at various temperatures, under standard atmospheric conditions[9] (20·9% oxygen in air by volume, total atmospheric pressure 29·9 in (760 mm)). The amount which can dissolve in the presence of pure oxygen at 29·9 in (760 mm) pressure is, therefore, about five times greater.

The oxygen deficit indicates the amount by which the dissolved oxygen content is less than saturation. It may be expressed simply as so many p.p.m. or, otherwise, as percentage below saturation. The greater the deficit, the greater the rate of reaeration (p. 59). *Percentage of saturation* is also commonly reported.

Oxygen demand

Volatile matter content (p. 30) is a parameter, e.g. for treatment works process loadings (pp. 276, 418), usually close to the proportion of total organic matter. Although the test does not distinguish readily biodegradable from highly condensed matter such as humus or from coal (essentially mineral, inorganic), for most water-borne wastes there is satisfactory correlation of ignition loss with unstable organic matter.

Other ways of measuring and expressing such loadings are in terms of *oxygen demand*, determinations[10] of which may use biological or purely chemical methods, and of direct determination of the *organic carbon* content. The *Chemical Oxygen Demand* (C.O.D.) and *organic carbon* tests, like the test for volatile matter, do not distinguish between stable and unstable contents. However, for any particular case, they can usually establish useful correlations, and since they can be completed within a few hours of taking samples, they have advantages, for example, in trade-waste surveys, over the *Biochemical Oxygen Demand* (B.O.D.) test which takes several days.

When water-borne wastes such as domestic sewage are discharged into natural waters containing dissolved atmospheric oxygen (free dissolved oxygen, see pp. 55, 57), not only is the sewage diluted in proportion to the respective volumes, but the microbiological flora and fauna of the water are supplemented by the many varieties (and greater numbers, pp. 58, 64) of the microbiota of the sewage. Aerobic decomposition of the organic matter ensues and, as a consequence, some free dissolved oxygen is taken up (pp. 62–69). If the overall demand for dissolved oxygen is greater than that available, the water becomes anaerobic and foul conditions result.

The oxygen demand of water-borne wastes is, therefore, of the utmost significance. The importance of the relationship of oxygen supply and consumption in natural waters was established in the middle of last century. Sir Edward Frankland[11] first used laboratory tests (before 1870) in which appropriate proportions of sewage or effluent were stored in the presence of an excess of dissolved oxygen, and the oxygen demand was used as a measure of the content of polluting organic matter. Work by Gérardin, Dupré, Spitta, Adeney and others helped to develop the test, which was finally accepted as a standard technique in most countries. In his work for the Royal Commission on Sewage Disposal, Adeney distinguished two stages of oxidation of organic matter, the carbonaceous and the nitrogenous, and believed at first that the former must be completed before the second could proceed. It was recognized that the demand for oxygen arose from decomposition processes due to biochemical (biological) activity, but the details of these processes are not completely understood even now, and the many facets of the test are a continuing source of inspiration for research and investigation.[12]

Biochemical Oxygen Demand (B.O.D.) tests have come to be used for two distinct though related purposes. The first, as above, is to provide an indirect measure of the total amount of unstable organic (polluting) matter contained in a waste.[13] The second is to provide a basis for assessing the effect of the discharge of the waste on the oxygen balance of the natural waters receiving it. Two techniques are available, the *dilution method*[14, 17] and the *direct method*,[15, 21] the former being used much more. Details are given later (p. 39).

During studies of pollution and natural purification of the Ohio River waters, Streeter and Phelps[16] established a law covering the first, or carbonaceous, stage of oxidation of organic matter in oxygenated waters: 'The rate of biochemical oxidation of organic matter is proportional to the remaining concentration of unoxidized substance, the concentration being measured in terms of oxidizability.'

Theriault[17] expressed this law mathematically by the equation

$$+\frac{d}{dt}(L_t) = -KL_t$$

the integrated form of which is

$$\frac{L_t}{L_0} = e^{-Kt} = 10^{-kt} \qquad 4.1$$

where L_0 is the total first stage B.O.D. (B.O.D.$_{20}$) and L_t the residual B.O.D. after t days, whence $L_0 - L_t$ is the B.O.D. exerted during t days

$$(\text{B.O.D.})_t = L_0 - L_t = L_0(1 - 10^{-kt}) \qquad 4.2$$

$k = 0{\cdot}4343 \times K$ being the deoxygenation rate constant.

Phelps established a value for $k = 0{\cdot}1$ at 68°F (20°C), confirmed by Theriault.[17] This value for the deoxygenation rate constant corresponds to a daily oxidation rate of 20·6% of the total first-stage biochemical oxygen demand measured from the beginning of the day. However, most later studies did not confirm $k = 0{\cdot}1$ even as a good mean working value for sewage. For industrial

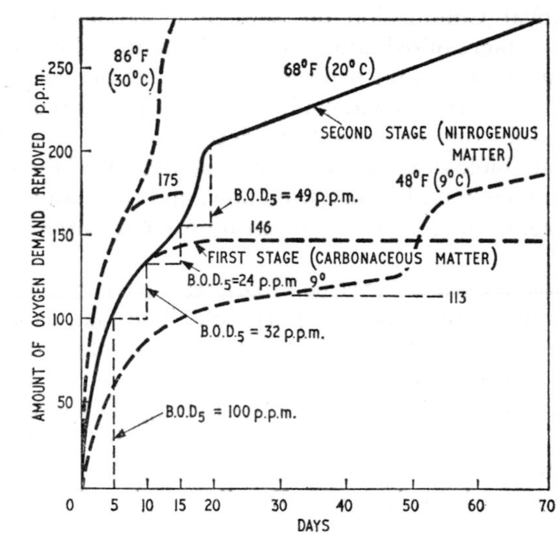

Fig. 4.2. *Rate of removal of oxygen demand from raw sewage, according to Theriault, taking the initial B.O.D.*$_5$ *at 100 p.p.m.*

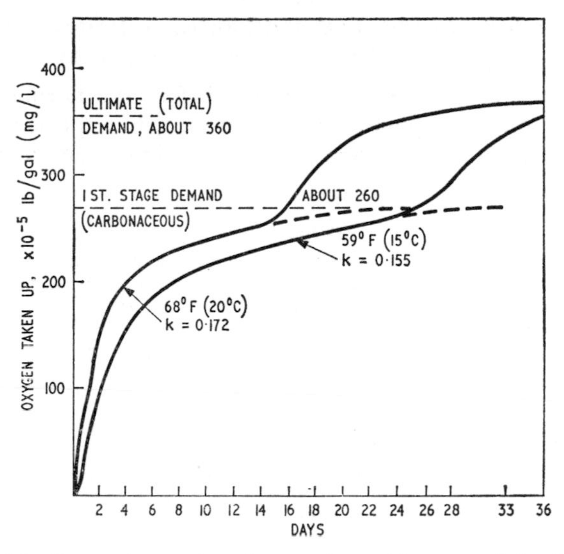

Fig. 4.3. *Progressive exertion of oxygen demand by sewage (after Gotaas)*

wastes, k values range from very low values around 0·01 to 0·30 or more. Gotaas[18] found $k_{68°F}$ for an ordinary sewage to be about 0·19. Also, whereas Theriault found the total first-stage B.O.D., L_0, to depend on temperature, Gotaas and others report that L_0 is not temperature-dependent but the same for all temperatures, as indeed would be expected. Streeter and Phelps[17] found the oxidation rate constant to vary with temperature according to the relationship

$$k_{T°C} = k_{20°C} \times 1{\cdot}047^{(T-20)}$$

Gotaas,[18] using a wider range of temperature, 40–85°F (5–30°C), observed that the following empirical equation fitted the data better

$$k_{T°C} = 0{\cdot}0115\ T^{0{\cdot}932}$$

(the corresponding $k_{20°C} = 0{\cdot}188$).

Figs 4.2 and 4.3 show plots of their data by Theriault and by Gotaas, and Tables 4.2a and 4.2b give the corresponding percentage oxidation rates from their findings in respect of oxidation rates at different temperatures.

These presentations of a rigid mathematical basis for the B.O.D. test are oversimplified. Many factors operate in controlling the rates of oxygen uptake, also in natural waters, and their interplay inevitably affects the chemical and biochemical rates of reaction.[12] Some of the consequences relevant for the laboratory test are:

1. There is commonly a 'lag' phase before the normal progress of the oxygen uptake sequence begins, but the lag varies in length from one or more hours (sometimes none at all) to as much as one or two days.

Table 4.2a

RELATIVE RATE OF CONSUMPTION OF OXYGEN IN B.O.D. TESTS (DILUTION METHOD), IN TERMS OF THE CORRESPONDING 5-DAY BIOCHEMICAL OXYGEN DEMAND (B.O.D.$_5$) AT 68°F (20°C)

Days	1	2	3	4	5	6	7	10	15	20
Relative oxygen consumption										
Theriault $k_{20} = 0{\cdot}1$	30	54	73	88	100	110	117	132	142	146
Gotaas $k_{20} = 0{\cdot}188$	51	83	106	120	129	135	139	144	145	146

Table 4.2b

DAILY OXIDATION RATE AS PERCENTAGE OF THE TOTAL BIOCHEMICAL OXYGEN DEMAND AT DIFFERENT TEMPERATURES (DILUTION METHOD)

Temperature		*Daily oxidation rate (% of total biochemical oxygen demand remaining at beginning of any one day)*	
°F	°C	Theriault	Gotaas
41	5	11	11
50	10	14	21
59	15	17	28
68	20	21	35
77	25	25	41
86	30	31	47
Values of k		$k_{T°C} = 0{\cdot}1 \times 1{\cdot}047^{(T°-20)}$	$k_{T°C} = 0{\cdot}0115 \times T^{0{\cdot}932}$

2. The rate of oxidation, first reaching a peak after the initial 'lag', then decreases progressively. The overall measured rates of oxygen demand represent the aggregate result of individual reaction rates which may vary widely among themselves and from time to time (differences in substances, in biological populations, etc.). In some cases, the overall rate may approach that of a first-order reaction, in others a second-order but in others neither model may be satisfactory.[12, 19–21]

3. The carbonaceous and nitrogenous oxidation stages do not necessarily proceed separately (whether in the laboratory or in the field) but may take place concurrently although not paralleled in respect of lag periods or relative oxidation rates.
4. When sewage samples are under test, there is usually an adequate supply of auxiliary and nutrient substances, including nitrogen and phosphorus. In testing industrial or other such water-borne wastes, however, special attention may have to be given to such factors, as well as to the possibilities of toxic effects.
5. The rates of oxidation are often quite sensitive to the ratio of waste to dilution water.

The *immediate or simple oxygen demand* may be important in particular for certain industrial wastes or mixed, including septic, sewages. An immediate or relatively rapid uptake of oxygen occurs in the presence of a variety of chemical substances, such as ferrous or manganous compounds, sulphides, some dyestuffs, etc. This may be measured by making dilutions as in the B.O.D. test, but using higher proportions of the waste. The depletion of oxygen is measured after a suitable period which may be as little as 5 or 10 min up to no more than, say, one hour.

Of the two techniques in use for the determination of the B.O.D., the *dilution method* is commonly preferred because it simulates the conditions applying where sewage is discharged into natural waters. In this procedure, a sample portion of the sewage, effluent or other aqueous waste to be tested is diluted with a suitable measured quantity of specially prepared air-saturated 'dilution water'. The resultant mixture is carefully dispensed into a number of airtight glass vessels, some of which are stored at 68°F (20°C) for the test period (usually five days). Two or more are set aside for a short period (see above), after which the concentration of dissolved oxygen is determined chemically. Determination of the residual dissolved oxygen after storage completes the test, the B.O.D. being calculated from the amount of oxygen depletion and the degree of dilution. Usually two or three different dilutions are necessary to ensure satisfactory results. About 100-fold dilutions are required with domestic sewage, about tenfold for trickling-filter effluents.

In the second technique, the *direct method*, a suitable quantity of waste is stored in direct contact with a definite volume of air or oxygen. Usually a special vessel is used (e.g. Sierp apparatus or Warburg respirometer) with provision for the absorption of the carbon dioxide formed, the oxygen consumption being measured manometrically. This procedure allows a progressive measurement of the B.O.D. at intervals during storage and has been receiving much more attention in recent years. It is the basis of most of the methods for automatic B.O.D. determination now in use or under development.

In the dilution method particularly, the aim is usually to measure only the first-stage (carbonaceous oxidation) demand. For this purpose, correction for nitrification may be necessary. For both techniques, it is usually assumed that sludge solids also are fully oxidized, as would be the case where sludge is kept in suspension by high velocities or turbulence in streams or shallow lakes. The sludge often contains complex nitrogenous matter, and the second, or nitrification, stage must also be taken into account in such cases. However, in treatment plants where sludge is first removed by sedimentation and nitrification is not established or barely initiated (as is the case with high-rate filters and many activated-sludge plants), the first-stage B.O.D. may be an adequate measure of overall purification and corresponding 'purification efficiencies'.

The *relative stability test* is used as an indication of the amount of available oxygen contained in an effluent or aqueous waste relative to the (first-stage) oxygen demand. 'Available oxygen' in this sense includes oxygen dissolved and present as nitrite or nitrate, all of which may be used up in satisfying the oxygen demand. The blue dye, methylene blue, is used as an indicator, changing to a colourless compound when the aqueous mixture is deoxygenated. The test is useful for routine checking especially of effluents from low-rate biological treatment processes.

The *putrescibility test* is another simple assay of the biological decomposition of sewages, effluents, etc. A sample portion is dispensed into a glass vessel in which it is stored undiluted and in the virtual absence of air, observations being made at intervals to determine after what period of storage (preferably at a given temperature such as 68°F (20°C)) a detectable amount of hydrogen sulphide is evolved. The test has no quantitative significance.

Chemical Oxygen Demand (C.O.D.) tests—Many water-borne wastes are not suited for B.O.D. tests in the same way as domestic or municipal sewages, perhaps because they cannot be utilized normally by bacteria until other nutrients are added, or because they contain high concentrations of growth-inhibiting substances and are not oxidized biologically unless they have first been diluted further than the B.O.D. test requirements permit. However, it is often possible to make indirect determinations of what their natural oxygen demand would be when diluted; e.g. when considering the proportional loading on a municipal sewage treatment works or their oxygen demand, say, when added to an already polluted body of water, and this can as a rule be correlated with C.O.D. tests. Another possible problem with the ordinary biochemical demand test, the relatively long period required before the result can be obtained, is avoided by using purely chemical 'oxidizability' tests, which have been in use for a century or more. These can provide quick and simple means for measuring the relative levels of the concentration of oxidizable organic matter in water-borne wastes.

Table 4.3

COMPARISON OF TESTS FOR ORGANIC MATTER

Results of analyses[22] of samples containing 1 000 p.p.m. of the substance tested

Substance tested	*Loss on ignition test,* p.p.m.	p.p.m. (*by weight*) *of oxygen consumed*		
		from permanganate	B.O.D.$_5$ *test*	*during complete combustion*
Lactic acid	1 000	260	540	1 067
Dextrose		600	580	1 067
Lactose		400	580	1 123
Dextrin		300	520	1 186
Starch		120	680	1 186
Phenol		2 400	1 700	2 383

For rough estimation of B.O.D., or vice versa, 1 lb (0·45 kg) of dry organic matter content of municipal sewage solids corresponds to about 1 lb (0·45 kg) of B.O.D.$_{20}$ or 0·7 lb (0·32 kg) of B.O.D.$_5$. River sludge deposits may have a B.O.D. of roughly 0·4 lb (0·18 kg) B.O.D.$_5$ per lb (0·45 kg) of volatile solids ('organic matter') content (see Table 6.4, p. 66).

Of these, the chemical tests most used are the *oxygen absorbed from permanganate* test, commonly referred to as oxygen absorbed or consumed, the dichromate oxidation test, usually called dichromate value, or, directly, the chemical oxygen demand. The test results may be calculated in the ordinary way in terms of the amounts of oxygen consumed in chemical oxidation of the organic matter, but the procedures of the permanganate tests used in England and Germany are different so that the various values cannot be readily compared. Whereas in English procedure standardized amounts of acid and potassium permanganate added to the sample are allowed to react for a period of 4 h at 80°F (27°C), in Germany the reaction is carried out at a temperature of 212°F (100°C) for 10 min. In the U.S.A. the dichromate reflux method is preferred.

Three different estimates of the amount of organic matter have been mentioned above, all involving combination of the organic matter with oxygen: loss on ignition (combustion), biochemical oxygen demand (biological oxidation) and the permanganate tests (chemical oxidation). Table 4.3 shows how different may be the relative amounts of these estimates for specific chemical compounds.[22]

In addition to these more common chemical oxidation tests, others have been used including *alkaline* permanganate, hypochlorite solutions and other oxidants such as ceric salts or periodate.

Other chemical tests

Organic carbon tests provide another means of determining organic matter, used also for assessing carbon : nitrogen ratios. Practically all the carbon in a waste or polluted water can be converted to carbon dioxide under conditions made strongly oxidizing with hot chromic acid and the carbon dioxide evolved is absorbed and measured (after purification to remove chlorine, sulphur dioxide, etc.).

The *nitrogen content* of sewage may be present in complex compounds such as proteins, in simpler ones like urea, and in inorganic compounds such as ammonium salts, nitrites and nitrates. Tests for nitrogen content, therefore, may include the determinations of organic nitrogen, albuminoid ammonia (from readily decomposed organic compounds), free ammonia or ammoniacal nitrogen and oxidized nitrogen which includes nitrite and nitrate nitrogen. When sewage is decomposed aerobically, organic and ammoniacal nitrogen is oxidized partly or wholly to nitrate nitrogen. Organic nitrogen (Kjeldahl nitrogen) is determined by digestion, with or without catalysts, with boiling sulphuric acid, and final measurement as ammonia. The *organic* nitrogen is the 'Kjeldahl nitrogen', obtained by subtracting the *ammoniacal nitrogen* from the *total Kjeldahl and ammoniacal nitrogen.*

The *chlorine demand* of a sewage is usually defined as that proportion of chlorine which must be added to it so that 0·3 mg/l of free residual (active) chlorine is present exactly 10 min after admixture. Fresh domestic sewage normally has a chlorine demand which may be taken as around 2–3 g *per capita* daily, corresponding to about 10–20 mg/l; that of septic sewage is higher.

The *oxidation–reduction (redox) potential*[23] of sewage or other wastes is sometimes measured. It is the potential of a bright platinum electrode immersed in the sewage relative to that of the normal hydrogen electrode and gives an indication of the 'state' of the sewage, dependent on the presence of oxidizing or reducing biological enzyme systems or of chemical substances such as sulphides, ferric salts, oxides of manganese, etc. Thus aqueous wastes containing sulphides might have a redox potential of 0 or -100 mV, whereas that of wastes containing free chlorine might be $+800$ mV or more. Continuous redox-potential measurements are being used in some activated-sludge plants for control of aeration.

The *grease content* is usually determined by solvent extraction, using either the whole waste or the concentrated sludge or scum formed by coagulation and precipitation or by heating. Grease measured in this way includes true oils and greases as well as fats and fatty acids and some alcohols and waxes but not normally volatile oils such as petrol, naphtha, etc. The extracts may be tested chemically for characterization (acid value, etc.). Proper sampling and test apportionment is often specially difficult.

The *volatile acid* content is another generalized test measuring formates, acetates, propionates, etc., used for control of sludge digestion. The 'acid' is separated by acidification and distillation and the equivalent acidity reported in terms of acetic acid.

Tests for *detergent concentration* have become important. The effective concentration is measured in terms of the capacity of a waste or treated effluent to capture dyestuff, compared with a standard dilution of a known detergent compound.

Tests for *biodegradability* may be applied to particular compounds or wastes, including detergent compounds, and are based on special laboratory cultures of activated sludges.

Radioactivity—Both in respect of waste disposal and water-pollution control, there is a growing need for tests for radioactivity, including total and β-activity and specific radioisotopes. Techniques have not yet been standardized but are well developed. Apparatus includes specialized equipment which has been assembled specifically for testing concentrations at the relatively low levels (from, say, 10^{-3} down to 10^{-9} μc/ml) appropriate for waste waters.

Specific chemical tests[24–26]—In addition, the examination of water-borne wastes requires other tests which are more specific in character or purpose, including reaction (pH) and the concentrations of particular elements or molecular groups, such as toxic metals, phosphorus or boron compounds, etc., chloride, sulphate, sulphide. Studies of receiving waters need to include the determination of dissolved ('free') oxygen concentrations. Gas analysis is required for proper control of sludge digestion.

Some of these tests should be carried out 'on the spot' or at least initiated at the time the sample is taken. Others may require lengthy preparation before the final measurement of concentrations can be made. A great variety of procedures and apparatus has been developed. Other special tests, arising from the growing problems of drinking-water pollution by, for example, insecticide residuals, are demanding more and more attention.

The *reaction* of aqueous wastes may be described as acid, neutral or alkaline, depending on the concentration of hydrogen ions. Pure water is neutral in reaction and contains 1 g-ion of hydrogen in 10^7 l, i.e. 10^{-7} g-ions/l. The pH scale of Sørenson is based on the hydrogen-ion concentration, neutrality corresponding to pH $= 7$, pH < 7 to acidic reactions and pH > 7 to alkaline. Colorimetric indicators are commonly used to determine the pH, with an accuracy of about 0·2 units. Electrometric apparatus is more useful and more accurate. For a rough qualitative indication, litmus paper is used, changing to pink if the pH is about 6 or less, and to blue if about 8 or more. Other 'universal' indicator papers can be used for approximate estimates of pH between the values of 1 and 12. In addition to the reaction, the *alkalinity* and *acidity* of wastes are measured by direct titration, with acid using different indicators to distinguish *total* and *caustic* alkalinity; with alkali, varying indicators again distinguishing *total* and *mineral* acidity.

Chlorinity, the concentration of the *chloride* ion (not the free chlorine or chloramine content used for chlorination control), is often important. That of water-borne wastes alone is sometimes high enough to limit the possibilities of water re-use, owing to problems of corrosion or taste.

Phosphate concentration may be important especially for studies of treatment and/or disposal possibilities.

BIOLOGICAL INVESTIGATIONS

As with chemical tests, *bacteriological* testing may be either general or specific in character. Because of the large numbers of bacteria occurring in wastes such as sewage or sullage, bacteriological tests provide specially delicate measures of the presence of pollution from such sources. General tests include nutrient agar plate counts, made at room temperature or at specific temperatures such as 95°F (35°C), or simple tests for the presence or absence of bacillus spores (e.g. *Clostridium welchii*) or for organisms of typical groups. Thus common tests are for actual numbers of individual organisms of the coliform group (lactose fermenters growing in the presence of bile salt with acid and gas formation) or enterococci, etc.

On the other hand, specific test procedures are available whereby it is feasible to grow and separate particular species of bacteria. Among the coliforms, for example, *Escherichia coli* (*E. coli*) types may be distinguished from *Aerobacter aerogenes*. *Streptococcus faecalis* numbers may be estimated separately and specifically as indicators of the degree of recent faecal pollution, or perhaps even bacteria of the typhoid or paratyphoid groups can be identified. These procedures may involve the use of special media, of biochemical reactions, specific serum or phage titrations, differential staining and microscopic examinations, etc., which require expert knowledge and practice.

For pollution control, nutrient agar total counts and coliform tests are usually sufficient, but additional tests for particular species or for particular serum and phage types can provide valuable supplementary data where required for distinguishing different sources of pollution or in the assessment of particular problems or dangers.

Counts of micro-organisms are made either directly by diluting the wastes until the numbers are low enough for separate colonies to be grown and counted, or indirectly by using multiple-tube dilution series. The colony counts employ gelatine or agar plates or the membrane filter technique.

Biological examinations are of considerable significance in the study of natural waters, because observations of plant and animal associations provide useful indications of the degree and nature of pollution, particularly if routine examinations are carried out.[27] Ecological studies of different waters by many workers have provided a sufficient basis for classification according to the degree of pollution. The saprobic system of Kolkwitz and Marsson, elaborated by Liebmann, classifies waters ecologically as follows:

	Organisms
1. Pure natural water	oligosaprobic
2. Slightly polluted natural waters	β-mesosaprobic
3. Highly polluted natural waters	α-mesosaprobic
4. Extremely polluted natural waters	polysaprobic

No single organisms or species but their associations or communities are found to indicate clearly the relative degree of pollution from place to place.

Liebmann has called more recently for a revision of the saprobic system groupings of organisms. For river waters he has suggested a system which includes seven Quality Classes ('Güteklassen'); (1) to

(4) generally as above, plus three intermediate classes 1–2, 2–3 and 3–4. Corresponding expansions of the saprobic system have been proposed, providing for subdivision of the polysaprobic group into β-polysaprobic organisms characteristic for waters rather more highly polluted than α-mesosaprobic, and α-polysaprobes, providing a much narrower biological spectrum, dominated by bacterial species which occur only in extremely polluted waters, very low in dissolved oxygen and with a high concentration of decomposable organic matter (B.O.D.$_5$, say, over 20 mg/l).

Some workers have proposed additional groupings such as hyperpolysaprobic, covering degrees of pollution approaching that of settled sewage; and katharobic, quite the opposite, providing for those organisms which are to be found only in unpolluted natural waters of the highest purity.

In the investigation of streams, lakes, reservoirs, etc., it is necessary not only to take samples of the waters for sieving, centrifuging and microscopic examination, but bottom deposits, plant growths, cavities under stones, attached animal forms, etc., must also be taken or examined to obtain broad estimates of the numbers of types, species and individuals. A series of such investigations reveals the general character of the aquatic environment and shows any permanent changes.

These broader classifications of the ecology in relation to natural or unnatural conditions may be supplemented by more specific characterizations. Liebmann, for example, has reported on the significance of pollution by intestinal parasites and on the effects of waste-treatment processes in the control of water pollution by the ova, larvae or adults of such species.[28]

Biological examinations require considerable specialized knowledge and experience to achieve valid assessments of the diversity of the organisms and ecological conditions which occur, and usually need to be combined with some physical and chemical investigations. Temperature and dissolved oxygen content are extremely significant in this respect.

Liebmann states that his seven 'Gewässergüte' classes are weighted equally according to chemical qualities on the one hand and biological composition on the other. For this, he allocates the waters to a grouping according to the 'saprobic spectrum' of the organisms observed, taking account of their numbers and frequency, and fits the waters into the appropriate place in the seven-class system on the further basis of (*a*) per cent dissolved oxygen saturation and (*b*) oxygen demand (two-day demand and B.O.D.$_5$).

The fish life also may provide a good index of the relative purity of particular waters and the 'sewage fungus', *Sphaerotilus natans*, a typical indicator of polluted streams, is commonly found to grow below discharges of incompletely purified wastes.[29]

BIBLIOGRAPHY

American Public Health Association *et al.* *Standard Methods for the Examination of Water and Wastewater* 12th Edn (1965) New York

Association of British Chemical Manufacturers and Society of Analytical Chemistry. *Recommended Methods of Analysis of Trade Effluents* (1958) Heffer, London

JENKINS, S. H., HARKNESS, N., HEWITT, P. J., SNADDON, M. X. V. *et al.* 'Some Analytical Methods used in the Examination of Sewage and Trade Wastes', *J. Proc. Inst. Sew. Purif.* (1965) Pt 6, 533

KLEIN, L. *River Pollution, I. Chemical Analysis* (1959) Butterworths, London

SAWYER, C. N. *Chemistry for Sanitary Engineers* (1960) McGraw-Hill, New York

TARZWELL, C. M. (Ed) *Biological Problems in Water Pollution* (1957) R.A. Taft Centre, U.S. Dept. Hlth Educ. and Welfare

REFERENCES

1. THISTLETHWAYTE, D. K. B. 'Aspects of Sampling and Testing for Water Quality and Pollution Control', *Proc. 3rd Fed. Conv. Aust. Wat. Wastewater Assoc. Brisbane* (1968) 107
2. ROSKOPF, R. F. 'A Composite-Grab of Water Pollution Control Sampling', *J. Wat. Pollut. Control Fed.* **40** (1968) 492
3. PRICE, D. H. A. 'Some Comments on Sewage Works Records', *J. Proc. Inst. Sew. Purif.* (1943) 108
4. LULEY, H. G. *et al.* 'Industrial Wastes Automatic Sampling', *J. Wat. Pollut. Control Fed.* **37** (1965) 508
5. WOOD, L. B. and STANBRIDGE, H. H. 'Automatic Samplers', *Wat. Pollut. Control* **67** (1968) 495
6. Water Pollution Research Laboratory, Watford. 'Some notes on the determination of dissolved oxygen', *J. Proc. Inst. Sew. Purif.* (1953) Pt 1, 15
7. *Notes Wat. Pollut.* No. 26 (1964); cf. *J. Proc. Inst. Sew. Purif.* (1965) Pt 2, 183
8. GOBE, J. H. and PHILLIPS, T. 'Dissolved-oxygen electrode', *Biotechnol. & Bioengng* **6** (1964) 491
9. TRUESDALE, G. A. and DOWNING, A. L. 'The solubility of oxygen in water', *Nature, Lond.* **173** (1954) 1236
10. INGOLS, R. S., HILDEBRAND, J. C. and RIDENOUR, G. M. 'Measuring the Strength of Sewages and Trade Wastes—B.O.D. or O.C.', *Wat. Sewage Wks* **97** (1950) 21

11. Frankland, Sir Edward. *Report of the Rivers Pollution Commission of Great Britain* (1870)
12. Busch, A. W. 'B.O.D. Progression in Soluble Substrates', *Proc. 13th Ind. Wastes Conf., Purdue Univ.* (1958) 54
13. Buswell, A. M., Mueller, H. F. and van Motor, I. 'The B.O.D. Test and Total Load', *Sewage ind. Wastes* **26** (1954) 276; **27** (1955) 1297
14. Eye, J. D. and Ritchie, C. C. 'Measuring B.O.D. with a membrane electrode system', *J. Wat. Pollut. Control Fed.* **38** (1966) 1430
15. Montgomery, H. A. C. 'Determination of Biochemical Oxygen Demand by Respirometric Methods', *Wat. Res.* **1** (1967) 631
16. Streeter, H. W. and Phelps, E. B. 'A study of the pollution and natural purification of the Ohio River. III. Factors concerned in the phenomena of oxidation and re-aeration', *U.S. Publ. Hlth Serv. Bull.* No. 146 (1925)
17. Theriault, E. J. 'The rate of deoxygenation of polluted waters', *Trans. Am. Soc. civ. Engrs* **89** (1926) 1341; cf. *U.S. Publ. Hlth Serv. Bull.* No. 173 (1927)
18. Gotaas, H. B. 'Effect of temperature on the biochemical oxidation of sewage', *Sewage Wks J.* **20** (1948) 441; also 'The Effect of Seawater' **21** (1949) 818
19. Stones, T. 'A study of the Kinetics of the Biochemical Oxidation of Settled Domestic Sewage', *J. Proc. Inst. Sew. Purif.* (1963) Pt 3, 285
20. Young, J. C. and Clark, J. W. 'Second-order equation for B.O.D.', *Proc. Am. Soc. civ Engrs* **91** SA(1) (1965) 43, SA(3) (1965) 136
21. Simpson, J. R. 'A Second-order Equation in Biochemical Oxygen Demand Reactions', *Wat. Pollut. Control* **67** (1968) 433
22. Schulze-Forster, A. 'Der biochemische Sauerstoffbedarf (BSB) und der Kaliumpermanganatverbrauch organischer Verbindungen', *Kleine Mitt. Ver. Wass.-Boden- u. Lufthyg.* **14** (1938) 214
23. Stumm, W. 'Redox Potential as an environmental parameter: conceptual significance and operational limitation', *Proc. 3rd Int. Conf. Wat. Pollut. Res., Munich* **3** (1966) 283
24. Christie, A. A., Kerr, J. R. W., Knowles, G. and Lowden, G. F. 'Methods of determining small amounts of metals in sewage and trade waste waters', *Analyst, Lond.* **82** (1957) 336
25. Kahn, L. and Wayman, C. 'Amino acids in Raw Sewage and Sewage Effluents', *J. Wat. Pollut. Control Fed.* **36** (1964) 1368
26. Stanwick, J. D. and Davidson, M. F. 'Determination of specific resistance to filtration', *Wat. Waste Treat. J.* **8** (1961) 386; cf. *Wat. Pollut. Control* **66** (1967) Pt 6, 622
27. Hawkes, H. A. 'Biological detection and assessment of river pollution', *Effl. & Wat. Treat. Jnl* **3** (1963) 651
28. Liebmann, H. 'Parasites in Sewage and the Possibilities of their Extinction', *Proc. 2nd Int. Conf. Wat. Pollut. Res., Tokyo* **2** (1964) 269
29. Baalsrud, K. 'Polluting Material and Polluting Effect', *Wat. Pollut. Control* **66** (1967) 97

PART B

METHODS OF DISPOSAL

5

GENERAL CONSIDERATIONS

SEWAGE and other water-borne wastes can be disposed of readily and without nuisance if the volume of waste water is small and relatively large volumes of receiving waters or areas of land are available for direct disposal. The demands of proper sanitation are then limited to local problems of hygiene, the control of bacterial and other infectious diseases involving *inter alia* that of dipterous insects, and of odours and aesthetic nuisance, which can be achieved by proper house drainage and sewerage.

The denser the population of an area and the higher the standard of living and the development of industry, the more important is the proper disposal of its wastes. The public nuisance which may otherwise develop may interfere with or cause damage to industries, water supply and conservation, waterways, harbours and beaches, to residential and recreational areas, to agriculture, pastures, fisheries and so on. Sewerage and drainage of part of an area, though satisfactory from the local viewpoint, may result in unsatisfactory conditions elsewhere, and in general the sanitation of a country can only be regarded as soundly and safely established if all units of all communities in a river-catchment area are sewered and drained to the same standard. The problem of disposal of sewage and other wastes from any one community may, therefore, involve technical disposal problems of much wider extent.

Administratively, it is always desirable that standards of sanitation be established by law, by Acts of government and by regulations which both local authorities and other bodies must adopt. In densely populated countries such regulation is essential. Where disposal problems may tend to clash with the intensive use of water and watercourses for other purposes, the establishment of special overriding authorities such as river boards has proved necessary to ensure a proper overall review of all associated problems. Some such boards, including the Thames Conservancy Board (1857), the Birmingham Tame and Rea District Drainage Board (1877) and the Emschergenossenschaft (1904), have been established for many years in England, Germany, the U.S.A. and other countries,[1, 2] where the problems of water supply, sewerage, drainage and general river usage had become critical. The need for similar organizations is now more widely appreciated and new administrative action is being taken to re-establish the responsibilities and powers of the controlling authorities, and to set up new authorities.[3–5] In the U.K., for example, the Water Resources Act of 1963 consolidates and extends the previous relevant Acts, such as of 1948 and 1961, which constituted the various river board authorities in particular to administer water-pollution control throughout the country. The new Act is aimed at a wider control, of *all* aspects of water conservation, as to quantity, maintenance of an appropriate quality, proper sanitation and drainage, related to water usage and consumption, including waste disposal, and sets out a new and fuller range of duties for the River Boards, prescribing their relationship and responsibilities to a Water Resources Board and to the Ministry of Housing and Local Government. In the Federal Republic of Germany, similar action was taken in the Federal Water Management Law (Wasserhaushaltsgesetz) of 1957; within its general framework the individual State water laws were established. On this basis, the existing State water authorities had to develop in particular river systems appropriate water-management planning schemes covering all kinds of water uses and including water-pollution control. The standard of sanitation achieved by such measures depends not only on direct questions of health and aesthetics but must take into account matters of finance and economy, not only of the water but also of fertilizing constituents and useful chemicals available from the wastes. The disposal of sewage and other such wastes will result in different cycles of use and re-use of the various constituents.[6] The waste of one area becomes wholly or in part reclaimed afresh by its neighbour. Generally, there has developed a tendency to re-use waste material over and over again, as far as may be economical and practicable, but here designers must not lose sight of the principles of proper sanitation. The ideal design and construction should provide for the present and yet permit future development.

Water-borne wastes may be considered to be composed of:

1. The water content which belongs to the natural cycle of water (see p. 55);
2. Utilizable constituents, which may be reclaimed for repeated use (see p. 109);
3. Non-utilizable or useless constituents, including harmful (toxic) or nuisance-causing matter.

Some thought should always be given to the ultimate fate of the various components.

The following general methods of disposal of sewage and other water-borne wastes may be applicable and should be considered carefully in the first stage of planning of drainage and sewerage systems prior to the formulation of preliminary lay-outs.

Disposal by dispersal

Disposal by dispersal into natural waters, or onto the land, was adopted in past ages as a natural thing to do with all waste or unwanted materials. However, the natural processes which result can lead to the many nuisances mentioned. The concentration of population in large communities has produced intolerable conditions in natural environments, for example, early last century the drainage of London fouled the waters and banks of the River Thames. This along with similar circumstances in many other parts of the world, eventually forced the development of modern sewerage technology. Many of the complexities of these apparently simple natural processes are now well understood, so that scientific design of such disposal systems is now practicable. The techniques of disposal into water or onto land differ considerably and are considered in detail in Chapters 6 and 7.

Disposal by reclamation

Either the whole waste may be re-used or it may be partly reclaimed, as when the water is used again or applied to some other purpose[7] or when some constituent such as phenol or lanoline is recovered. It is seldom, probably never, possible to reclaim or re-use all of a waste. Practical techniques are discussed in Chapters 7 and 8.

Disposal with solid wastes

When disposal along with solid wastes, such as garbage, is employed, the aqueous wastes lose their liquid character.

Disposal by special techniques

An example of these is disposal by incineration including evaporation. Some special applications are dealt with in later chapters, for example incineration applied to screenings or sludge.

Such techniques may also be utilized for other purposes, e.g. by raising steam during the combustion of otherwise intractable water-borne organic wastes.

Some utilization (partial reclamation) may also result from fishing industries, etc., following disposal into rivers or into the ocean. On the other hand, waste constituents which cannot serve even as land-fill, including certain types of inorganic and organic matter and some radioactive wastes must be dispersed finally in the oceans or buried deep underground to obviate undesirable accumulations or concentrations at their source.

BIBLIOGRAPHY

WISDOM, A. S. *The Law of the Pollution of Waters*, 2 Edn (1966) Shaw, London

'Aspects of Water Pollution Control' Conf. on Wat. Pollut. Problems in Europe, 1961, *Publ. Hlth Pap. W.H.O.* **13** (1962) Geneva

Waste Management and Control Report to Federal Council for Science and Technology by Committee Nat. Acad. Science and Technology and Nat. Res. Council. *Publication No. 1400, NAS–NRC*, Washington D.C.

Water Pollution Control (1966) Report on of W.H.O. Expert Committee (W.H.O. tech. Rep. Ser. 318)

REFERENCES

1. *Royal Commission on Sewage Disposal*, Reports 1–9 with appendices, and Final Report (1915) H.M.S.O., London
2. *Ohio River Committee*, Report upon Survey of The Ohio River and its Tributaries for Pollution Control, Pt I (U.S. War Dept.) and Pt II with supplement (U.S. Public Health Service) (1944) Washington
3. LESTER, W. F. 'The Water Resources Act 1963 and River Water Quality Management', *J. Proc. Inst. Sew. Purif.* (1966) Pt 2, 103
4. BERRY, A. E. 'The Ontario Plan in the Administration of Water Resources', *J. New Engl. Wat. Wks Ass.* **75** (1961) 16
5. CANHAM, R. A. 'Status of Federal Water Pollution Control Legislation', *J. Wat. Pollut. Control Fed.* **38** (1966) 1
6. MÜLLER, W. J. 'Landwirtschaftlische Abwasserverwertung und Abwasserbehandlung', *Wass. Boden* **5** (1953) 5
7. STANBRIDGE, H. H. 'From Pollution Prevention to Effluent Re-use', *J. Proc. Inst. Sew. Purif.* (1965) Pt 1, 20

6

DISPOSAL INTO NATURAL WATERS

NATURAL waters may be taken broadly to include those of creeks, rivulets, rivers, lakes, estuaries and oceans, also groundwater streams and other underground bodies of water. All these have some interconnection, at least within particular areas.

Attention is drawn in the first place to the inevitable existence of a water cycle.

THE WATER CYCLE

The organized use of water naturally depends upon a water cycle which involves towns or municipalities, by reason of their water supply and sewage disposal. The water-sewage cycle is practically always, at least in some degree, to be found associated with water supply. The presence in drinking water of some part of the sewage from upstream is unavoidable in a community or factory using river water. It is a fallacy to imagine that uncontaminated water can be obtained anywhere and that sewage can be so disposed of that its water content and some impurities would never again be used by man. On the contrary, all water, even from springs and other groundwater sources, has derived from surface water at some time past and has been more or less polluted, and sewage and sullage water must finally in practically every case reach a watercourse or body of groundwater and may be used again, even for drinking. Thus the cycle is, with few exceptions, inevitable. The important conclusion is that the path of the cycle should be as long as practicable and short-path connections avoided and guarded against.

Observations of such a cycle were first reported by Imhoff,[1] who discovered that the greater part of the water of the River Ruhr in Germany, the source of drinking and domestic water supplies for the district, completed three full cycles of usage under the dry-summer conditions experienced in 1929, and showed how harm or nuisance were avoided. Since that time the re-use of water in similar cycles has been noted more and more.[2–5] It has also been reported that during the dry summer of 1948 the waters of the River Verdigris in Kansas, U.S.A., were completing 17 full water–sewage–water cycles.[4] In Chanute (Kansas), however, an eightfold cycle of usage resulted in poor-quality drinking water,[5] and similar observations were made in the Ruhr River area in Germany during the very dry summer of 1959. A different cycle of re-use is exemplified by the sale of treatment works effluent to a steelworks[6] at Baltimore (U.S.A.) and another by artificial groundwater recharging for agricultural usage,[7] using highly purified sewage from the Whittier Narrows Water Reclamation Plant, Los Angeles County, California.

During each cycle of use some of the water is lost. According to Schröder,[8] average losses are approximately as shown in Table 6.1.

Table 6.1

CYCLICAL LOSSES OF WATER, %

Cooling water	Very little
Sewerage	20
Rural water supply without sewerage	70
Sewage disposal by	
broad irrigation	40
spray irrigation	100

Problems of waste disposal are simplified where relatively unpolluted river catchment areas are available for water supply and another river, the waters of which are used for less critical purposes or are of such volume that the discharges are relatively very small, is available to take waste discharges.

When a river must be used for both purposes—water supply and waste disposal—the required degree of sewage purification will be determined by considerations of water-supply quality downstream. Such cases occur along industrial river areas, and it often happens that the total water consumption exceeds the dry-weather flow in the river. This means, of course, that the demand cannot be met without re-use of the water in a water–sewage–water cycle, using 'sewage' broadly to cover any water-borne waste from town or industry. Each user of the river water should then, as a matter of duty, provide adequate treatment to ensure a proper degree of purification of his waste waters before discharge into the river. No objection to the re-use of the water in this way, even for town water supply, can be sustained if suitable and adequate artificial and natural treatment processes, especially the filtration of water through natural soils, are properly applied and used. In more densely populated and highly industrialized countries the prevention of river pollution by adequate sewage treatment is the best insurance for a safe and cheap water supply to people and industries.

Water supply by re-usage cycles is also involved in the many cases of 'artificial' groundwater supplies described in the literature.[9]

Certain salts and other end products of decomposition are added to the water during each cycle of usage, and their relative concentration becomes progressively greater after each cycle. Thus the salinity of the water is increased; it may become 'aggressive' (corrosive) owing to a relative increase in carbon dioxide or even oxygen (p. 242) and may develop excessive algal growths because of the build-up of nitrates and other plant nutrients (pp. 189, 242). These disadvantages are inherent and may impose limits to the number of re-use cycles, depending upon the actual uses to which the water is applied. Except in isolated areas of naturally high salinity, permissible salinities for drinking waters normally limit re-use in the water–sewage cycle to perhaps five or six cycles.

Some consideration must, therefore, also be given to such special qualities of recycled waters as may result from the accumulation of particular constituents such as carbon dioxide, nitrogen compounds, chlorides, sulphates, phenols and other substances derived from industrial wastes.

SELF-PURIFICATION OF NATURAL WATERS

Left to itself, any polluted water is subject to various processes of decomposition by natural biological agencies, and some degree of purification results, especially from the breaking down and partial reconstitution of organic matter: this is called self-purification. Many different associations of organisms, including both plant and animal forms but especially bacteria, fungi and algae, and protozoa (with more highly organized forms such as insects, worms and even fish), may play their part in these processes. It is important to distinguish between *anaerobiosis*, namely biological decomposition in the absence of atmospheric (free, dissolved) oxygen and *aerobiosis*, in the presence of air or oxygen. Some organisms are strictly anaerobes, i.e. cannot grow normally if any free oxygen is present; others are facultative, adaptable to the presence or absence of free oxygen, but most are aerobes, growing normally only if free oxygen is continually present.

Anaerobic or *septic* decomposition is not suitable for the self-purification of polluted river water because normal plant and animal life, such as fish, cannot survive anaerobic conditions, also because odour nuisance results and colour and general appearance are aesthetically repugnant.

However, anaerobic conditions normally prevail in the mud, silt and organic sludge deposits on the beds of lakes and sluggish and medium-velocity streams and estuaries, and in the muds of river flats.

Aerobic decomposition usually proceeds without nuisance, provided that sufficient dissolved oxygen is held in solution in the water and replenished as fast as it is consumed. If oxygen is always available, then purification will be accomplished aerobically as a consequence of the natural activity of many species of plants and animals, including especially bacteria. The rate of self-purification depends on the overall levels of biological activity maintained from place to place as purification proceeds and, in turn, on the number of organisms which is generally governed by the amount and concentration of appropriate nutrients and food available to them.

In a pure river, the numbers of organisms are relatively low, but after the introduction of nutrients and food in the form of organic wastes, their concentration increases, usually relatively rapidly and progressively until a maximum level of biological activity (and self-purification) is reached. Thus a polluted river or lake has a higher capacity of self-purification and is able to deal with further pollution —and likewise to withstand harmful substances such as acids, toxic substances and mineral oils—more quickly and easily than less polluted rivers with low and more sensitive levels of biological activity.

The following discussion of the self-purification of natural waters refers to the maintenance of aerobiosis at least in the free water zones. The basic consideration in such a discussion is the oxygen economy of the water in which the oxygen consumption ('expenditure') must be controlled to balance the oxygen intake ('receipts'). If oxygen is always available, then nature itself will accomplish the purification aerobically.

SUPPLY AND INTAKE OF OXYGEN

Natural waters anywhere may contain dissolved oxygen derived from any of three sources: (1) already present in solution, including nitrites and nitrates; (2) entering solution through the surface from the atmosphere–water interface; (3) released into solution, under high partial pressure, by the photosynthesis of green aquatic plants.[10, 11]

Generally the oxygen in nitrites and nitrates becomes available only when the dissolved oxygen is nearly exhausted and should not be taken into account for practical cases, since exhaustion of the oxygen is not compatible with fish life. The amount of oxygen which can dissolve from the atmosphere depends on temperature and partial pressure, so that the saturation concentration decreases as the temperature increases and the partial pressure decreases (see pp. 32–33). Thus streams can dissolve more oxygen in winter than in summer because water temperatures are lower, and those at sea level more than mountain streams because the atmospheric pressure is greater. Sometimes natural waters may be found to contain more than the corresponding saturation concentration, that is, they are supersaturated with respect to the atmospheric oxygen pressure, this being due to the liberation of oxygen by the photosynthetic activity of plants under or within the body of water. Since the effective partial pressure of the pure oxygen (produced some feet below surface in strong sunlight) may exceed the total atmospheric pressure, oxygen concentrations up to five times the normal saturation limits of Table 4.1 are possible. Field studies have shown that the contributions of dissolved oxygen by photosynthesis are highly variable, as would be expected, and that under normal conditions the total net contribution resulting from algal activity, for example, is not large, even where such concentrations are undesirably high. Therefore, although the contribution from photosynthesis alone, especially without taking plant respiration into account, may be substantial and should not be overlooked in particular for quietly flowing or stationary bodies of water, it is commonly unsafe to take photosynthesis into account in calculating the oxygen balance.

The rate of oxygen uptake is usually calculated on the basis of stream flow. The rate of solution from the atmosphere varies in proportion to the difference (oxygen deficit) between the actual dissolved oxygen content and that at saturation corresponding to prevailing conditions of temperature and pressure. Apart from this important potential effect, the rate of oxygen uptake also greatly depends upon mechanical factors,[12–14] especially relative velocities of air and water, including turbulence effects and wave formation, which alter the rate by increasing the effective contact surface of air and water; in rough seas and in surf the effective area is very greatly increased.

The problem has been simplified by Fair[15] who considered natural surface waters under a classification into six broad types, including ponds, lakes and streams, and calculated values of oxygen uptake from the atmosphere, for a temperature of 68°F (20°C), in terms of the oxygen deficit (Table 6.2).

Column (*a*) applies only to the limiting case of conservation of all oxygen taken up, none being consumed or lost in any way, the content of dissolved oxygen steadily approaching saturation. Such a case might be that of a stream generated from an underground source, the water being highly purified (without residual oxygen demand) but originally devoid of free dissolved oxygen.

Column (*b*) applies to the case of a polluted stream the dissolved oxygen content of which has fallen to the level at which the rate of its removal (absorbed by the B.O.D. of the water) is just equal to that of reoxygenation. The corresponding level for any class of water depends, of course, on the relative concentration of polluting matter. This is then the general case. Should the pollution be so

Table 6.2

RATE OF OXYGEN UPTAKE BY NATURAL WATERS OF VARIOUS CLASSES AT 68°F (20°C) (AFTER FAIR)

	Daily uptake of oxygen as percentage of oxygen deficit		
	Effective oxygen consumption of water		*Ratio b : a*
	equal to zero (*a*)	*exactly equal to rate of uptake* (*b*)	
1. Small ponds and backwaters, 3 ft (1 m) deep	10·9–20·6	11·5–23·0	1·08
2. Large lakes and impounding reservoirs	20·6–29·2	23·0–34·5	1·15
3. Large streams at low velocity, 6 ft (2 m) deep	29·2–36·9	34·5–46·0	1·22
4. Large streams of average velocity, 6 ft (2 m) deep	36·9–49·9	46·0–69·0	1·31
5. Swift-flowing streams, 6 ft (2 m) deep	49·9–68·4	69·0–115	1·55
6. Rapids	>68·4	>115	>1·55

great as to result in complete deoxygenation, an entirely new regime of decomposition may ensue; the rate of oxygen uptake is then governed by the same factors, but that of oxygen demand will be different and may vary widely.

By way of example, take a stream of 1 000 ft³/s (28 m³/s), found to contain 2 p.p.m. of oxygen, with water temperature averaging 68°F (20°C), over a range of 50 miles (80 km). How much oxygen does it absorb each day on this basis? Saturation at 68°F (20°C), near sea level, is 9·2 p.p.m.; the deficit is then 7·2 p.p.m., and taking the stream at, say, 2 miles/h (1 m/s), Group 4 in Table 6.2 under (*b*) gives us 46–69, say 57%, or an uptake of $7{\cdot}2 \times 0{\cdot}57$ p.p.m. daily $= 4{\cdot}1$ p.p.m. or $7{\cdot}2 \times 0{\cdot}57 \times 1\,000 \times 62{\cdot}4 \times 10^{-6}$ lb/s of oxygen corresponding to $7{\cdot}2 \times 0{\cdot}57 \times 62{\cdot}4 \times 60 \times 1\,440 \times 10^{-3}$ lb $= 22\,000$ lb of oxygen daily. (An uptake of 4·1 p.p.m. $= 4{\cdot}1 \times 10^{-6} \times 28 \times 10^{3}$ kg/s $= 4{\cdot}1 \times 10^{-1} \times 28 \times 864$ kg daily $= 9\,900$ kg daily.)

Such calculations take no account of surface area which is an important factor; Table 6.2 may agree satisfactorily with observations in the average case, but surface area must be taken into account for shallow or very deep waters. The rate of absorption depends

Table 6.3

VARIATION OF RATE OF OXYGEN UPTAKE BY NATURAL WATERS ACCORDING TO OXYGEN DEFICIT

(pressure 29·9 in (760 mm); temperature 68°F (20°C)

P.p.m. dissolved oxygen	9·2	7·4	5·5	3·7	1·8	0
Percentage oxygen deficit	0	20	40	60	80	100
Percentage saturation	100	80	60	40	20	0
Class of water body	lb of oxygen taken up/1 000 ft² d (kg/100 m² d)					
1. Small ponds and backwaters	0 (0)	0·06 (0·03)	0·12 (0·06)	0·18 (0·09)	0·25 (0·12)	0·31 (0·15)
2. Large lakes and impounding reservoirs	0 (0)	0·20 (0·10)	0·39 (0·19)	0·59 (0·29)	0·78 (0·38)	0·98 (0·47)
3. Large streams of low velocity	0 (0)	0·27 (0·13)	0·55 (0·27)	0·82 (0·40)	1·10 (0·53)	1·37 (0·66)
4. Large streams of average velocity	0 (0)	0·39 (0·19)	0·78 (0·38)	1·19 (0·57)	1·56 (0·75)	1·96 (0·95)
5. Swiftly flowing streams	0 (0)	0·63 (0·30)	1·25 (0·60)	1·90 (0·92)	2·54 (1·23)	3·17 (1·53)
6. Rapids	0 (0)	1·96 (0.95)	3·93 (1·90)	5·85 (2·83)	7·85 (3·80)	9·82 (4·75)

mainly upon the oxygen deficit in the surface layer of the water and, as before, on the relative velocity of air and water. Average values are given in Table 6.3, also derived from Fair.

Simple calculations based on broad classifications and generalized data such as these may in many cases be sufficient to give a good general idea of what is to be expected and are always valuable for preliminary investigations. However, it is often desirable or necessary to make more precise assessments of reaeration rates.

The rate of solution of atmospheric oxygen in rivers and other moving waters can be expressed in terms of an exponential rate constant. This is because the rate at which the oxygen deficit is decreasing (or, oxygen concentration increasing) at any time, due to reaeration alone, is proportional to the oxygen deficit at that time. Thus

$$-\mathrm{d}D/\mathrm{d}t = KD$$

or

$$D_{t_2} = D_{t^1} . \mathrm{e}^{-K(t_2 - t_1)} \qquad 6.1$$

where D_{t_1} and D_{t_2} are the oxygen deficits at time t_1 and t_2 and K is the rate constant, usually now referred to as the overall absorption coefficient. Alternatively

$$D_{t_2} = D_{t_1} . 10^{-k_2(t_2 - t_1)}$$

where k_2 is the rate constant to base 10.

Many attempts have been made to develop theoretical or empirical relationships of at least the most important factors which determine the value of the rate constants K and k_2 for different natural environments. The staff of the Water Pollution Research Laboratory at Stevenage, England, have conducted studies in which rates of oxygen uptake measured in laboratory tests were correlated with field observations, assessing in this way the overall effects of various impurities such as detergents on actual rates of uptake in various natural streams and other water bodies.[12] Churchill, Elmore and Buckingham[16] of the Tennessee Valley Authority in the U.S.A. used naturally deoxygenated waters discharged through a power station from the hypolimnion of a reservoir to derive empirical relationships between reaeration rates and various parameters of river geometry and hydraulics.

Based on their own mathematical and practical studies and those of many other workers to whom they refer, O'Connor and Dobbins, and later Dobbins developed mathematical models which appear to be more widely applicable and reliable. These and other treatments of such problems now permit predictions of reaeration rates which, at least for a number of cases, have been observed to approach closely to what actually does occur.[17] Owens, Edwards and Gibbs[12] of the Water Pollution Research Laboratory correlated their own observations with the Tennessee Valley observations by the equation

$$k_2^{20°\mathrm{C}} = 9{\cdot}41\ U^{0{\cdot}67}\ H^{-1{\cdot}85}$$

where
k_2 = reaeration coefficient, unit per day
U = mean stream velocity, ft s^{-1}
H = mean depth of stream, ft

OXYGEN CONSUMPTION WITHIN NATURAL WATERS

Generally, as noted above, the amount of self-purification by aerobic decomposition of organic polluting matters is related quantitatively to that of oxygen consumed by the metabolic activity of animals and plants, more especially bacteria. It was from their observations in the laboratory of the rate of oxygen consumption within the waters of the Ohio River system that Streeter and Phelps, Theriault[18] and others developed the concept of the natural biochemical oxygen consumption rates as a simple first-order reaction, following the law referred to (pp. 36–38). This simplified concept has proved a valuable tool for water-pollution studies, although it has been recognized that the actual rates of oxidation, i.e. of consumption of dissolved oxygen, are often very different from those indicated by Streeter and Phelps.

The idea of a single-order reaction (with $k = 0{\cdot}1$) applied to the original averaged data of the Ohio River studies corresponds to a daily consumption of dissolved oxygen in the river equal to 20·6% of the B.O.D., that is, the total degradable organic matter content of the river water at the beginning of the day. Thus, if the water contained organic matter equal to 100 lb (45 kg) of oxygen demand at time zero, it would have 79·4 lb (35·7 kg) remaining at the end of the first day, $79{\cdot}4 \times (79{\cdot}4/100) = 60{\cdot}3$ lb (27.1 kg) at the end of the second day, and so on, and so would consume progressively $100 \times 0{\cdot}206 = 20{\cdot}6$; $79{\cdot}4 \times 0{\cdot}206 = 16{\cdot}3$; $60{\cdot}3 \times 0{\cdot}206 = 12{\cdot}4$ lb (9·3, 7·3, 5·6 kg respectively) etc., of dissolved oxygen in successive days.

However, observations of polluted waters indicate that the actual rates at which dissolved oxygen is used up may be very different. It has been shown[19, 20] that at least in relatively polluted waters the oxidation rate, k, is usually very much higher than 0·1, e.g. 0·2–0·3, sometimes even more. Therefore, instead of 20·6%, daily rates may under natural conditions reach 30 ($k = 0{\cdot}15$) or 50% ($k = 0{\cdot}3$). Lower rates of oxidation are applicable particularly for large streams of relatively low velocity maintaining a low degree of biological activity and carrying smaller concentrations of organic pollution. Small or swift streams or shallow ponds, where relatively more surface is available on which an active plant and animal life may develop, or where concentrations of pollutants are greater, are usually characterized by higher oxidation rates. Even under the most favourable circumstances, however, the natural rates effective in surface waters which are highly diluted compared with settled sewage or other such wastes, usually are much less than is achieved in artificial works such as percolating-filter beds (p. 196) or activated-sludge tanks (p. 217). The amount of oxidation which occurs during 5–10 days' travel in a large stream is accomplished in about 5 h in the aeration tank of an activated-sludge plant or after about 1 h contact with the zoogloeal slimes of a trickling filter.

In highly saline waters, oxidation rates tend to be lower, down to about one-half those in fresh waters, but again depend on the level of biological activity; in highly polluted saline waters, they also may reach relatively high levels. Temperature affects natural rates of decomposition, which may increase by 4% per °C rise in temperature. However, adjustments to temperature are not always direct or immediate, and the effective increase in oxidation rates is seldom more than 3%. There is also a slight increase in rates of uptake as the temperature increases, and it is probably safe to ignore temperature changes, unless they are large or need special consideration, as where hot waters are discharged.

It is also important to take account of the nature and composition of the polluting matter in the particular case under review. The rates of oxidation referred to above are applicable, for example, to heterogeneous types of wastes such as municipal sewages. If the wastes in question were more complex or otherwise less susceptible to microbial degradation, perhaps requiring also the development of more highly specialized groups of organisms, the rates might be much less or, in fact, the wastes might not be degradable to any significant degree until supplied with special nutrients or mixed with other types of wastes. With such a change of circumstances, they might thereupon suddenly develop a high proportional oxygen demand. Another attribute of wastes which must be taken into account is the degree of dispersion of the polluting solids in the aqueous phase. They may be completely soluble or may be insoluble (suspended) but comprised of very small particles not readily settleable; or may contain larger particles of such density as to be readily settleable.

In this connection it is to be noted that pollution control authorities in some countries are becoming more and more concerned with the direct and indirect pollution of waterways arising from earth-moving operations involved in construction works such as roadway preparations, foundations for buildings, etc., or in clearing natural sites for urban development. Rainfall run-off often carries relatively large volumes of solids into lake, stream or estuary, and the resulting deposits, even where practically inert and perhaps exerting no significant oxygen demand, may alter profoundly the character of the bottom and, hence, of the overlying waters.

A considerable proportion of solids may remain suspended in the water while undergoing decomposition, some being too highly dispersed (colloidal or microscopically fine) to settle readily even in still water, some being kept in suspension by motion within the fluid body. The limiting velocities for sedimentation have been determined by sewage treatment practice. Thus it is known that fine solids will still settle out from streams flowing at velocities up to 10 ft/min (say about 0·16 ft/s or 0·05 m/s). Model experiments by Streeter[18] indicate that, at velocities around 10 ft/min (0·05 m/s) laminar flow merges into turbulent. Operation of grit chambers (channels and detritus tanks) shows that a stream velocity around 1 ft/s (0·3 m/s) allows sedimentation of sand but carries forward organic solids. It can be taken that stream velocities determine sedimentation phenomena as follows:

$<$ 0·16 ft/s (0·05 m/s): settleable solids remain suspended only for so long as is required for particles to reach the bottom where they remain:

0·16–1 ft/s (0·05–0·3 m/s): more and more solids remain suspended, only larger and denser particles reaching and remaining on the bottom:

$>$ 1 ft/s (0·3 m/s): most organic matter remains suspended in the stream and the suspended matter is oxidized along with the dissolved.

Suspended matter settling and remaining on the bottom, undisturbed by the stream, collectively constitutes a sludge deposit or sludge bed.[21] Only the uppermost stratum (according to Fair, to a depth of only 0·16 in (4 mm)) is subject to oxidation at the expense of the dissolved oxygen content of the stream above.[22] Below this superficial layer, conditions are always anaerobic, and anaerobic digestion similar to sludge digestion (p. 260) proceeds, with evolution of methane, carbon dioxide, etc., (in a case quoted by Fair, 69% N_2, 17% CH_4, 14% CO_2), but the particles remain stationary in the deposits so that decomposition is slower and digestion continues over a period of many years.

The progress of decomposition of the sludge can be followed by carrying out determinations of B.O.D. from time to time which serve as a measure of the residual decomposable organic content. Fair, Mohlman, Rudolfs and others have thus followed the progress of digestion during long periods; typical results are taken from Mohlman[23] (Table 6.4).

Calculating these results as the proportion of decomposition completed, and plotting against time of digestion, yields the curve of Fig. 6.1, similar in shape to those depicting the normal progress of digestion of sewage sludge in typical digestion tanks (see Fig. 13.2, p. 262) but showing the decomposition rate for ordinary sludge digestion to be about 20 times greater. According to Fair, the rate of digestion (anaerobic decomposition) will increase about 2·7%

if the temperature of digestion is increased by 1°C. This may be compared with a temperature effect of about the same order for aerobic decomposition.

Table 6.4

RIVER SLUDGE DIGESTING AT 68°F (20°C)

No. of days digesting	0	40	80	120	200	400	600	700
B.O.D.$_5$ (lb $\times$ 10^{-3} O_2/lb sludge, or mg/g)	380	270	210	170	140	105	97	96

Particularly in colder climates, winter temperatures may be so low as to interrupt the digestion of sludge deposits, so that decomposition is virtually halted for at least several months every year, depending upon duration and severity of the winter. The net effect of this in simple cases is shown in Fig. 6.2 which indicates no effective digestion during each winter, the seasons comprising four months of summer, averaging 68°F (20°C), followed by eight months

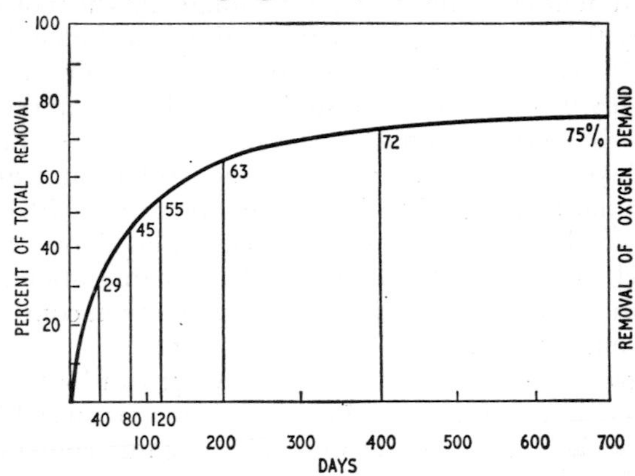

Fig. 6.1. *Relative rate of removal of B.O.D.$_5$ of river sludge deposits at 68°F (20°C) according to Mohlman*

of winter with temperatures of about 41°F (5°C) or less. 55% of the B.O.D.$_5$ is removed during the first summer then only 11% and 5% respectively, in following summers, a total of 71% in three years. Further decomposition will ensue at reducing rates, but it should be noted that a considerable proportion of the organic matter remains practically indigestible under these circumstances (about 30% of the ignition loss).

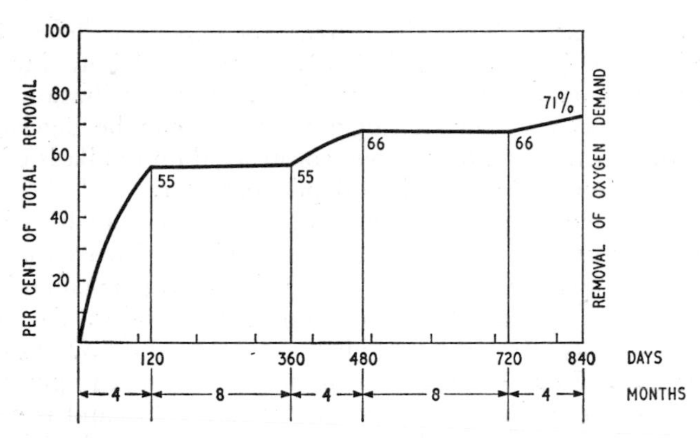

Fig. 6.2. *Relative rate of removal of B.O.D.$_5$ of river sludge deposits at* 68°*F* (20°*C*) *allowing* 8 *months' interruption during winter periods*

So long as the sludge lies undisturbed, its effect on the waters of the stream itself remains small and may be negligible. The only oxygen demands exerted are those of the uppermost layer and those resulting from the diffusion or escape of decomposition products (such as sulphuretted hydrogen) into the water. Fair, Moore and Thomas[22] have also investigated this aspect, measuring the oxygen-consuming substances of the sludge in terms of organic solids (ignition loss) and determining the rates of oxygen consumption from day to day on the basis of quantity per unit area. Their results are reproduced in Table 6.5.

These figures show clearly that the deeper the solids lie in the deposits, the less they affect the rate of oxygen consumption in the

Table 6.5

OXYGEN CONSUMPTION OF SLUDGE DEPOSITS UNDER NATURAL WATERS

Organic solids content of sludges, lb/1 000 ft² (kg/100 m²)	780 (380)	290 (140)	100 (50)	40 (20)
Calculated depth of sludge,* in (mm)	5·00 (127)	1·85 (47)	0·67 (17)	0·28 (7)
	Oxygen consumed, lb/1 000 ft² d (kg/100 m² d)			
1st day	0·96 (0·47)	0·63 (0·31)	0·35 (0·17)	0·23 (0·11)
100th day	0·45 (0·22)	0·25 (0·12)	0·14 (0·07)	0·08 (0·04)
300th day	0·25 (0·12)	0·12 (0·06)	0·06 (0·03)	0·04 (0·02)
400th day	0·08 (0·04)	0·04 (0·02)	0·02 (0·01)	0·014 (0·007)

* Calculation based on freshly formed deposits containing 90% water and 7% fixed residue.

water above them. Thus an increase of deposition from 40 to 780 lb/1 000 ft^2 (20 to 380 kg/100 m^2), i.e. about 20 times, increases the oxygen demand only about five times. The distribution of the deposits behind weirs or in sluggish streams can be determined approximately from settling curves constructed on the basis of rates of sedimentation and of flow and depths of water.

Entirely different are the circumstances resulting from disturbance of the sludge deposits which causes these to be redispersed into the body of the water. An immediate (chemical) oxygen demand must then be satisfied, using dissolved oxygen from the water. According to Fair, this may amount to as much as 28% of the B.O.D.$_5$ of the sludge solids. In addition to this immediate effect, an appreciable proportion of the finer particles remains suspended and is carried away, exerting a normal biochemical oxygen demand and so increasing the overall demand. If, however, the deposits are disturbed by large flood flows, no damage to the oxygen balance need be feared, because flood waters are well oxygenated and so carry a very large oxygen supply; floods are usually beneficial in that oxygen-consuming deposits are scoured away, thereby temporarily reducing the oxygen demand of the stream. On the other hand, lifting of sludge deposits into the water as a result of summer conditions may endanger the oxygen economy, especially when the water temperature first reaches 60–68°F (15–20°C). Gasification in the deposits then becomes so rapid as to disturb them and lift sludge towards the surface. If the river flow or the amount of water in the reservoir is also low at this time, as is commonly the case, serious depletion of the oxygen content may result and conditions may become unpleasant or even dangerous. It might be noted here that similar circumstances occur in lakes or ponds into which organic sludge-forming wastes are discharged. Artificial ponds such as oxidation ponds (p. 242) commonly also show similar phenomena of sudden deoxygenation owing to rapid seasonal changes of temperature or to disturbance of sludge deposits by wind or wave action. In some cases, the deoxygenation effects may not be sufficient to deplete all the oxygen, but the increase in septic-sludge decomposition rates may produce sludge gases in amounts sufficient to cause odour nuisance even at considerable distances from the water surfaces.

In ponds, behind dams, etc., sludge deposits will inevitably be formed because, apart from solids carried by stormwater discharges, natural purification itself produces some settleable solids which can undergo further decomposition. Therefore, some allowance must always be made for their oxygen demand. Technically it may be possible to scour deposits away or remove them by dredging prior to the onset of summer weather or before low-water conditions prevail. The spring floods which are a regular feature of many rivers may happily achieve the same result. Alternatively, conditions during critical periods might be controlled sufficiently by the technique of artificially increasing the flow through the ponded section so as to scour the deposits away.

THE DISSOLVED OXYGEN BALANCE IN POLLUTED NATURAL WATERS

The basic determinants of the dissolved oxygen balance in polluted natural waters[24] are: (1) the consumption of dissolved oxygen due

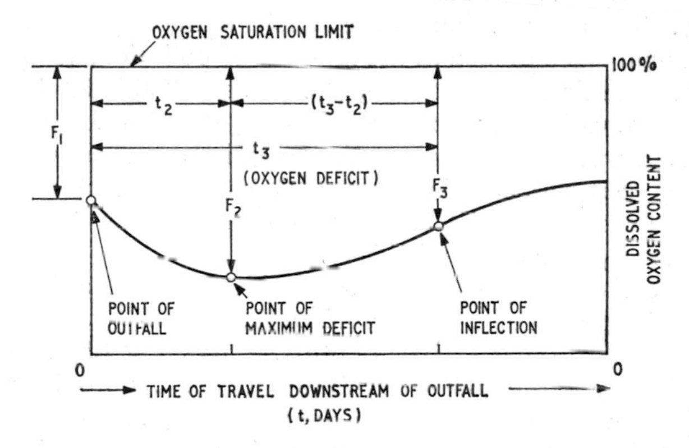

Fig. 6.3. *The oxygen sag curve, after Fair, showing the relative dissolved oxygen content along the course of a river polluted by a sewer outfall discharge*

to the oxygen demand of the natural processes of self-purification of the water (debit), and (2)—credit—the total available dissolved oxygen made up of the original content, the oxygen supply by reaeration—the gradual uptake into solution of atmospheric oxygen —and by photosynthesis. Data are available of the phenomena involved[26] except that the production of oxygen by photosynthesis is highly variable, depending not only on the amount of light available but also on the numbers and kinds of photosynthetic organisms which are vegetatively active at the time. If photosynthesis is neglected, it is feasible to calculate the oxygen balance to be expected under any given circumstances relevant to a natural body of water, and hence to make approximate estimates of the probable contents of dissolved oxygen at various locations in it. These estimates may be plotted graphically, e.g. as the dissolved-oxygen line along the course of a natural stream into which polluting water-borne wastes are to be discharged. Such a line can provide a basis for the design of waste-water disposal problems.[26]

An example of such an oxygen line, termed an *oxygen sag curve*, is shown in Fig. 6.3. The initial deficit, F_1, indicates some residual pollution, but the sewage discharge results in a comparatively rapid increase in oxygen consumption, the maximum deficit, F_2, being reached after t_2 days. Following this there is a progressive improvement during a further t_3–t_2 days, and thereafter a gradually lessening rate of recovery, oxygen saturation being again approached only at long distances below the outfall. The times t_2 and t_3–t_2 variously range from a few hours in the case of small, rapid and active streams to several days in large streams of low velocity and small biological activity. The actual values of the deficits F_1 and F_2 depend in each case on many factors such as the amount of pollution, temperature, physical condition of the stream, etc. (see also Table 6.6 below).

There are various methods of estimating the probable pollutional effect of a water-borne waste discharge into natural waters; the following examples are based on (1) calculation of the oxygen balance (Examples 1–3) and (2) the method of load factors, established by Fair (Examples 4–7).

Example 1—A city of 100 000 inhabitants with a high living standard is situated on a medium-sized river of relatively pure clean water used for fishing. River depth averages 7 ft (2 m), breadth 50 ft (15 m), and the flow 350 ft^3/s (9·9 m^3/s). Approximately 16½ miles (26·5 km) downstream the river discharges into a very large stream which has a high capacity of self-purification. It is to be decided whether primary treatment only is sufficient to avoid harm to the fish life.

Fish life requires a dissolved oxygen content of not less than 3–4 p.p.m. (Table 6.7, p. 80). Taking the B.O.D.$_5$ of settled sewage at 0·10 lb (45 g)/hd d (Table 3.2, p. 19), the total is 10 000 lb (4 500 kg) per 100 000 people. In a river 7 × 50 ft = 350 ft^2 (32 m^2) cross section carrying 350 ft^3/s (9·9 m^3/s), velocity is 1 ft/s (0·3 m/s), equal to about 16½ miles (26·5 km) per day. Taking the daily oxidation rate at 30% of the total (37% of B.O.D.$_5$), the 16½ mile stretch will use up 3 700 lb (1 665 kg) of dissolved oxygen. The surface area of the 16½ mile stretch is about 4 × 10^6 ft^2 (37 × 10^4 m^2), hence the daily oxygen load is 0·9 lb/1 000 ft^2 (0·44 kg/100 m^2). Values of Table 6.3 show such an uptake if the oxygen deficit is around 50–60%. Taking the worse figure, 60%, and the corresponding saturation value for 70°F (21°C) at 9 p.p.m., we might expect the dissolved oxygen to fall as low as about 3½ p.p.m. This is barely sufficient, and primary treatment alone could prove inadequate. Before deciding the question, therefore, careful investigations would be required to establish the basic data of probable rates of biochemical oxidation more reliably for the particular river stretch and for several possible degrees of purification. Also, of course, the contributions of wastes from the particular population should be investigated and compared with the generalized data of Tables 3.1–3.3. The rate of reaeration to be expected should be investigated, using the more specialized models mentioned above (p. 62).

This example illustrates the influence of a single discharge of sewage from a city into an otherwise pure stream. In practice, this is an unusual case, especially in the more densely populated countries where pollution loads are imposed one after another. Their combined effect is illustrated by a later example (p. 83 et seq.).

Example 2—A large impounding reservoir is provided in a heavily polluted stream carrying fish life, having an area of 350 acres (140 ha) and a depth of 7 ft (2 m). Sewage discharged into the river from residential areas with a moderate living standard is treated by plain sedimentation only. How much sewage, expressed as population loadings, can be treated biologically in this reservoir?

(*a*) *Neglecting sludge deposits*, that is, taking no account of oxygen demand from the bottom sludge deposits, the $B.O.D._5$ of settled sewage will in this case be about 0·08 lb (36 g)/hd d. Again from Table 6.3 the uptake of oxygen for, say, 60% saturation is 0·39 lb/1 000 ft^2 = 17 lb/acre (19 kg/ha) daily, and the reservoir is, therefore, able to purify the settled sewage from $(350 \times 17)/0{\cdot}08$ = about 74 000 persons, but the sewage has to remain in the reservoir for about 5 days. If a greater degree of purification is desired, a longer detention period is necessary, and the loading must be lowered. Thus to achieve a 90% removal of $B.O.D._{20}$ (say, $k = 0{\cdot}1$, $B.O.D._{20} = 1{\cdot}46\ B.O.D._5$), about 10 days' detention at 68°F (20°C) is required, and the load reduces to $(350 \times 17)/(0{\cdot}9 \times 1{\cdot}46 \times 0{\cdot}08)$ = about 57 000 persons. On the other hand, a smaller detention period can only achieve partial purification. Thus one day's detention will remove only 30% of the $B.O.D._{20}$, and the impounding reservoir may partially treat this type of settled sewage from 74 000/0·3 = about 247 000 persons.

(*b*) *Taking sludge deposits also into account*—Of the total $B.O.D._5$ of the raw sewage (in this case, say 0·12 lb (54 g)/hd d from Table 3.2), only about 0·04 lb (18 g) is settleable, the settled sewage containing the remaining 0·08 lb (36 g). Of this 0·08 lb, the non-settleable, very fine or colloidal particulate matter contains 0·03 lb (14 g) of $B.O.D._5$. During self-purification, some of the non-settleable and dissolved solids of the settled sewage give rise to humus-like solids. Sludge deposits during a detention period of 5 days in such a reservoir, therefore, may be expected to amount to about 0·04 lb (18 g)/hd d of B.O.D.$_5$, that is, about equal to the quantity of the settleable solids in the raw sewage. Again according to Table 3.2, the settleable solids amount to 0·13 lb (59 g)/hd d, containing 0·09 lb (40 g) of organic matter. It may be taken, therefore, that the latter figure is about what can be expected in the sludge deposits. It may be assumed that winter floodwaters flush away all deposits, and hence we may reckon on six months' accumulation; this amounts to 16 200 lb (7 290 kg) of organic solids per 1 000 of population. If the deposits are spread over 350 acres (140 ha) from 74 000 persons, we have then 78 lb of organic solids/1 000 ft^2 (38 kg/100 m^2). From Table 6.5 we may take an average oxygen consumption for the sludge deposits of about 0·10 lb/1 000 ft^2 (0·05 kg/100 m^2). The consumption from the sewage is 0·39 lb/1 000 ft^2 d (0·19 kg/100 m^2 d) (from above), and the total therefore is 0·49 (0·24). An uptake of 0·49 lb/1 000 ft^2 d (0·24 kg/100 m^2 d) would correspond (Table 6.3) to only about 50% saturation.

(c) *Additional allowance for rapid decomposition of deposits in summer*—Under certain circumstances, the solids in sludge deposits may undergo such rapid decomposition that sludge rises with gas bubbles from the bottom and floats as scum or is again carried along in suspension before settling in some other place. This may happen, for example, during the first hot days of spring or summer when the temperature of the water suddenly increases, and it commonly occurs especially where the deposits accumulate in greater proportions, such as near the sewer outfalls or in deeper holes. Assuming an accumulation of six months' deposits, taking again a population loading of 74 000 persons, the accumulation of organic matter amounts to 16 200 × 74 = say, 1 200 000 lb (540 000 kg) as calculated above. Assuming further that one-half of the six months' summer deposits remains accumulated at the top end of the reservoir, and one-fifth of this part is raised by gasification following a rise in temperature early in the summer period, the total additional burden of organic solids derived from the deposits will amount to 120 000 lb (54 000 kg) with a B.O.D.$_5$ of 0·4 lb per lb (Table 6.4), totalling 48 000 lb (21 600 kg). However, it is found that only about one-half of the oxygen demand of such solids is satisfied during the first summer, and we may take their demand as 24 000 lb (10 800 kg), say 3 000 lb/d (1 350 kg) reckoned on the first hot summer days. Five days' detention in this reservoir (350 × 7 acre ft) corresponds to a river flow of $(6{\cdot}24 \times 350 \times 7 \times 43\,460)/5 = 134 \times 10^6$ gal

(610 × 10⁶ l)/d, and the demand of 3 000 lb (1 350 kg) amounts to a significant debit of 2·2 p.p.m.

The amount of 3 000 lb/d is also equal to 3 000/(350 × 43 560) = 0·20 lb/1 000 ft² (0·10 kg/100 m²). The total oxygen consumption during these 8 days due to the sewage, sludge deposits and rising sludge, as calculated in the above Example 2 (*a*)–(*c*), is, therefore, 0·39 + 0·10 + 0·20 = 0·69 lb/1 000 ft² d (0·33 kg/100 m² d). According to Table 6.3, this corresponds to a saturation deficit of about 70% or, at 68°F (20°C), a dissolved oxygen content of only 0·30 × 9·2 = 2·8 p.p.m. It is apparent that fish life can be seriously endangered by the additional oxygen demands due to sludge deposits. It must also be noted that conditions of some particular regions of the reservoir would probably be much worse than these figures indicate, because uniformity cannot be expected. The early removal of sludge deposits may be advisable in such cases.

This last calculation shows clearly that the effects of sludge deposits cannot be ignored. Similar methods of investigation were applied in a fish-pond used as an artificially controlled biological method of sewage treatment.

Example 3—Take the case of a fish-pond for sewage treatment (see p. 244) providing 1 acre (0·4 ha) per 800 persons, depth 18 in (0·46 m), the sewage flow amounting to 40 gal (180 l)/hd d, the sewage being treated by plain sedimentation as preliminary treatment.

The requisite flow of dilution water amounts to 5 × 40 = 200 gal (900 l)/hd d, whence the effective detention period in the pond is then calculated, taking 1 acre 18 in deep = 43 560 × 1·5 × 6·24 = 408 000 gal (1 840 000 l), loaded by (200 + 40) × 800 = 192 000 gal (860 000 l) of diluted sewage daily, corresponding to 408/192 = say, 2 days' detention. In 2 days, 54% of the B.O.D.$_5$ of the sewage may be satisfied (Table 4.1 (*b*)) or perhaps more, because the pond is only 18 in deep. Oxygen consumption is calculated as at least 0·08 × 0·54 × 800 = 35 lb (16 kg) per 800 persons, or 35 lb/acre d (40 kg/ha), equal to 35/43·56 = 0·8 lb/1 000 ft² d (0·4 kg/100 m² d) on the basis of a moderate living standard of population or as at least 0·10 × 0·54 × 800 = 43 lb (19 kg) per 800 persons, or 43 lb/acre (47 kg/ha) daily, equal to 43/43·56 = say, 1·0 lb/1 000 ft² d (0·5 kg/100 m² d) if a high living standard has been assumed. Taking the minimum dissolved oxygen content for selected fish life to be 3 p.p.m. and assuming a temperature of 68°F (20°C), we have 3/9·2 = about 33% saturation, and from Table 6.3, an uptake of oxygen at the rate of about 0·21 lb/1 000 ft² d (0·10 kg/100 m² d). (For such a shallow pond the rate of uptake may be considerably greater.) The dilution water carries an oxygen excess of 6·2 p.p.m., which at 200 gal (900 l)/hd and 800 persons/acre corresponds to an excess supply of $(200 \times 800 \times 6{\cdot}2 \times 10^{-6})/43{\cdot}56 = 0{\cdot}23$ lb/1 000 ft² (0·12 kg/100 m²). The total available oxygen supply (credit) from dilution water and reaeration is, therefore, at least $0{\cdot}21 + 0{\cdot}23 = 0{\cdot}44$ lb/1 000 ft² (0·21 kg/100 m²), while the oxygen demand as calculated above (debit) is 0·8 or 1·0 lb, respectively, per 1 000 ft² (0·4 or 0·5 kg/100 m²). A further 0·36 or 0·56 lb (0·16 or 0·25 kg) is required for balance if 33% saturation is to be maintained. This might be gained by photosynthetic activity, especially in such shallow ponds if settled sewage pollution is low.

Again, no account has been taken of sludge deposits. It may be concluded that the oxygen demands of a fish-pond require shallow water to increase the relative rate of reaeration and, in addition, a considerable net supply of oxygen from photosynthesis. During periods of higher temperatures particularly, and then especially at night, such a supply sufficient to maintain a satisfactory balance may be endangered, and in such cases the load may be reduced or the dilution-water flow increased or, again, the pond waters may be aerated mechanically. In this last case, some additional sludge oxidation must be allowed for.

Another method of determining the oxygen balance by means of *load factors* and using the oxygen sag curve, due to Fair,[15] takes account of variations in the oxygen balance along the course of a polluted stream, as shown in Table 6.6.

Take the simple equation, first put forward by Streeter and Phelps,

$$\frac{dD}{dt} = k_1 L - k_2 D \qquad 6.2$$

where D = oxygen deficit (at prevailing temperatures) at time t
L = biochemical oxygen demand at time t
k_1 = biochemical oxidation rate constant (under the conditions prevailing during exertion of the B.O.D. at, and subsequent to, time t)
k_2 = the reaeration rate constant, again the actual rate prevailing.

From this Fair has calculated a series of load factors, $f = k_2/k_1$, appropriate to different circumstances (cf. Tables 6.3 and 6.6) and classes of ponds or streams. These of course, grade continuously from one to another. The extremes of circumstances are characterized by the two different 'oxygen lines', Cases A and B of Table 6.6,

Table 6.6

LOAD FACTORS AND DISSOLVED OXYGEN CURVE

$B = F \times Z$ where B = allowable B.O.D.$_5$ of mixed water below sewage outfall
F = oxygen deficit allowed in mixed water at minimum oxygen concentration, p.p.m.
Z = load factor, k_2/k_1 (Case A); $Z = (k_2/k_1).10^{k_1 t_c}$ (Case B)

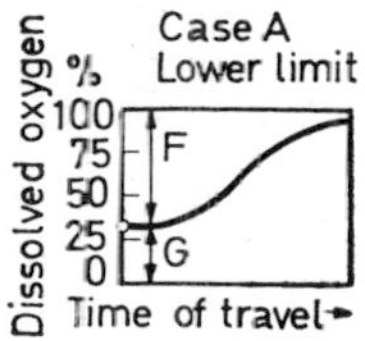
Case A
Lower limit
Dissolved oxygen %
100
75
50
25
0
F
G
Time of travel

Maximum oxygen deficit reached immediately below outfall point

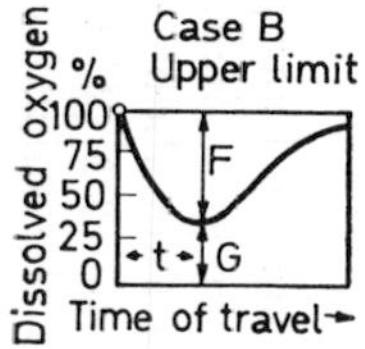
Case B
Upper limit
Dissolved oxygen %
100
75
50
25
0
F
t
G
Time of travel

Maximum oxygen deficit reached after critical time (t_c days) of travel by the mixed water downstream of outfall

Type of water body	*Values of Z* (Case A)			*Values of Z* (Case B)			*Values of t_c days* (Case B)		
	59°F (15°C)	68°F (20°C)	77°F (25°C)	59°F (15°C)	68°F (20°C)	77°F (25°C)	59°F (15°C)	68°F (20°C)	77°F (25°C)
1. Small ponds and back-waters	0·6	0·5	0·4	2·1	1·6	1·3	5·9	5·0	4·3
2. Large lakes and reservoirs	1·1	0·9	0·7	2·7	2·1	1·6	4·5	3·9	3·3
3. Sluggish streams	2·6	1·2	0·9	3·2	2·5	2·0	3·8	3·2	2·8
4. Large rivers	2·2	1·7	1·3	4·0	3·2	2·5	3·0	2·6	2·3
5. Swiftly flowing streams	3·5	2·7	2·1	5·4	4·3	3·3	2·3	2·0	1·8
6. Rapids and waterfalls	22·0	17·0	13·0	25·0	20·0	15·0	0·6	0·6	0·5

but again any particular experience may fall into a category similar to A or, alternatively, rather similar to B but different from either. For example, there might be a case where a typical sag occurs, as in Case B, but the dissolved oxygen at time zero may be, say, only 60% saturated, a consequence of residual pollution from upstream, instead of 100% saturation as shown (unadulterated or fully recovered stream conditions). Case A, by contrast, represents the lower extreme case where the initial dissolved oxygen concentration is already at the lowest permissible level (for this example taken to be 4 mg/l).

Application of this method is illustrated in Examples 4–7.

Example 4—The dissolved oxygen content of the waters of a large river of normal type should not be less than about 4 p.p.m. Given the maximum effective temperature to be 77°F (25°C), what may be the maximum allowable $B.O.D._5$ of the mixed waters below a sewage outfall?

From Table 6.6 for Type 4 the minimum and maximum values for Z are given as 1·3 and 2·5, respectively. The dissolved oxygen at saturation from Table 4.1 is 8·4 p.p.m., whence the allowable deficit, F, is 4·4, and hence the average $B.O.D._5$ below the outfall may be $4{\cdot}4 \times 1{\cdot}3 = 5{\cdot}7$ p.p.m. if the river water from upstream contains itself only 4 p.p.m., and $4{\cdot}4 \times 2{\cdot}5 = 11{\cdot}0$ p.p.m. if it is saturated, that is, before mixing with the sewage discharge.

Example 5—Taking a similar case in which the river-water temperature reaches only 68°F (20°C), what dilution is required for settled sewage of $B.O.D._5$ 230 p.p.m. (a) if the river water is saturated with oxygen and has itself no oxygen demand, and (b) if it contains only 4 p.p.m. dissolved oxygen and has a $B.O.D._5$ of half that allowable?

Case (a): From Table 6.6, $Z = 3{\cdot}2$, and using Table 4.1(a), $F = 9{\cdot}2 - 4 = 5{\cdot}2$ p.p.m., whence $B = 5{\cdot}2 \times 3{\cdot}2 = 16{\cdot}6$ p.p.m. $B.O.D._5$ and the dilution required is $230/16{\cdot}6 =$ about 14 times the sewage flow.

Case (b): From Table 6.6, $Z = 1{\cdot}7$, and hence $B = 5{\cdot}2 \times 1{\cdot}7 = 8{\cdot}8$ p.p.m. The $B.O.D._5$ of the river water is then $8{\cdot}8/2 = 4{\cdot}4$ p.p.m. and the required dilution $230/(8{\cdot}8 - 4{\cdot}4) =$ about 52 times the sewage flow.

If the $B.O.D._5$ of the river water from upstream itself amounted to the allowance of 8·8 p.p.m., no sewage could be discharged without disturbing the oxygen balance.

Example 6—Continuing Example 5, what is the limit of the load allowance in countries with a high living standard, expressed as persons/ft^3 s (or /m^3 s) of river flow?

The allowable B.O.D.$_5$ was 8·8 p.p.m.; 1 ft^3 (0·028 m^3)/s = 0·539 × 10^6 gal (0·245 × 10^6 l)/d, equivalent to 5·39 × 10^6 lb (2.45 × 10^6 kg) water/d, which at 8·8 p.p.m. allows a total of 5·39 × 8·8 = 47·4 lb of B.O.D.$_5$/ft^3 s per day (760 kg/m^3 s per day). Since the B.O.D.$_5$ of raw sewage averages 0·15 lb (0·07 kg)/hd d (Table 3.2), the maximum allowance would be 47·4/0·15 = nearly 320 persons/ft^3 s (11/l s) of river flow.

In countries with a more moderate living standard, the load allowance would be greater because the average pollution per head of population is lower, 0·12 lb (55 g) B.O.D.$_5$/hd d (Table 3.2) or maybe less. In such cases, the maximum load allowance may be taken as 47·4/0·12 = nearly 400 persons/ft^3 s (14/l s) of river flow.

Example 7—The reservoir of Example 2 is reconsidered as a further example to compare the calculations from Table 6.6. Taking, as before, the allowable deficit as 40% and temperature as 68°F (20°C), the deficit, $F = 9{\cdot}2 \times 0{\cdot}4 = 3{\cdot}7$ p.p.m., and the Z values are 0·9 and 2·1 for cases (*a*) and (*b*), respectively.

Case (*a*): $B = F \times Z = 3{\cdot}7 \times 0{\cdot}9 = 3{\cdot}4$ p.p.m. From Table 3.3 (basis 40 gal (180 l)/hd) the B.O.D.$_5$ of settled sewage is 300 − 100 = 200 p.p.m. (average standard), so that a dilution of 200/3·4 = say, 60 times the sewage flow is required. The reservoir capacity (350 acres (140 ha), 7 ft (2 m) deep) is 665 × 10^6 gal (3 020 × 10^6 l and the sewage flow 40 gal (180 l)/hd, giving a total river flow of 60 × 40 gal/hd, so that design for 1 day's detention allows (665 × 10^6)/(60 × 40) = say, 280 000 persons, and for 5 days' detention (665 × 10^6)/(60 × 40 × 5) = say, 55 000 persons (cf. Example 2 values of 247 000 and 74 000 respectively).

Case (*b*): $F \times Z = 3{\cdot}7 \times 2{\cdot}1 = 7{\cdot}8$ p.p.m., $t = 3{\cdot}9$ days. Dilution required is then calculated as 200/7·8 = 26 times the sewage flow. This gives a total daily flow of 26 × 40 = 1 040 gal (4 700 l)/hd d. Now, since the critical time is 3·9 days, the loading should be adjusted for an average detention of not more than 3·9 days, corresponding to a sewage flow of 40 gal (180 l)/hd from (665 × 10^6)/(26 × 40 × 3·9) = 164 000 persons.

The influence of temperature on the rate-load factors of Table 6.6 was determined by laboratory experiments to be about 3% increase in rate of self-purification per °C rise in temperature. In practice, the differences may not be nearly so great. Viehl[27] reports observations on natural waters which show clearly that the effective bacteria become acclimatized gradually to different temperatures, so that after some time the practical differences between self-purification at different temperatures, between 45° and 86°F (7° and 30°C), are not as great as might be expected.

These examples demonstrate that water-pollution control programmes have to be based not only on the number of population living in the area of the particular river but also on the average rate of pollution contributed per head of population, a figure which may vary substantially in different areas of one country or of different countries.

The deoxygenation rate constant, k, from Eqn 4.2 $X_t = L_0 (1 - 10^{-kt})$, must be assessed reliably for the calculation of self-purification rates. Originally $k = 0{\cdot}1$ was commonly assumed to be a good average value for all cases but, as mentioned before, this has been found to be an unjustified simplification. In his original presentation of the analysis of the dissolved oxygen sag curve and the limiting cases of the dissolved oxygen lines of Table 6.6, Fair assumed $k(k_1) = 0{\cdot}1$, and his values of k_2/k_1 (Z of Table 6.6) were arrived at accordingly. In a recent re-presentation of the matter, he concluded that the factors affecting deoxygenation rate (k_1) for different natural classes of streams tended to affect the corresponding reaeration rate (k_2) likewise, and it appears that the original tabulation of the k_2/k_1 ratios is at least of the right order.

These simplified systems of assessing and predicting the oxygen balance in waters to which additional pollution is to be added or for which corrective measures are to be devised, should as a rule be used merely for preliminary investigations. Only when these indicate positively that there is no danger of oxygen balance problems along a watercourse, etc., should they be regarded as sufficient. Otherwise, more rigorous methods should be used which require reliable estimates of local values appropriate for the rates of deoxygenation and reaeration.[10–27]

Classical theory leads to an equation which gives the overall effect on the oxygen balance for a lake or stream in terms of the oxygen deficit

$$D_t = \frac{k_1 L_0}{k_2 - k_1}(10^{-k_1 t} - 10^{-k_2 t}) + D_0\, 10^{-k_2 t} \qquad 6.3$$

where

D = oxygen deficit, at times t and 0, respectively
k_1, k_2 = deoxygenation and reaeration coefficients
L_0 = (first-stage) B.O.D. (total) present at time 0, i.e. in the case of a stream, at the upstream end.

Especially in shallow waters and swift-flowing streams, both k_1 and k_2 will be high, but other local factors and conditions may be especially favourable—or unfavourable—for either or both. If, for example, the k_1 value is high, then the time of flow required for a given degree of purification is shortened, but the area available to balance the corresponding oxygen consumption by oxygen uptake is thus reduced, perhaps very considerably, and unless the effective k_2 value is correspondingly high, serious oxygen depletion may become inevitable. Some practical data from local investigations are always desirable, if not essential, for reliable predictions.

In many cases, load allowances will be close to the average values given above for large, slow streams at 68°F (20°C). However, deoxygenation or recovery rates can be different enough to result in very real and costly nuisance, unless the possibilities are adequately realized and bad estimates avoided. In critical cases the experienced judgment of waste-disposal experts, backed by the data of local investigations, may well be required.

WATER-POLLUTION CONTROL

Water-pollution control should be related to the uses of the waters by the community, taking into account the common and special problems of waste disposal. Recently, however, the simple preservation of waters as far as possible in a naturally balanced state has begun to be adopted as a basic principle of water-pollution control. As pointed out earlier, the many fundamental aspects of water reclamation, including pollution control in the cycles of re-use, are as a consequence now receiving more and more attention by research workers and regulating authorities throughout the world.

Regulation of stream usage and control of pollution are already common practice in the U.S.A. and some countries of Europe but administrative procedures differ considerably and implementation has been achieved in relatively few localities. Table 6.7 is based upon a classification of waters according to usage and can be regarded as a summary which is generally in accord with the views of administrative authorities in most civilized countries, even where no strictly legal regulations have yet been promulgated.[28, 29] Proposals consistent with this table would probably be accepted in any country.

The table indicates suitable minimum requirements for the quality of the water during times of minimum flow, except that no provision is included for temperature (see p. 98), especially under summer or hot-weather conditions. In cases *C* and *D*, the required degree of treatment for sewage or other wastes discharged into the stream depends largely on the oxygen balance and is determined by the permissible minimum oxygen content of the water of the stream,

Table 6.7

CLASSIFICATION OF STREAMS ACCORDING TO WATER QUALITY AND USAGE

Class of stream	*Appropriate water usage*	*Minimum quality requisites**	*Treatment required of most wastes before discharge*	*Additional provision† for pollution control*
D (lowest class)	Surface water drainage, irrigation, transport, minor industrial usages	absence of odours, not too corrosive, not toxic to plants; general appearance tolerable	primary sedimentation plus specific chemical treatments, e.g. for acid or toxic wastes	storage; plant for intermittent chemical treatment, etc., according to circumstances
C	Boating, fishing (plus Class *D* usage)	as *D*; also absence of scum and excessive vegetation, algae, etc.; moderate hygienic standard; strict toxicity limitations; D.O. $\nless$ 3–4 p.p.m., B.O.D.$_5$ $\ngtr$ 5 p.p.m.	as *D* plus partial or full biological treatment according to quantitative relationships or special circumstances	as *D* plus provision for aeration or supplementation of stream flow with clean or purified water, etc.
B	As for *C* and *D*, but also bathing, aqua-planing, etc.	as *C* and *D* but stricter hygienic standards; bacterial pollution limits generally specified (e.g. coliforms average 500, max. 2 000/100 ml; *E. coli* average 200, max. 500/100 ml)	as *C* and *D* but more efficient purification; disinfection a specific requirement in some countries	as *C* and *D* plus adequate control of sewage or other waste-water overflows
A (first class)	Water supply	as *B–D*, plus limitations for salinity (e.g. chlorinity $\ngtr$ 250 mg/l), residual pollutants, etc. Stricter hygienic standards—absence appreciable colour, turbidity, suspended matter; limits for algal contamination	as *B–D* but greater degrees of purification (by treatment or self-purification); removal of taste or odour-producing compounds, etc.	as *B–D* with stricter control of overflows; preferably also river-monitoring systems with automatic shut-down of water-supply intakes, etc.

* Aesthetic aspects are also very important, in different degrees for the different class of usage.

† For example, to cover abnormal circumstances such as might occur seasonally during low stream flow for short periods only or, more irregularly, as a consequence of prolonged dry weather (droughts), etc.; temporary plant breakdowns must also be considered.

but it should be noted that undesirable or harmful concentrations of toxic substances must also be avoided. For classes *A* and *B*, personal hygiene and public health protection involve more than the oxygen balance, and adequate control of water-borne diseases from bacteria, viruses, intestinal parasites, etc., must also be provided for by appropriate additional measures. For class *A*, moreover, the presence of chemicals such as detergents, chlorides, nitrates or phenolic compounds in undesirable or intolerable concentrations may be important, because they may not be sufficiently removable by common water-treatment methods. Phenols even in very low concentrations may yield objectionable tastes and odours after chlorination (see p. 97). Acids may result in excessive corrosion, and even alkalis in excess can seriously affect the domestic, industrial or other common uses of public water supply systems.

However, the most important engineering problem is to provide for such treatment of sewage and other water-borne wastes as is necessary to ensure the maintenance, along any particular water-course, of a sufficient content of dissolved oxygen for the continuation of natural plant and animal life and to obviate nuisance. Originally it was assumed that no nuisance would occur if the waste were automatically diluted by a river flow at least a certain number of times greater than that of the waste. In Germany, when a river flow of more than 20 times the sewage flow was available, treatment by primary sedimentation only was often regarded as adequate. In Great Britain, the Royal Commission on Sewage Disposal recommended that full sewage treatment was sufficient if it produced a well-nitrified effluent containing less than 20 p.p.m. B.O.D.$_5$ and 30 p.p.m. suspended matter, provided at least an eightfold dilution was assured.

All the rules with regard to individual outfalls requiring specific degrees of purification as independent problems only, have failed to prevent river pollution in almost every country today. This is so mainly because the *combined* effect of all the discharges from place to place along a stream determines the degree to which it is polluted. It was shown when considering the oxygen balance along water-courses that the influence of waste discharges is not confined to the river section just below the outfalls but extends downstream for very considerable distances, up to many miles, depending on the rate of travel of the river waters. The effects of one discharge are added to those of others travelling from upstream as well as downstream, and often the latter are under the control of different authorities who may be responsible for limited sections of the stream only. It is, therefore, important to consider the actual effective load on the stream resulting from all sources of pollution, natural and artificial, along its whole course, or at least down to that receiving water which can be regarded as the ultimate, which may be a very large stream, a large lake or the ocean. In some cases, it is necessary or desirable to continue investigations into the ultimate receiving water, even where this is the ocean itself, for they have to be extended as far as the influence of the water discharges can be observed and may possibly be objectionable. The development of new plans for water supply, sewerage or drainage might also make it necessary to extend or renew investigations on a larger scale to take account of all sources of pollution within a wider area including, for example, other portions of the main river system which is being used for waste disposal.

A convenient and rational term for the purposes of such investigations is that of the sewage load,[26] usually expressed as the number of persons or equivalent of sewered population/ft^3 s of river flow. However, this flow varies throughout the year, from the minimum dry-weather flow (D.W.F.) to the maximum flood flow (F.F.) and also shows annual variations. The flow records of a river are usually evaluated to yield the following distinctive data:

Average minimum flow = average of yearly minimum flows
Mean flow = average daily yield (from total flow over all days of many years)
Average maximum flow = average of yearly maximum flood flows.

In addition, the lowest minimum flow and the highest maximum flow also are recorded during observations extended over as long a period as is feasible.

The worst conditions usually result from pollution during periods of minimum flow. In practice, in European and such other countries where streams are perennial, the average minimum flow is taken as the basis for the determination of the sewage load of streams, so that this is expressed as the number of persons/ft^3 s of average minimum flow. Industrial wastes are taken into account by calculating population equivalents (p. 309). In countries where streams are not perennial, the direct use of such data may not be appropriate (see pp. 87–96). It is possible to investigate rivers or streams along their whole course using sewage load units, a continuous diagram representing the variations in conditions.[30] This is shown as a curve summarizing the various polluting loads discharged but taking into account their progressive decrease as the mixed waters travel downstream and are subjected to the cleansing effects of self-purification.

The mathematical expression of the rate of self-purification, measured in terms of the B.O.D., is identical with that of the rate of biochemical oxidation in the laboratory B.O.D. tests, but the actual values of the oxidation rate constant, k_1, may be very different from the laboratory values. Using Eqn 4.2 (see pp. 35–39),

$$X_t = L_0(1 - 10^{-k_1 t})$$

where X_t = the B.O.D. exerted during t days (after discharge of the waste into the river)

L_0 = total B.O.D. of mixed waste and river water immediately below the outfall

t = time (days)

k_1 = the oxidation rate constant which depends on rate of stream flow, depth and gradient of the river and roughness of the river bed, on temperature and biological activity (d^{-1}).

The value of k_1 for the laboratory B.O.D. test at 68°F (20°C) is $k_1 = 0{\cdot}1$, but as we have seen, in many rivers, especially in shallow and swift-flowing ones, observed values of $k_1 = 0{\cdot}2$ and $0{\cdot}3$ have been found and for swift and turbulent waters even higher. The value $k_1 = 0{\cdot}1$ corresponds to an oxidation per day of 30% of the B.O.D.$_5$ remaining, 0·2 to 54 and 0·3 to 73% per day (see Fig. 6.4).

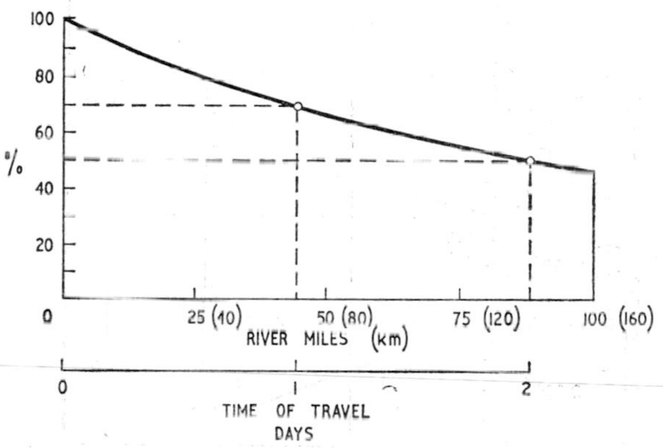

Fig. 6.4. *Self-purification in a flowing stream assuming 30% removal of B.O.D. daily*

The sewage load must be kept down to such a level as will permit the maintenance of a dissolved oxygen content of not less than the value desired.[31] In Example 6 it was calculated that, in order to maintain dissolved oxygen sufficient for fish life, the (raw) sewage load should not exceed about 300–400 persons equivalent/ft^3 s

(11–14/l s), according to the relative level of the living standards of population. In this case the sewage load must be kept below this figure along the whole course of the river. The following example[30] illustrates the calculations necessary for such a case.

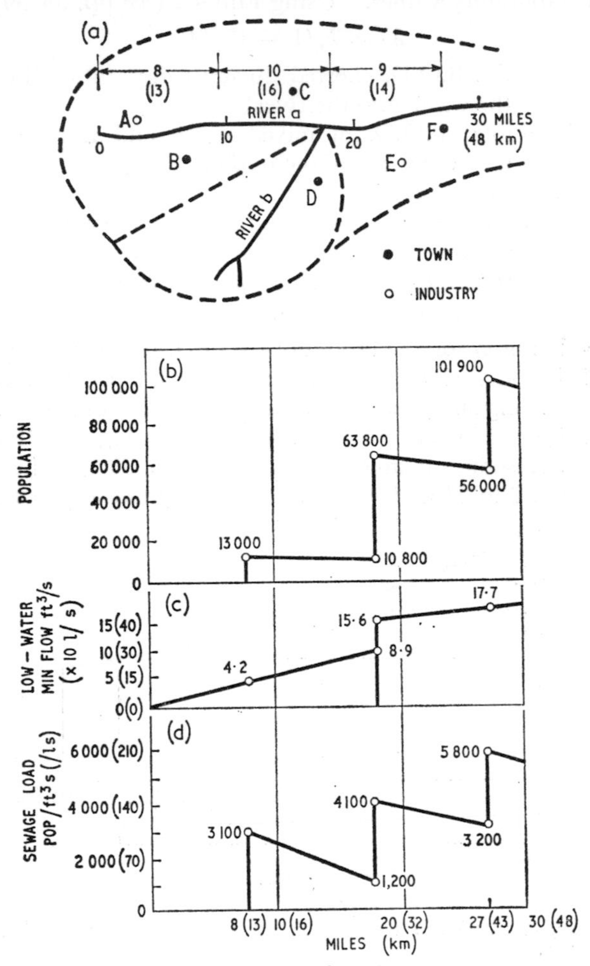

Fig. 6.5. *Sewage load diagrams*

Example 8—In a country with moderate living standard, a river 30 miles long must carry the sewage of four towns and two industrial plants located according to the drainage area sketch plan shown in Fig. 6.5 (*a*). The towns *B*, *C*, *D* and *F* have populations of 10 000, 35 000, 18 000 and 45 000 respectively, totalling 108 000. The factories *A* and *E* discharge wastes with a population equivalent of 3 000 and 900 persons. At low flow, the river flow varies from practically nothing upstream at 0 miles to 17·7 ft³ (500 l)/s at 30 miles (48 km). Taking the stream velocity as 2·5 ft (0·77 m)/s = 40·9 miles (67 km) avge/day, and a temperature of 68°F (20°C), the curve of self-purification is first set up as shown in Fig. 6.4. Under these circumstances, the rate of purification is assumed to be 30% per day of B.O.D.$_5$, which corresponds to $k_1 = 0{\cdot}10$. From Figs 6.4 and 6.5, progressive sewage loads are calculated as shown in Table 6.8. The values of sewage load are plotted as Fig. 6.5 (*d*), and from this may be found the degree of treatment required for each of the discharges. In Example 6 it was found that, for

Table 6.8

(1) *Length of section*, miles (km)	(2) *River discharge*, ft³/s (l³/s)	(3) *Population loadings, town or factory population equivalent*	(4) *Residual load from upstream (reduced by self-purification, see Fig. 6.1) population equivalent*	(5) (=(3) + (4)) *Progressive total load*	(6) (=(5) ÷ (2)) *Sewage load*, persons/ft³s (/l s)
8·0 (13)	4·2 (120)	Factory *A* = 3 000 Town *B* = 10 000 13 000	nil	13 000	3 100 (109)
18·0 (29)			83% of 13 000	10 800	1 200 (43 400)
upstream	8·9 (250)	Town *C* = 35 000 from stream *b* ex			
18·0 (29) downstream	15·6 (0·44)	Town *D* = 18 000 53 000	10 800	63 800	4 100 (145)
27·0 (43)			88% of 63 800	56 000	3 200 (113)
upstream	17·7 (500)	Factory *E* = 900			
27·0 (43) downstream	17·7 (500)	Town *F* = 45 000 45 900	56 000	101 900	5 800 (205)

preservation of fish life, the sewage load should not exceed about 400 persons/ft^3 s (14/l s) (D.O. 4 p.p.m.). Fig. 6.5 indicates that complete treatment is required for all sewage discharged, the necessary degree of purification being at least about 93% for all cases. Settling tanks only will not suffice.

Parts of the river may have to carry additional intermittent loads, especially at low-water periods during summer, and the oxygen demand of sludge deposits which occur particularly in ponded sections and behind weirs or dams and consume dissolved oxygen from the overlying water, must be allowed as equivalent additional oxygen demand (pp. 66–67). In any case, the possible formation of sludge deposits and the consequences which may result should always be considered.

So far, attention has been confined to the sewage load in perennial streams, but in arid zones and even in temperate or subtropical regions, the minimum flow in some streams may be virtually zero or they may dry up entirely, especially during drought periods. For example, near Sydney, Australia, the Warragamba river drains a catchment area of 3 400 $mile^2$ (8 800 km^2), over which the mean annual rainfall varies between about 80 and 30 in (2 030 and 760 mm), averaging about 40 in/year (1 020 mm/year); allowing a 25% run-off, the average daily yield is thus about 3 000 ft^3/s (85 m^3/s) but the maximum flood flow is about 500 000 ft^3/s (14 150 m^3/s), and the minimum flow fell during a prolonged drought to less than 10 ft^3/s (0·3 m^3/s) daily for some months and to less than 2 ft^3/s (0·06 m^3/s) for many days; in ordinary seasons, the mean D.W.F. of this river is about 400 ft^3 (11 m^3)/s.

Where insufficient dilution is available naturally during low-flow or dry periods, it is usually feasible to construct storage reservoirs along the streams to ensure permanent supplies of dilution water. Where sewage must be disposed of finally by discharging into such natural streams, even as highly purified effluents from biological treatment plants, these have to become part of the sewage disposal works, unless low-rate irrigation on land (p. 105) is used. Alternatively, groundwater may be used for dilution by withdrawing it from the ground and using it to supplement the stream flow (pre-aerated if necessary), or a similar dilution achieved by the return of purified water from downstream. Where inadequate dilution is available, discharge to open watercourses necessitates a relatively high degree of treatment, perhaps also including disinfection or other special measures to overcome public health dangers.

Even complete biological treatment by processes such as trickling filters, activated sludge, intermittent sand infiltration and land treatment may result in secondary pollution of the receiving waters as a result of unusually high concentrations of fertilizing constituents such as nitrates, phosphates, etc., and this may be intolerable. Such secondary pollution or contamination may also be caused by excessive growth of algae and fungi and subsequent accumulation and decay of large quantities of organic matter in reservoirs, locks, harbours, etc. As a result of photosynthetic activity stimulated by the fertilizing substances, the amount of organic matter accumulated may exceed that of the original pollution. Removal of fertilizing constituents also is necessary in such cases, and to achieve this, an additional or tertiary treatment of the effluents from normal biological treatment processes is required.[32–37] Final disposal onto land may provide one solution; another method is to absorb the nitrates, phosphates, etc., by means of irrigation with crops, or in specially prepared ponds or lakes in which water plants grow and are then removed mechanically and composted to produce agricultural fertilizers (see pp. 102, 112). In such cases, the proper treatment of the wastes leaves naturally balanced waters for final disposal. The removal of phosphates may be especially important if the effluent is to be discharged into lakes or similar impounded sections of river systems. Regarding nitrates, see also pp. 189, 237.

CONTROL OF SELF-PURIFICATION IN NATURAL WATERS

It has been shown that for each distinct region of any natural body of water, there is a definite capacity for self-purification which may be estimated with reasonable accuracy. Provided that the amount of pollution added to the water does not exceed this capacity, discharges of water-borne wastes will be absorbed and disposed of by the waters without nuisance. In fact, some gain may be anticipated because of the consequent supply of foodstuffs and fertilizers, directly and indirectly, the biological growth cycle leading in the end to an increased fish population or greater development of plant life and wildfowl. It is possible to control, and to increase to some extent, the self-purification capacity of natural waters by engineering means designed to alter physical factors and to modify the environmental conditions governing biological activities. If this is done, it will offset the need for purification of wastes before discharge. However, such control methods may have different purposes, which may be characterized broadly as follows:

1. To control the self-purification capacity directly by
   (*a*) *Increasing* either the physical capacity for, or the effective rate of, self-purification, or both;

(*b*) Delaying or postponing normal decomposition of polluting matter (i.e. *reducing* the rate of self-purification) so that high loadings may pass on without nuisance or harm to regions of adequate purification capacity.

2. To control the pollution loading by

(*a*) Providing mechanical or other treatment processes which remove polluting matter from the water body (i.e. by reducing the total pollution load) or

(*b*) Providing engineering means for spreading the pollution load more uniformly or in appropriate proportions to the self-purification capacity from place to place within or along a stream or other water body.

These two different purposes usually show a natural interplay, but it is useful to consider the possibilities separately as far as is feasible. The following are typical techniques which are effective for case 1.

River activation[38] is a technique similar to the activated-sludge process used in sewage treatment, whereby appropriate quantities of 'activated-sludge cultures' are maintained, for example by some recirculation of river waters. Water-borne wastes are introduced with regulation of the quantities of water and impurities discharged from place to place, in such a way that an optimum level of biological activity is maintained at each stage. Artificial aeration may be required (see below).

Creation of impounding reservoirs[39] provides a second technique, whereby the surface area for reaeration is enlarged and the time of flow lengthened; polluting matter, settling or being changed to form sludge, is thus prevented from reaching downstream sections of the river which may otherwise be overloaded. Such reservoirs or ponds are simply biological treatment works and may be efficient and economical enough by comparison with artificial works to replace them partly or wholly. However, the accumulation of deposits in such bodies of water should be limited so as to avoid danger of unduly thick sludge deposits which might cause seasonal or even continual nuisance owing to excessive gasification and sludge disturbance. Therefore, it is usually necessary or at least desirable to treat sewage and similar wastes before discharge in efficient sedimentation plants with provision for separate sludge disposal. Impounding reservoirs may only be loaded with polluting matter so far as will allow maintenance of an adequate content of dissolved oxygen at all times. Otherwise, if the oxygen consumption should exceed the supply and the oxygen be exhausted, the whole system will become septic and all aerobic life will die. The reservoir will then become a serious nuisance and instead of alleviating will increase the burden of disposal.

A third technique involves *an artificial increase of the minimum dry-weather river flow*. The pollution of a stream is usually most critical in times of minimum flow at summer temperatures. The capacity for purification may be increased by the addition of more natural water of good quality, thus diluting the polluted water and providing a greater capacity for dissolved oxygen, resulting in greater biological activity. Such additional water may be taken from other river systems, from the same system using previously stored floodwaters, or by recirculating water from downstream after recovery from pollution.

Artificial aeration[40] may be applied more particularly to avoid oxygen depletion to levels dangerous to fish life (less than 3–4 p.p.m.). The water may be aerated by spraying adequate volumes into the air above the river surface, using, for example, a rotating disc. Spraying water low in oxygen content up to 3–4 ft (about 1 m) into the air is calculated to result in about 3 p.p.m. oxygen uptake. River overflow weirs also afford a means of reaeration, at a rate of the same order. Oxygen may be supplied to the water of hydraulic power stations by introducing air into the intake of the turbines. Aeration by blowing compressed air into overloaded river water has been applied successfully. Other mechanical techniques for activated-sludge plant aeration systems such as rotating brushes or aeration cones are also being developed for aerating streams or ponds (foam formation is sometimes a problem).

Oxygen may be supplied chemically by the addition of nitrates (usually sodium nitrate). However, nitrate oxygen does not become available until the dissolved oxygen content is practically exhausted, so that this technique is useful only as a means of avoiding general septicization and its attendant nuisances. The oxygen becomes available by reduction of nitrate to nitrite and ammonia, and it might be noted that both these compounds are objectionable at relatively low concentrations. In alkaline waters, a concentration of more than 2 p.p.m. of ammonia is toxic to fish life. The quantity of sodium nitrate added to sewage ponds is usually calculated to provide oxygen equivalent to 20% of the B.O.D.$_5$. It must also be realized that the ammonia formed by reduction must ultimately be re-oxidized to nitrate elsewhere, so that the overall oxygen requirement is not reduced by the addition of nitrate. Other chemicals have likewise been used for control of pollution problems.[41]

In 1(*b*) the aim is to transfer the main burden of purification downstream or, alternatively, to distribute it over greater stretches of a river or more widely within a lake or reservoir, and these aims may be achieved by decreasing the time of travel or by reducing the level of biological activity. By *reconstruction of the river* into more direct channels, with better hydraulic characteristics (smooth banks and bed and good uniform grades), the time of travel and the total amount of active slimes are reduced, so that less decomposition is effected over the reconstructed stretch and the pollution load is carried further downstream where circumstances may permit more self-purification. Otherwise, the water may be *chlorinated*, reducing biological activity to a level dependent upon the relative proportion of chlorine added. The chlorine may oxidize some of the organic matter and accordingly eliminate part of the oxygen demand. Hydrogen sulphide, ferrous iron and other chemical impurities may also be oxidized and the water rendered odourless and more stable. However, chlorine and many chlorination reaction products are relatively toxic to fish and other higher organisms; their use in rivers must be restricted. *Copper sulphate* and other compounds may be used to control the development of algae or the growth of aquatic vascular plants. Small dosing rates, of the order of 0·2–0·5 p.p.m. of copper sulphate, may be sufficient, but larger doses are usually not permissible, being unduly toxic to many species of fish. *Cooling of the river water* can also be used as a temporary measure to delay or reduce the rate of decomposition and to improve the oxygen balance.

For method (2), control of pollution loadings, purely physical techniques are envisaged.

Scouring of the river bed or total flushing of the stream from time to time provides a means of reducing the total amount of purification required. Natural scouring results during flood flows. If floods are prevented artificially, scouring may still be provided, for example, by the opening of sluice gates so as to maintain scouring velocities for just as long as is necessary to carry away deposits and flush back-waters, etc. In general, this technique transfers the polluting matter further downstream and is, therefore, used only where circumstances permit this. As mentioned earlier, flushing is preferably carried out according to a programme determined by seasonal factors. On the other hand, sludge deposits may be removed by dredging, thereby eliminating the equivalent oxygen demand load entirely from the stream. Dredging programmes should also be arranged according to seasonal factors.

CLASSIFICATION OF NATURAL WATERS IN RELATION TO THE DISPOSAL OF SEWAGE AND OTHER WATER-BORNE WASTES

Natural waters were classified broadly earlier and may for the present purpose be considered under the general headings of creeks, permanent streams, estuaries, oceans, lakes and groundwaters.

Creeks

The term creeks as used here includes watercourses which sometimes have no dry-weather flow. The worst conditions for waste disposal are then to be expected at times of zero flow when no water is available to dilute discharges. Waste waters must then be stable enough in themselves to pass along the watercourse without causing harm or nuisance, until finally reaching some relatively large body of water or until self-purification is completed. If such watercourses traverse only uninhabited lands to which direct access of people and grazing animals may be prevented, it may be sufficient if the wastes remain in a fresh condition during their passage so that odour nuisance is avoided and aerobic decomposition of organic matter proceeds steadily. However, if the adjoining lands are public property or part of residential or rural (grazing) areas, wastes must first be purified sufficiently for Class *B* usage (Table 6.7), so that people, animals and even plants may be adequately protected from harm due to bacteria, intestinal parasites or direct toxicity. Although the need for such protection may appear of greatest importance in the case of sewages or house drainage discharges, many cases of trade-waste disposal bring special problems and are also potentially dangerous for man or beast. Apart from this danger, development of septic conditions will almost certainly result in odour and aesthetic nuisance generally, which may be intolerable. In less extreme cases, it may be sufficient to construct suitable fencing along the banks where the creek is not used for agriculture, stock watering or other purposes. Its reconstruction as an artificial open channel (see pp. 8, 90) may also be helpful.

Relatively large 'waterholes' along creek beds must be considered separately. Other special cases occur particularly in areas of relatively low rainfall. Some such watercourses have low average gradients and in dry periods may consist of a series of relatively large waterholes separated by barriers of river gravel and sand through which a steady but invisible stream of water flows. Such streams, which are actually permanent, may have a considerable capacity for self-purification. Waterholes may be naturally subject to organic pollution, particularly by leaf-fall.

Permanent streams

Permanent streams always have some flow but, of course, this varies between a minimum and maximum (flood flow). The first occurs usually in summer time when temperatures may be at a maximum so that the solubility of oxygen is low and biological activity high. These circumstances may be worsened, for example, by discharge of appreciable volumes of water from factory cooling plants with relatively high temperatures. Thus it is usually during minimum flows that the least favourable conditions occur with regard to pollution loadings, and the common rule has been to determine only these and to provide for sewage treatment accordingly. It is proper, however, to give consideration to all stages of river flow, at least to representative minimum, average and maximum stages. In some cases, streams flowing in well-graded natural channels may have adequate self-purification capacity during minimum flows, whereas during higher flows the reduced travel time and the flushing and scouring by the higher velocities may transfer pollution loads downstream to sections where the purification capacity is critical or barely adequate. Occasionally it may be economical to vary the extent of pretreatment of wastes before discharge according to the purification capacity of the stream; this would, of course, be significant for the design and operation of treatment works.

Estuaries

Estuaries comprise bodies of water affected both by stream flow from surface and underground waters and by tidal flow into and out of the ocean. Problems of disposal of waste into tidal waters are more complicated than discharge into simple streams, so more detailed investigations of purification are necessary.[42, 43]

Self-purification of salt water is different from that of fresh water.[27] In sea-water, bacteriological self-purification[44] is much more rapid, to approximately 90% within one day at summer temperature and two days in winter periods, compared with several days in fresh water. However, in terms of B.O.D., self-purification of sea-water from pollution by organic matter requires about twice the time experienced in fresh water. In sea-water, and in mixtures of fresh and sea-water within estuaries, the formation of deposits is favoured generally; some organic compounds are precipitated, some are coagulated, others may tend to be more highly dispersed.

The limit of salt-water influence in an estuary is determined by the volume of river flow; it is located farther downstream with larger river flows and farther upstream with smaller flows. This limit is called the 'upper end of the estuary'. The salt content of estuary water gradually increases from the upper end to the mouth of the estuary. River water containing discharged sewage and wastes is moved through the estuary with the upstream and downstream tidal movements and is thereby diluted more and more by salt water. It is possible to calculate the time of travel of the river water from the determination of the salinity of the water of the estuary. For instance, in the Tees estuary the average travel of river water was found to be 1–3 mile (1·6–5 km)/d, in the Mersey about 1·65 mile (2·65 km)/d.[45]

Since in estuaries the river waters are diluted further by the ocean water, higher loadings of organic pollution may be permissible. It is possible to determine the rate of dilution of river water by salt water from a calculation of the tidal prism which is formed by the difference between flood- and low-water levels upstream of the particular cross section of the estuary. However, it is to be assumed that some of the water which escapes during ebb tide will return during flood, resulting in a tidal volume available for dilution of only a portion of the total, say one-third or less, depending on local conditions. As the tidal volume is greater near the mouth than upstream, greater dilutions are obtained if the polluting wastes are discharged as near the mouth as possible. The discharge of sewage into tidal waters may be regulated, for example, by equipping the end of the outfall line with a storage tank and tide-flap in order to retain the sewage while the tide is rising and to discharge only when it is falling, thus confining deposits to the channels of the lower estuary and avoiding deposition and exposure of solids on tidal banks, etc. Septic conditions, however, should be avoided.

The Ocean

The ocean itself has a practically unlimited capacity to absorb pollution from sewage or, for that matter, any kind of wastes. However, wastes must be discharged at such distances from the shore—especially from beaches, baths or other recreation areas on foreshores—that the waste is sufficiently diluted and significant pollution of waters, sands,[46] rocks, vegetation and, indeed, the atmosphere (spray contamination) does not occur.[47–51] The discharge into the ocean of water-borne wastes, which in practically all cases have a significantly lower specific gravity than ocean waters, results in the production of a 'sewage field' or 'Sleek field'. This always occurs and may be visible as a distinct area, generally circular, elliptic or parabolic according to ocean currents and the direction and strength of the wind. The actual appearance of such a field formed by the discharge of raw or settled sewage varies with the strength and composition of the latter. The area depends primarily on the total amount of pollution, especially the suspended (particulate) matter discharged and the relative concentration ('strength') of the waste, but it is also modified, according to the location of the outfall, by local, mainly physical factors. From observations made by the California State Board of Health, the Sleek field area produced by raw sewage discharges can be calculated as

$$A = 11{\cdot}5 - 3{\cdot}5 \times \log N \qquad 6.4$$

where A = visible Sleek field area in acres per 1 000 population
N = population in thousands.

For sewage or similar wastes, the amount of organic pollution and of particulate matter is best expressed in terms of the population contributing plus equivalents. This provides a useful basis for estimating the probable area of sewage field. However, it should be noted that greasy surface scums may travel long distances under wind influences and are thus spread far beyond Sleek field limits. It may be feasible to transport wastes or sludge from sedimentation plants by specially built vessels to deeper water, say 15 miles from shore or more, thereby obviating further treatment or otherwise simplifying problems of disposal.

Lakes

Lakes have a capacity for self-purification depending upon their size and on special characteristics. Ponds and impounding reservoirs are special cases in this sense. The basic characteristics of lakes are determined by the volumes of fresh water flowing through and by other factors depending on the nature of the surrounding country, including vegetation, climate and geology, and on the composition of the waters, especially with regard to inorganic constituents, including trace elements. The types of biological activity are controlled by these and other ecological factors. Algal blooms and other nuisances may result from over-fertilization of lake water by sewage or other wastes carrying nitrogen and phosphorous compounds into it. Lakes may be classified as eutrophic (well nourished), oligotrophic (poorly nourished) or dystrophic (abnormally nourished).

Eutrophic lakes generally receive natural waters from areas rich in vegetation so that both waters and lake bottom contain optimum proportions of mineral and organic matter and can support a vigorous aquatic community of microscopic and macroscopic forms. Self-purification capacity is determined by the intake and output of fresh waters from streams and waste discharges and the effective storage or detention period, a high level of biological activity being characteristic.

Oligotrophic lakes, on the other hand, are naturally poor in nutrients such as phosphates and essential trace elements; biological activity is accordingly much reduced and purification capacity may be very limited. The introduction of wastes must be undertaken very cautiously because the conditions prevailing may be completely altered thereby, with unpredictable consequences. The discharge of polluting wastes into oligotrophic lakes requires consideration not only of the dissolved oxygen economy but also of the biological equilibria and of material balances (intake and output), the more so if the outflow of water is relatively small.

Dystrophic lakes are usually closed, i.e. without effluent, and characterized by a high, gradually increasing organic content which is unbalanced so that normal conditions do not develop. The waters are usually dark in colour, penetration of sunlight being restricted to the surface; generally, the water is markedly acidic in reaction, and relatively few species of bacteria and other forms of life can proliferate. For most types of wastes, the capacity for purification is relatively very low, and careful consideration of prevailing conditions and of material balances is required before disposal plans are prepared.

Swamps

Swamps, as well as shallow or old lakes in the final stage of filling, are usually not suitable for disposal of water-borne wastes. They may take the tertiary stages of purification of well-purified effluents.

Groundwaters

Groundwaters tend to be freed of living matter as they percolate through soils or other subterranean structures. It may be noted that soils, sands and gravels effect filtration, essentially as a physical process, particularly at greater depths. Water flowing through favourable strata may thereby be rapidly purified, filtration, precipitation or adsorption of colloidal matter being important features of the process. In some strata, for example, in limestone areas and in disturbed strata, very little purification may be effected in this way even after long distances and caution must be exercised in such areas.[52] In any case, however, most of the soluble constituents, for example, phenols and salts, travel long distances, subject only perhaps to some dilution in other waters, and may contaminate wells or surface waters elsewhere. Contamination of underground waters by wastes carrying mineral oils can make the water unsuitable for drinking purposes, even if highly diluted, for a long period.

As noted earlier, groundwaters always have a connection with surface waters and, especially in dry periods, rivers and lakes may derive a large part, if not all, of their water from groundwaters. Consequently, groundwater pollution, chemical or otherwise, can result in pollution of the water of rivers or lakes, restricting its use for water supplies or for waste disposal. On the other hand, reservoirs or streams of groundwaters may be supplied or supplemented from surface waters which may be polluted, so that polluting matter may penetrate subterranean structures and so contaminate supply water drawn from such groundwater bodies. In particular cases, it may be economical to employ engineering means to improve the capacity of groundwaters for waste disposal or to divert intakes, e.g. by constructing subsurface dams, by physical or chemical alteration of subterranean structures or in other such ways. It should also be noted that it may be very difficult to overcome groundwater pollution once it has occurred. In some cases, even several decades or longer after cessation of pollution, contamination of groundwaters is still observable and has limited their otherwise possible usefulness. Dissolved oxygen is usually absent.

SPECIAL ASPECTS OF DISPOSAL INTO NATURAL WATERS

The disposal of waste waters is less troublesome if the wastes are kept in fresh condition, because septic wastes, especially septic sewage, are toxic to fish and have a much higher immediate oxygen demand. Of course, relatively small volumes of septic waste may cause little harm in well-oxygenated waters.

Where different kinds of wastes arise, particularly industrial, it is generally wrong to discharge them separately into common waters. They should be mixed together and allowed to react together before discharge through one or more outfalls. The separation of industrial wastes from municipal sewage frequently results in a higher degree of nuisance in natural receiving waters than common discharges. Toxic effects of certain constituents may be neutralized or buffered by interaction, by dilution and consequent better distribution into the receiving waters, and by balancing of variations in concentration. This is better achieved if the various wastes are collected by a common sewerage system.

Pretreatment of wastes before discharge may be necessary in order to reduce the amounts of toxic substances which could destroy fish life or otherwise affect the natural purification capacity of the receiving water, or for other such reasons, apart from the ruling consideration of the oxygen balance in the receiving waters. General requirements to be met in practically all cases, involving at least a minimum of treatment, are as follows:

1. The discharged wastes must be freed of floating solid or liquid matters including especially oils, which may spoil appearances, interfere with normal oxygen uptake or, indeed, increase health dangers resulting from flotation of polluting matter and delay in its purification.
2. The wastes must likewise be freed from any heavy suspended matter such as silt which may accumulate near the outlet or elsewhere and interfere with normal use of the waterways.
3. The wastes must not carry sizeable amounts of any biologically inert or otherwise toxic substances which may unduly delay decomposition or reduce biological activity. Such substances would be, for example, free chlorine, acids, alkalis, sulphides, metal salts, cyanogen compounds, chromates and phenolic compounds.

Allowable concentration limits for many such substances have been the subject of numerous investigations, especially in the U.S.A. and Germany, and laboratory data are being reviewed constantly in the light of actual experience (see Table 14.2, p. 308). However, all such problems are inherently complex and cannot be dealt with in a few words or summarized adequately by simple tabulation. Taking phenols, for example, experience has shown that phenol itself is quite readily oxidized in rivers if sufficiently diluted, being completely decomposed within three or four days. The process is hastened in streams already polluted by other organic wastes such as sewage, in which biological activity is well established, and is very rapid in waters previously 'seeded' with phenols; that is, rivers into which phenols are discharged from point to point may develop a considerable capacity for their oxidation (including cresols, xylenols and other more complex phenolic compounds). The oxidation proceeds actively at normal temperatures but is markedly lessened below 50°F (10°C). Relatively high concentrations of phenols (100 p.p.m. and more) are oxidized successfully by trickling filters (see p. 315) and could, therefore, be presumably oxidized likewise in natural streams at such concentrations; but a concentration of only 5 p.p.m. is toxic to fish, while even lower degrees of it may impart disagreeable flavours to the cooked fish. Furthermore, concentrations as small as 0·005 mg/l in water used for water supplies produce marked odours during chlorination, rendering the water unpalatable and objectionable. Thus it may be seen that even in the relatively simple case of phenols, which are readily oxidizable organic substances with a definite B.O.D., many complications may arise.[53]

The wastes must not be so hot nor so cold, nor so variable in temperature as to endanger the natural plant and animal life of the water or to render the aquatic environment unattractive, i.e. upset the natural balance of life. It has been observed, for example, that trout are likely to die when temperatures reach 77°F (25°C), pike at 86°F (30°C), goldfish at 95°F (35°C), but sudden changes are already fatal at lower temperatures. Of course, warming up of natural waters could be beneficial in winter or at times when temperatures are very low, but in such cases it is even more important to provide rapid local mixing. Consideration of temperature effects has a direct bearing on cooling-water discharges.

The *design and distribution of outfalls*[54] are of considerable importance. The principal aim is rapid dispersion of the discharged wastes into as great a relative volume of the receiving waters as possible, particularly of wastes with high oxygen demand and into a sluggish stream or sheltered lake. In some of these cases, it may be necessary or desirable to use multiple outlets suitably spaced along outfalls extended from the shore or along the shoreline. Alternatively, the waste might be mixed before discharge with dilution-water taken from the stream or lake, or discharged through outlets located at the centre of the stream.

Discharge of wastes according to *time schedules* is also a possibility which should not be overlooked, especially where water supply offtakes must be considered. Then adequate storage provision, both for the water-carried wastes and for the water supply demand, allows for the discharge of wastes and the operation of water supply works according to a timetable arranged to avoid any contamination. In the same manner, the use of storage ponds might permit retention of part or all of the water-borne wastes until periods of higher flow occur or because of some other circumstances. Of course, very large ponds might then be required and must be considered as engineering works for which there will be suitable designs and proper methods of maintenance, operation and control. Water supply works might likewise provide reservoirs filled from the stream when no pollution is present but isolated at times of discharge of wastes or when the stream is otherwise polluted.

BIBLIOGRAPHY

Hynes, H. B. N. *The Biology of Polluted Waters* (1960) Univ. Liverpool

Isaac, P. C. G. (Ed.) 'River Management', *Proc. Symposium Univ. Newcastle-on-Tyne 1966* (1967) Gordon Press, Letchworth

BIBLIOGRAPHY

KLEIN, L. *River Pollution*, Vol. II *Causes and Effects* (1962) Vol. III *Control* (1966) Butterworths, London

McGAUCHEY, P. H. *Engineering Management of Water Quality* (1968) McGraw Hill, New York

McKEE, J. E. and WOLF, H. W. *Water Quality Criteria*, 2nd Edn (1963) Publ. 3A, California State Water Quality Control Board and Supplt

SOUTHGATE, B. A. *Water—Pollution and Conservation* (1969) Thunderbird, London

'Advances in Water Pollution Research', *Proc. Int. Conferences* (published biennially 1962 to date) Pergamon, London

'Effects of Polluting Discharges on the Thames Estuary', *Wat. Pollut. Res. Bd Tech. Pap. 11* (1964) H.M.S.O.

Water Quality Criteria, Fed. Wat. Pollut. Control Admin. (U.S. Dept of Interior) (1968) Govt Printer, Washington D.C.

REFERENCES

1. IMHOFF, K. and HYDE, C. G. 'Possibilities and Limits of the Water Sewage–Water Cycle', *Engng News Rec.* **106** (1931) 883
2. TEBBUTT, T. H. Y. 'Sewage Effluents as a Source of Water', *Effl. & Wat. Treat. Jnl* **5** (1965) 565
3. Min. of Technology Gt. Brit., *Notes Wat. Pollut. No. 31* (1965); *J. Proc. Inst. Sew. Purif.* (1966) Pt 6, 591
4. VEATCH, N. T. 'Industrial uses of reclaimed sewage effluents', *Sewage Wks J.* **20** (1948) 3
5. METZLER, D. F., CULP, R. L. *et al.* 'Emergency use of reclaimed water for potable supply at Chanute, Kans.', *J. Am. Wat. Wks Ass.* **50** (1958) 1021
6. KEEFER, C. E. 'Improvements and operation at Baltimore's Back River Sewage Works', *J. Wat. Pollut. Control Fed.* **33** (1961) 22
7. PARKHURST, J. D. and GARRISON, W. E. 'Water Reclamation at Whittier Narrows', *J. Wat. Pollut. Control Fed.* **35** (1963) 1094
8. SCHRÖDER, G. 'Aus dem Tätigkeitsgebiet der Wasserwirtschaftstellen', *Dt. Wass Wirt.* **36** (1941) 329
9. KAZMAN, R. G. 'River infiltration as a source of ground water supply' *Proc. Am. Soc. civ. Engrs* **73** (1947) 837
10. 'Oxygen Relationships in Streams', *Tech. Rep. Rt A. Taft sanit. Engng Center W58–2* (1958)
11. O'CONNELL, R. L. and THOMAS, N. A. 'Effect of Benthic Algae on Stream-Dissolved Oxygen', *Proc. Am. Soc. civ. Engrs* **91** (SA3) (1965) 1
12. OWENS, M., EDWARDS, R. W. and GIBBS, J. W. 'Some re-aeration studies in streams', *Int. J. Air Pollut.* **8** (1964) 469
13. DOBBINS, W. E. 'B.O.D. and Oxygen Relationship in Streams', *Proc. Am. Soc. civ. Engrs* **90** (SA3) (1964) 53, (SA6) 131
14. HANYA, T. and HIRAYAMA, M. 'Factors influencing gas exchange between fresh water and air' (with discussion) *Proc. 2nd Int. Conf. Wat. Pollut. Res., Tokyo* **1** (1964) 259
15. FAIR, G. M. 'The dissolved oxygen sag—an analysis', *Sewage Wks J.* **11** (1939) 445

16. CHURCHILL, M. A., ELMORE, H. L. and BUCKINGHAM, R. A. 'The Prediction of Stream Re-aeration Rates', *Proc. Am. Soc. civ. Engrs* **88** (SA4) (1962) 1; cf. GUNNERSON, C. G. and BAILEY, T. E. **89** (SA4) (1963) 95; **90** (SA1) (1964) 175
17. NEGULESCU, M. and ROJANSKI V. 'Recent Research to Determine Reaeration Coefficient', *Wat. Res.* **3** (1969) 189
18. STREETER, H. W. 'Measures of Natural Oxidation in polluted streams', Pts I–III (with WRIGHT, C. T. and KEHR, R. W.) *Sewage Wks J.* **7** (1935) 251, 534; **8** (1936) 282
19. MEINCK, F. 'Kritische Betrachtungen zur Belastbarkeit der Gewässer', *Gesundheitsingenieur* **71** (1950) 216
20. NEJEDLY, A. 'An explanation of the difference between the rate of the B.O.D. progression under laboratory and stream conditions', *3rd Int. Conf. Wat. Pollut. Control, Munich* **1** (1966) 23
21. IMHOFF, K. 'Der Bodenschlamm in verschmutzten Gewässern', *Gesundheitsingenieur* **65** (1942) 154
22. FAIR, G. M., MOORE, E. W. and THOMAS, H. A., Jr. 'The natural purification of river muds and pollutional sediments', *Sewage Wks J.* **13** (1941) 270, 756, 1209; cf. EDWARDS, R. W. and ROLLEY, H. L. J. *J. Ecol.* **53** (1965) 1
23. MOHLMAN, F. W. 'Oxygen demand of sludge deposits', *Sewage Wks J.* **10** (1938) 613
24. INKSTER, J. E. 'The oxygen balance in polluted water', *J. Proc. Inst. Sew. Purif.* (1943) 123
25. CAMP, T. R. 'Field estimates of Oxygen Balance Parameters', *Proc. Am. Soc. civ. Engrs* **91** (SA5) (1965) 1; **92** (SA3) (1966) 14
26. IMHOFF, K. 'Die Sauerstofflinie in verschmutzten Gewässern', *Dt. Wass Wirt.* **36** (1941) 338
27. VIEHL, K. 'Uber den Einfluss der Temperatur und der Jahreszeit auf die biologische Abwasserreinigung', *Zentr. Bakt Parasitkde* **91** (II) (1934) 14; cf. STOLTENBERG, D. H. and SOBEL, M. J. 'Effect of temperature on the Deoxygenation of a Polluted Estuary', *J. Wat. Pollut. Control Fed.* **37** (1965) 1705
28. HERBERT, D. W. M. 'Pollution and Fisheries', *5th Symp. Br. Ecol. Soc.* (1965) 173, Blackwell, Oxford
29. JACOBS, H. L. *et al.* 'Water Quality Criteria—Stream *v* Effluent Standards', *J. Wat. Pollut. Control Fed.* **37** (1965) 292
30. IMHOFF, K. 'Graphical representation of load on polluted streams', *Sewage ind. Wastes* **22** (1950) 1614
31. MÜLLER, W. J. 'Selbstreinigung und zulässige Abwasserlast', *Gesundheitsingenieur* **74** (1953) 15
32. DIETRICH, K. R. 'Die dritte Reinigungstufe für den Vorfluter Bodensee', *Gesundheitsingenieur* **84** (1963) 305
33. TRUESDALE, G. A. and BIRKBECK, A. E. 'Tertiary Treatment Processes for Sewage Works Effluents', *Wat. Pollut. Control* **66** (1967) 371
34. SAWYER, C. N. 'The Need for Nutrient Control', *J. Wat. Pollut. Control Fed.* **40** (1968) 363

35. OGLESBY, R. T. and EDMONDSON, W. T. 'Control of Entrophication', *J. Wat. Pollut. Control Fed.* **38** (1966) 1452
36. IMHOFF, K. 'The final step in sewage treatment', *Sewage ind. Wastes* **27** (1955) 332
37. ROHLICH, G. A. 'Methods for the Removal of Phosphorus and Nitrogen from Sewage Plant Effluents', *Proc. 1st Int. Conf. Wat. Pollut. Res., London* **2** (1962) 207
38. MÜLLER, W. J. 'Steigerung der Selbstreinigung im verschmutzten Fluss durch Flussbelebung', *Gesundheitsingenieur* **72** (1951) 299
39. KRENKEL, P. A., CAWLEY, W. A. and MINCH, V. A. 'The Effect of Impounding Reservoirs on River Water Assimilative Capacity', *J. Wat. Pollut. Control Fed.* **37** (1965) 1203
40. SUSAG, R. H., POLTA, R. C. and SCHROEPFER, G. J. 'Mechanical Surface Aeration of Receiving Waters', *J. Wat. Pollut. Control Fed.* **38** (1966) 53
41. LANG, M. 'Chemical Control of Water Quality in a Tidal Estuary', *J. Wat. Pollut. Control Fed.* **38** (1966) 1410
42. HEANEY, F. L. 'Design, Construction and Operation of sewer outfalls in estuarine and tidal waters', *J. Wat. Pollut. Control Fed.* **32** (1960) 610
43. Institute of Sewage Purification, 'Memorandum on discharge of sewage and trade effluents into tidal waters', *J. Proc. Inst. Sew. Purif.* (1964) Pt 2, 119
44. MITCHELL, R. 'Factors affecting the decline of non-marine microorganisms in seawater', *Wat. Res.* **2** (1968) 535
45. SOUTHGATE, B. A. and PREDDY, W. S. 'Discharge of sewage and industrial wastes to estuaries', *J. R. sanit. Inst.* **22** (1952) 424
46. FLYNN, M. J. and THISTLETHWAYTE, D. K. B. 'Sewage Pollution and Sea Bathing', *Proc. 2nd Int. Conf. Wat. Pollut. Res., Tokyo* **3** (1964) 1
47. OAKLEY, H. R. 'A Study of Beach Pollution in Tidal Waters', *Proc. 2nd Int. Conf. Wat. Pollut. Res., Tokyo* **3** (1964) 85
48. GAMESON, A. L. H., BUFTON, A. W. J. and GOULD, D. J. 'Studies of the Coastal Distribution of Coliform Bacteria in the vicinity of a Sea Outfall', *Wat. Pollut. Control* **66** (1967) 501
49. MÜLLER, W. J. 'Die Kläranlage Subiaco in Perth, West-Australien', *Gas-u. Wassfach* **106** (1965) 1334
50. PEARSON, E. A. 'Some developments in marine waste disposal', *J. Proc. Inst. Sew. Purif.* (1966) Pt 3, 223
51. COVILL, R. W., DAVIES, A. W. and CHANDLER, J. R. 'Parameters of Marine Pollution in the Forth Estuary', *Wat. Pollut. Control* **69** (1970) 12
52. Proc. Symp. Ground Water Contamination, *Tech. Rep. Rt A. Taft sanit. Engng Center W61–5* (1961)
53. MEINCK, F. and SPALTENSTEIN, A. 'Abwässer der Kohleveredlungsindustrien und Trinkwasserversorgung', *Gesundheitsingenieur* **68** (1947) 7
54. MÜLLER, W. J. 'Die Einleitung von Abwasser in einen Flusslauf', *Gesundheitsingenieur* **57** (1934) 676

7

DISPOSAL ONTO LAND

INTRODUCTION

IN CHAPTER 6, the complexities involved in the disposal of water-borne wastes into natural waters were discussed with reference not only to surface but also to underground waters and to their interconnection. Normally natural waters and the land over or under which they move are also interconnected to some extent, and this must be considered when water-borne wastes are disposed of directly onto the land.

As with natural waters, distinctive regimes of aerobiosis and anaerobiosis (pp. 104–107) must be distinguished in relation to land disposal. Well-drained aerated soils under favourable circumstances could have a natural oxygen supply up to or even exceeding 10 times the daily rate available to an equivalent volume or area of water, but waterlogged soils far less. Moreover, once water has percolated deeper underground, the rate of oxygen supply to it becomes negligible. Hence, unless it has been relatively highly purified beforehand, opportunities for purification are confined to mechanical filtration, which in many cases would not be sufficient, and residual pollution may be present practically permanently wherever such water may travel.[1, 2] Land-disposal systems should be operated so as to ensure always an overall positive balance of oxygen supply over demand.

Land-disposal systems involve biological aspects in much the same way as disposal into water, many of the organisms involved being the same or of similar type. The different environment results in some important differences, however. The end product of waste assimilation in rivers is often found in the growth of fish. With soils, indigenous animals are relatively small, and as fish growth is perhaps most important for waters, so is plant growth for the soil. Even so, as with natural waters or biological waste-treatment processes, there must be an integrated system of micro-organisms, fungi, etc., with a succession of animal predators, to preserve a balanced ecology.

THE CYCLE OF FERTILITY

Some of the polluting matter of sewage and other wastes is of value as fertilizers, that is, as substances or elements essential to plant life.

The principal fertilizing constituents of domestic sewage are given by Sierp[3, 4] (Table 7.1).

The organic matter of the wastes includes humus and other compounds which finally produce more humus. In addition, many metallic salts, present perhaps in very small amounts ('trace elements'), are essential to pastures and agriculture. Some of the microscopic organisms and other more elementary living forms in the waste or stimulated by its constituents may also play a part in improving soil conditions.[5] It was Liebig who early last century first emphasized the idea that, in order to maintain the natural

Table 7.1

FERTILIZING CONSTITUENTS OF SEWAGE (after Sierp[3])

Fertilizing constituent	lb/hd d (g/hd d)		
	Raw sewage	*Biologically purified sewage*	*Digested sewage sludge*
Nitrogen	0·0282 (12·8)	0·0240 (10·9)	0·0029 (1·3)
Phosphate (P_2O_5)	0·0077 (3·5)	0·0062 (2·8)	0·0015 (0·7)
Potash (K_2O)	0·0154 (7·0)	0·0147 (6·7)	0·0004 (0·2)
Organic matter (loss on ignition)	0·121 (55)	0·042 (19)	0·044 (20)

cycle of the elements, it was essential to return to the soil of a country all the matter, such as is in sewage, which had been derived from it, thereby conserving its fertility.

As with the water–sewage cycle (pp. 55–57), there are natural cycles of these elements and compounds, and these may be likewise direct or indirect, with short or long return paths. The waste may be discharged directly onto the land, with or without pretreatment, or indirectly by discharging untreated or treated sewage into surface waters which are then used for irrigation. The cycle will be relatively short on direct discharge onto the land, longer where it leads into rivers, streams, lakes or oceans. It is pointed out that fertilization of plankton in surface waters—including oceans—finally increases fish life from which new foods can be won by fisheries. Disposal of wastes into the ocean, directly or indirectly, may nevertheless represent a relatively large loss of fertility to a country which has to be replaced by special means. Therefore, sewage disposal must be considered as part of the problem of soil fertility of a country.[5]

In practice, however, the return of fertilizing constituents in sewage and other water-borne wastes to the soils from which they derive is nearly always difficult and in many cases impossible.

It should be noted that by no means all the fertilizing elements and substances lost from the soil will eventually appear in the domestic and industrial water-borne wastes of the country concerned, including even garbage and other refuse. The soil itself may also lose significant amounts of important constituents directly by erosion, by solution into groundwaters and in other ways. Therefore, although its fertility may be maintained in any given area only by perfect cycles of soil–food–body and other wastes–soil, and although such cycles may be very helpful in obtaining a practical balance, the possibility of other losses must also be taken into account.

PURIFYING CAPACITY OF NATURAL SOIL

Natural soils have a capacity for absorption and decomposition of water-carried polluting matter analogous to the self-purification of natural waters, the polluting matter being decomposed as a result of the biological activity of the organisms living in the soil. As with natural waters, this activity may proceed either through anaerobic (septic) decomposition or aerobic processes (oxidation). The first is generally slower and not normally selected for waste disposal onto land. However, septic conditions develop naturally if ponding occurs and the supply of oxygen to the soil, either gaseous or dissolved, fails or becomes inadequate. The maintenance of aerobic (oxidizing) conditions requires that oxygen is always present in the zones of purification, and this usually necessitates drainage and ventilation of the active soil strata. For the complete disposal of wastes onto land, any residual matter other than gases or substances carried away by water must finally appear as vegetation, e.g. grasses, shrubs or trees, or as mobile animal life, such as adult insects which fly away, or earthworms taken by birds, grazing animals, etc., so that a practical balance is maintained within the soil. According to Dunbar and Calvert, in the absorption and oxidation of domestic sewage the utilization of the fertilizing constituents by vegetation sets the limits for disposal onto land, shown in Table 7.2.

In practice, the load allowance should be lower than these figures indicate, because the conditions may not be so favourable. As a general rule, for average conditions in moderate climates the sewage load should be limited to 12 persons/acre (30 persons/ha), and other wastes suitable for such disposal to the same population equivalent.

Up to such limits, the purifying capacity of the soil is not likely to be exceeded and disposal is virtually complete, so that no ancillary disposal works are needed.

If higher loading rates are used, the effluent waters draining from the land-disposal area will not be completely purified and some secondary purification will be necessary, usually by disposal into natural waters. The higher the specific rate of loading of the land-disposal area, the greater is the secondary load to be carried by such receiving waters. Standard sewage farming practices involve much

Table 7.2

LAND DISPOSAL OF DOMESTIC SEWAGE (MODERATE CLIMATES)

*Limiting fertilizing constituent**	*Limit of load allowance,* contributing persons /acre	/hectare		*Corresponding yearly loading rate* cwt/acre	kg/ha
Nitrogen	30	75	(as N)	3	380
Phosphorus	40	100	(as P_2O_5)	1	130
Potassium	80	200	(as K_2O)	1–2	130–250

* The amounts of potassium, sodium, calcium and magnesium largely depend on the composition of the original supply waters which carry the water-borne wastes. All figures are approximate.

higher loading rates (see p. 190), so that only partial disposal of polluting matter is achieved, the effluent from such works carrying some residual pollution which must finally be disposed of elsewhere.

LOW-RATE IRRIGATION OF WATER-BORNE WASTES

On the above basis, low-rate irrigation of water-borne wastes is essentially an agricultural problem. A load allowance of 12 persons/acre (30 persons/ha) for a flow of sewage of about 40 gal (180 l)/hd d corresponds to a rate of about 8 in (200 mm) of sewage effluent per year. Continuous operation of an irrigation process cannot be ensured because the agricultural demand for irrigation varies throughout the year, and in most areas effluents must be disposed of by other means for some part of the time, because irrigation is undesirable, e.g. in periods of adequate or excessive rainfall. Adequate alternative arrangements may include additional biological treatment processes such as intermittent sand filtration or ponding, or perhaps only primary treatment with discharge of the treated effluent into natural waters (which then may be flowing at a rate appreciably higher than normal).

Special attention has to be given to matters of public nuisance and possible dangers to health of man or beast[6–8] requiring adequate regulations which may include

1. Sewage to be irrigated must first be treated at least partially by biological processes which, in order to obviate odour nuisance, may if necessary be followed by chlorination.
2. Along public roads or other public areas, carriers must be closed. Where open channels are permissible, the flow must be contained by concrete or stoneware linings to prevent uncontrolled leakage.
3. Nuisances to settlements and associated areas of land, including orchards, gardens, parks, roads, etc., must be absolutely prevented, suitable space being provided to ensure that no sewage spray or other contamination may reach them under any conditions.
4. Irrigation onto, or adjacent to, areas used for water supply is forbidden.
5. The time schedule for irrigation of any area shall ensure that at least one week must elapse before any areas irrigated with sewage can be harvested or used for pasture or forage.
6. Potatoes and other root crops for the vegetable market must not be irrigated with sewage after flowering, and other vegetables not at any time during growth by other than fresh water. However, vegetables may be grown on soils which have previously been fertilized by sewage irrigation.
7. Any wastes which may possibly carry spores of anthrax may not be used for irrigation. This includes any water-borne wastes from tanneries, horsehair mills or other factories using raw material from anthrax areas or which for any reason cannot be guaranteed to be anthrax-free.

Either surface or spray irrigation systems may be used. Surface irrigation requires careful grading of surfaces and extensive systems of channels for distribution and ditches to collect effluent and surplus water. Spray irrigation, on the other hand, is virtually independent of the shape and contour of the area, and is more flexible accordingly, having also the advantage that it can be sprayed in any amount, even fractions of one inch, over the area, thus conserving water. However, odour nuisance is more noticeable because odorous gases escape more readily from solution when the liquid surface is spread in fine droplets. Moreover, the danger to health is greater because the spray carries germs of disease onto plants, and even long distances through the air in sewage mists.[9] In such cases, therefore, it is necessary to treat sewage not only by biological processes but by additional means such as chlorination or by adding oxygenated fresh diluting water before spraying.[10]

Costs of such irrigation schemes are generally higher than for other methods of disposal but may in some cases be tolerable in view of benefits derived from agricultural production. Irrigation methods have been successfully applied to many trade-waste disposal problems; spray irrigation has proved especially useful for disposal of milk-processing (p. 327) and canning wastes (p. 323). Spraying over forested lands is also practicable.

Sewage farming by broad (high-rate) irrigation is not a method of disposal onto land but of treatment before disposal, the rates of loading exceeding the purifying capacity of the soils, so that a substantial part of the pollutional load remains in the effluent. Accordingly, it is considered below as a biological treatment method.

BIBLIOGRAPHY

DUNBAR, W. P. and CALVERT, H. T. *Principles of Sewage Treatment* (1908) Griffin & Co., London

SCHRÖDER, G. *Landwirtschaftlicher Wasserbau*, 4th Edn (1968) Springer, Berlin

VAN VUUREN, J. P. J. *Soil Fertility and Sewage* (1949) Faber, London

REFERENCES

1. DEUTSCH, M. 'Incidence of chromium contamination of groundwater in Michigan', *Tech. Rep. Rt A. Taft sanit. Engng Center W61–5* (1961) 98
2. WALTON, G. 'Public health aspects of the contamination of ground water in the vicinity of Derby, Colorado', *Tech. Rep. Rt A. Taft sanit. Engng Center W61–5* (1961) 121
3. SIERP, F. 'Der derzeitige Stand der Abwasserforschung', *Zentbl. Bakt. Parasitkde* **155** (1950) 318
4. HEUKELEKIAN, H. 'Utilization of sewage for crop irrigation in Israel', *Sewage ind. Wastes* **29** (1957) 868
5. MÜLLER, W. J. 'Die landwirtschaftliche Verwertung der Abwässer', *Wass. Boden* **1** (1949) 125
6. ANTZE, H. H. 'Erfahrungen bei der gesundheitlichen Überwachung eines Rieselfeldes', *Schr. Reihe Ver. Wass. Boden-u. Lufthyg. No. 19* (1961) 82
7. HIRTE, W. 'Vergleichende microbiologische Untersuchungen an unterschiedlich rieselmüden Böden der Berliner Rieselfelder', *Zentbl. Bakt. Parasitkde* **114** (1961) 367, 490; **115** (1962) 606

8. VIEHL, K. 'Zur Frage der Abwasserverwertung', *Gesundheitsingenieur* **71** (1950) 126
9. REPLOH, H. and HANDLOSER, M. 'Untersuchungen über die Keimverschleppung bei der Abwasserverregnung', *Arch. Hyg. Berl.* **141** (1957) 632
10. MÜLLER, W. J. 'Landbewässerung mit verdünntem Abwasser', *Gesundheitsingenieur* **73** (1952) 17

8

RECLAMATION AND RE-USE

INTRODUCTION

IN CHAPTER 5, the possibilities of re-use and recovery of either the water or other constituent fractions of water-borne wastes were listed among the various methods of disposal. Technically, reclamation and re-use must be considered as part of the disposal problem, not only because of their inter-relationships with the techniques of handling and treatment involved in disposal, but also because of their importance for the overall economy of any wastes system, especially of many particular industrial installations or complexes.

Nevertheless, the possibilities of reclamation and re-use are not properly matters of waste disposal but aspects of the broader problems of *conservation* both of water resources (Chapter 1) and raw materials, within a factory, community or whole country.[1-3] Therefore, possibilities of recovery will be considered before the later stages of preparation for dispersal or destruction.

The re-use of sewage and other water-borne wastes and the reclamation of particular constituents fall under the general headings

1. Reclamation of water
2. Reclamation of water and fertilizing constituents for agricultural purposes
3. Utilization of waste nutrients
4. Reclamation or recovery of waste materials or specific substances such as textile waste, metals, phenols, lanoline and other raw materials and products of chemical, engineering or other industries
5. Re-use of waste water for industrial purposes.

Apart from the physical or chemical processes involved, any such re-uses also affect aspects of hygiene and aesthetics. It may fairly be said that, notwithstanding careful scientific control of purely artificial processes, natural means of disposal involving natural cycles of use and re-use provide the most complete and safest procedures where hygiene is the paramount consideration. As a first principle, therefore, where the reclamation is concerned with re-use of water or foodstuffs for human consumption, long natural cycles should be used as far as practicable.

In other cases, where economy and practicability predominate, short cycles of re-use are preferable, even if incomplete.

RECLAMATION OF WATER

Water for drinking and for bathing should not be taken directly from sewage or other waste effluents, no matter how elaborate the processes of treatment may be, since danger to health is too great. Use of the natural purifying agencies of natural soils and waters as additional barriers to the transmission of diseases greatly increases the degree of safety in re-use of the water. Thus the incorporation of surface water in the water–sewage–water cycle has proved practicable even for municipal water supplies (Chapter 6, pp. 55–57) where sufficient dilution is provided by the receiving water and adequate high-efficiency purification is ensured before re-use.[4–7] Such re-use is limited by the technical and economical possibilities to produce safe potable water from such sources. A higher degree of pollution of surface water, i.e. a lower dilution ratio, can be accepted if the path is lengthened by incorporating further natural purifying agencies in the water–sewage–water cycle. This is done by making proper use of the purifying capacities of natural soils, so that water for re-use is drawn from the river or other surface waters only indirectly, e.g. from wells or infiltration galleries separated from it and thus flowing first through a sufficient volume of ground. (Suitable geological formations are an essential part of any such system.) The effective detention period of the recycled water in the ground should be at least 1½–3 months, if possible more, because natural purification is a slow process. A similar procedure is the 'artificial' recharge of groundwater by introducing river water into the ground with the aid of infiltration wells or recharging basins. Here, too, the path of the water through the ground to the supply wells has to be long.

For continuous and satisfactory operation of the infiltration systems, the surface water must not be too polluted, satisfactory upper limits being about 5 mg/l for settleable suspended solids content and for total bacterial count no more than 100 000/ml. With higher degrees of river pollution it is usually necessary to pretreat the river water before infiltration, in some cases by sedimentation, in others also biologically, e.g. in reservoirs.

Artificial replenishment of groundwater supplies is standard practice in various countries, especially in Germany[1] where artificial groundwater is the most important source for water supplies in the Ruhr river area, as it accounts for more than 60% of the total. Re-use in the U.S.A. has been referred to above (pp. 55–57). Treatment of waste water, in particular municipal sewage, to a relatively high degree of purification in treatment plants, followed by groundwater recharging through infiltration basins and by abstraction as groundwaters for irrigation, or, finally, after further stages of treatment for industrial and domestic water supply, is now fully established as a means of extending the availability of water, e.g. at Whittier Narrows in California and for Long Island, New York.[8–11] In Holland and elsewhere, the replenishing of groundwaters in such ways is also playing an increasing part in the holding back or even reversal of advancing pollution from salt-water intrusion.[12]

In Israel, water reclamation is likewise of great importance. Based largely on systems of oxidation ponds, water usage is pursued through cycles of municipal water supply–sewerage–three stages of purification, the first suitable for agricultural use, the second for certain industrial applications, the final being effected by groundwater spreading and attendant natural purification processes.

It is clear that water reclamation is fast being recognized throughout the world as providing a logical means of expanding the effective capacity of local areas to meet growing supply requirements. A widening of the technologies of both waste-water disposal and water treatment forms part of this endeavour. The possibilities of use and re-use in extended and even repeated water–waste–water–treatment–dispersal– retreatment–reclamation cycles should always be kept in mind, during any investigation into water supply or waste-water disposal problems, be it for municipalities, regions or individual industrial installations.

Reclamation of municipal sewage for industrial purposes may follow a similar pattern as for public water supplies, except that the requirements of water quality vary considerably according to the industrial usage. The food industry, for example, requires water of high quality, as for municipal supplies. Other users, especially for ordinary cooling purposes, may tolerate water of poorer quality. On the other hand, boiler-feed waters have to meet particular quality standards different from those for drinking water. In many cases, especially where the industry cannot otherwise obtain sufficient cheap water of satisfactory quality, it has proved feasible to use reconditioned effluents from sewage treatment works directly and economically.

The reclamation of water from the water-borne wastes of municipalities for purposes of supply is of special importance in areas with inadequate natural sources or where permanent streams are few or widely separated. Under such circumstances, the costs of purification matter far less in comparison with the total costs of water supply and waste disposal, both for individual industries and their regional areas. The most economical means of reclamation are generally achieved by the formation of co-operative regional associations of communities and industries with specific responsibilities for water conservation and pollution control.

RECLAMATION OF WATER AND FERTILIZING CONSTITUENTS FOR AGRICULTURAL PURPOSES

The agricultural use (reclamation) of sewage or other water-borne wastes may be based partly on the water content, partly on the fertilizing constituents, or on either or both. Waste waters may be economically applied particularly to areas which need irrigation, i.e. where the amount of water available is limiting agricultural production. This demand depends on the characteristics of the soil, on rainfall, evaporation and other climatic factors, on groundwater conditions, etc., and on the type of crops to be grown.[13–15] It is the general experience that in most areas of the world the annual water requirements of irrigable lands suitable for agricultural utilization average about 10 in (250 mm) of water. This corresponds to 625 gal/acre d (7 100 l/ha d) or, on the basis of 40 gal (180 l)/hd d, to a sewage effluent of about 16 persons/acre (40/ha) (see also Table 7.2).

For most satisfactory operation, the wastes should preferably be treated, before irrigation, at least to remove settleable suspended solids by sedimentation (see Chapter 10). However, when the rates of irrigation are higher than the above, or the settled wastes contain much greater concentrations of polluting matter, it may also be necessary to apply first some biological treatment process to avoid anaerobic conditions (see Chapter 12). The sludges formed and separated in the sedimentation tanks can be reclaimed for agricultural purposes, on account of either nutrient or humus content, as described below.

The nutrient or fertilizing value of water-borne wastes, as discussed in Chapter 7, including the significance for plant growth of nitrogen, phosphorus and potassium contents of domestic sewage, is equally relevant for agricultural utilization.

UTILIZATION OF WASTE NUTRIENTS

Utilization as food

Nutrients in sewage cannot be re-used directly as human food. Higher animals, such as cattle, sheep, pigs and even fish, can also use such waste nutrients only indirectly, through the agencies of lower forms of life such as bacteria and other micro-organisms, and algae and other plants. On the other hand, some waste waters may contain organic matter which could conceivably be recovered and concentrated for use directly as food for beasts or for man. The proportions of such waste-food materials escaping from one series of industrial processes may not be economically significant to these but could still be so otherwise: waste waters from milk factories and sugar refineries might provide examples. However, once they reach the sewer, direct re-use is no longer feasible. They then become charged with pathogenic viruses, germs, protozoa, parasitic ova and, possibly, toxic substances, requiring more or less elaborate and careful treatment before recycling.

Generally, the waste water itself with all its constituents may be re-used, for example, as fertilizer for the land or in rivers, lakes or ponds, or the sludge may be separated by treatment. It is essential to ensure that pathogenic organisms are prevented from contact with food, including plants used for food supply. Irrigation of sewage must be regulated accordingly (see p. 106). The application of sludge also involves consideration of pathogenic contamination, and the effects of digestion, drying and direct disinfection must be studied accordingly.[16, 17]

Nevertheless, as mentioned before, sewage and other wastes contain useful concentrations of nutrients such as nitrogen, etc., also substantial quantities of organic matter which provide a possible basis for the direct growth of animal foods by means other than agricultural production. Such are, for example, the growth of fish in fish ponds (p. 244), a system in which waste purification is effected by bacteria and algae which grow by utilizing their organic content in the presence of oxygen and sunlight, thus providing food for growth in turn of protozoa, smaller polyzoa, crustaceans, etc., in a chain culminating in the growth of edible fish. Basically, the bacteria use the oxygen to oxidize the carbonaceous matter to carbon dioxide and water, and from this (which may be supplemented by 'inorganic' carbon dioxide from mineral bicarbonates in the water itself), the photosyntheses of the algae produce carbohydrates and other organic matter used to build more algae which, in turn, provide the food for animals small and large.

This algal–bacterial symbiosis[18] may be intensified in special high-rate ponds as a means of growing large amounts of microscopic algae which may be suitable for direct feeding to animals such as pigs or poultry, or for incorporation as high-protein supplements to their diets. Means of achieving suitable concentration of such algal cultures by sufficiently economical means are being sought now, and some measure of success is anticipated. In all such operations, the rules applying to hygiene and animal health must be observed.

Agricultural utilization

Agricultural utilization of sewage sludge and other waste solids is a topic of considerable interest and increasing importance in sewage disposal projects; it is, therefore, discussed here in some detail.

Compared with commercial inorganic fertilizers, sewage sludge (see p. 103) has only a low content of available nitrogen, phosphate and potash but, as with farmyard manure, it is valuable for its capacity to produce water-retaining humus.

From the agricultural viewpoint, the first question arising is whether such sludge should be used raw or only after digestion (p. 254). During digestion, its protein–nitrogen content decreases by about 40%, the nitrogen appearing as ammonia dissolved in the supernatant liquor which separates out during this process. It follows that, if the liquor is also used as fertilizer, there is little loss of nitrogen as a result of digestion. The case is different for composting and rotting where there is a real and substantial loss of nitrogen. Fresh sludge contains aggregates, of, for example, fats and paper, and carries many viable seeds of weeds and other plants, but must actually be decomposed before it is useful to plants. Digested sludge, however, is more nearly homogeneous. The gases formed during decomposition may be recovered as a valuable fuel from digestion tanks but are lost if sludge is used before digestion.

Before utilizing sludge as fertilizer for soils, it must be tested analytically and tried in pot experiments to make sure that it does not contain toxic or other harmful substances. Frequent problems are transport to the area where it is to be used and its distribution in the soil. These can be difficult and costly because the distances involved are often considerable; transport by pumping as a fluid usually affords the cheapest and simplest means. Sludge may be pumped readily and economically as far as 10 miles (16 km) or more, and discharged onto the land directly in liquid form. In most cases, however, it is necessary to dry it; before transport it should also be milled to facilitate spreading. It has to be noted that, during drying, most of the nitrogen contained in the supernatant liquor is lost with the water, and hence its nutritional value cannot then be reclaimed.

Composted sludge is also used as a fertilizer, and various composting processes are now receiving attention all over the world. In this treatment, the mixture must of course be wet enough for active chemical and biochemical decomposition but without any free water which might prevent adequate ventilation. Composting of sludge and garbage is the most promising of the processes available for large-scale use.[19, 20] Such mixtures are favourable as to moisture content, degree of porosity and carbon: nitrogen ratio (optimum about 30 : 1). Coal ash must not be present in the garbage in appreciable proportion, as it is detrimental to the process; it may, therefore, be necessary to eliminate it before preparing the mixture to be composted. In some cases it might be advisable to use sludge not after digestion but fresh from the biological treatment processes (excess activated sludge, trickling-filter humus sludge), partially dried by vacuum filtration and finally composted with garbage.

Other standard composting procedures have also been used. Kühnlenz introduced at Frankfurt[21] high-temperature composting of sewage sludge; dried raw sludge is composted alone, the temperature of the compost heap reaching 160°F (70°C). Seeds, parasitic ova and most pathogenic bacteria are destroyed, the final product after milling being a suitable powder of 15% moisture content. However, a high wastage of organic matter and some odour nuisance are inevitable in such high-temperature processes. Low-rate rotting in compressed heaps results in much lower maximum temperatures, usually not exceeding 90–100°F (32–38°C).

Artificial drying (p. 287) and milling produces a low-moisture fertilizer of a character generally suitable for spreading, and since temperatures may reach 200–210°F (93–99°C), sterilization is practically ensured. Prepared in this way from excess activated sludge, it contains below 10% moisture, 3–6% nitrogen and 2–3% phosphate.

Generally, digested sludge carries less dangers to health than fresh, and suitably composted sludge is relatively safe except for anthrax and some parasitic worm ova. Heat-dried sludge is mostly sterile. It has been found, for example, that the causative organisms of foot and mouth disease and of swine fever are killed by temperatures of 110°F (43°C), of tuberculosis at 130°F (54°C), but anthrax bacilli not below 160–165°F (70–74°C). Hook worm ova (European types) do not survive a temperature of 150°F (66°C). The full control of these organisms may require protracted periods of digestion and drying (at least six months)[16, 17, 22] or heating for short periods[23, 24] at 130°F (55°C) (two hours for raw or digested sludge).

The high cost of transport of sludge restricts its distribution so that it can only be utilized at relatively short distances from the treatment works. In large cities, these costs may be subsidized out of the additional expenditure otherwise incurred by alternative means of disposal. The sludge should be as dry as possible to reduce transport costs to a minimum and prepared for easy spreading onto land.

RECLAMATION OF SPECIFIC CONSTITUENTS FROM WATER-BORNE WASTES

In many cases, specific constituents of wastes can be partly or almost wholly recovered. This may be advantageous to industry, apart from relieving the load on receiving waters, since a reduction in the amount of waste increases the efficiency and economy of manufacturing processes and reduces the size of waste treatment works and, consequently, the costs of treatment and disposal.[2] The re-use or sale of reclaimed waste material may add appreciably to the profits of business.

Fats, greases and oils entrained in sewage or industrial waste waters may be recovered by means of separating tanks ('grease-traps') or by special processes. For example, the partial recovery of lanoline (wool grease) is common practice in wool-washing plants, but at Bradford sewage treatment works, England, it is also recovered, processed and sold as a valuable by-product. The raw sewage is treated with acid to improve separation of the grease. Revenue from sales substantially offsets the costs of purifying the sewage.[25]

Similarly, treatment of abattoir waste may yield up to 0·1% by weight of fats calculated on the weight of the animals killed, while blood and other refuse yield valuable fertilizers. Using various processes, mechanical, chemical and even biological, many such substances can be recovered for re-use or for other uses including, typically, recovery of fibres from dyehouse or paper mill wastes, of cyanogen and metallic compounds or elements from plating-room or pickling wastes (including partial recovery of sulphates), recovery (for re-use) of phenols and the use of waste sugar or cellulose wastes for growth of food yeasts or production of dilute acetic acid solutions.

Related to this aspect of reclamation is the recovery of combustible gases evolved during sludge digestion processes (see p. 279). The resultant gas mixture has a high calorific value and may partly or largely offset costs of sewage treatment, either as power for works operation, leading to higher efficiency of treatment processes, or by sale, after some treatment, for use outside the works, e.g. for power, as a substitute for petrol.

RE-USE OF WASTE WATER FOR INDUSTRIAL PURPOSES

A distinction should be made between the re-use of waste waters —that is, waters used in a particular factory and in some way soiled or altered in the process—within the same factory and partly or entirely outside it.[3]

The re-use of waste water within a factory is an effective means of reducing the volume of liquid wastes discharged from it and, therefore, has been applied in many industries. Waste water may be recycled where water is used for cooling, material transport, washing of raw materials or manufactured products, etc., where adequate treatment for reconditioning is practicable and not too costly and, in particular, where process water can be obtained only through costly treatment of surface or groundwater. Provision for recycling this water as many times as possible gains ready acceptance, subject only to holding its pollution at a satisfactory level. This need is usually met most easily with cooling-water systems which commonly are neither mechanically nor chemically polluted to any significant degree. However, even cooling waters may become infected with slimes or other biological growths and may require special disinfection processes. In other cases, in foundries for example, heated cooling water is used in sequences of increasing temperatures and thus several times, but perhaps without any recycling as such; in many cases it may even then be fit for further re-use as process water.

The recycling of process waters is possible in many industries, also their re-use from one operation to others. Naturally this is the more attractive the greater the overall usage of water in a particular industry. The cellulose and paper industries provide well-known examples, where large flows of water are used to transport relatively low concentrations of fibres, as in the paper-making process. Here the re-use of water is combined with the recovery of waste fibre, but care is required to control bacterial and fungal growths in the recycled water systems. Sugar factories, starch, fibreboard, coke quenching, vegetable-, coal- and wool-washing, dyeing and laundering and electroplating process plants provide other examples of possible water re-use, though varying considerably in ways and means and in the relative economies which can be effected.

The deliberate re-use of waste waters partly or wholly outside the factory in which they arise is also possible but uncommon except where it is controlled by responsible water resources authorities as part of a local or regional water conservation programme, where it is incidental to cycles of use and re-use of water involved in the waste disposal system, or, in the case of local industrial complexes operating under special contractual agreements. Generally speaking, waste waters dirtied in one process or factory require substantial purification before being suitable for use elsewhere.[3]

BIBLIOGRAPHY

GOTAAS, H. B. *Composting: sanitary disposal and reclamation of organic wastes* (1956) World Health Organization, Geneva

WYLIE, J. C. *Fertility from town wastes* (1955) Faber, London

American Institution of Chem. Engineers, *Water Re-use* (Ed. CECIL, L. K.) Chem. Engineering Progress Series 78/63 (1967)

International Union of Pure and Applied Chemistry, *Re-use of Water in Industry* (1963) Butterworths, London

REFERENCES

1. MÜLLER, W. J. 'Re-use of waste water in the Federal Republic of Germany' (1969) O.E.C.D., Paris
2. STANDER, G. J. 'By-product Recovery, Re-use of Water, and Clean House-keeping—Their importance in Industrial Effluent Disposal and Pollution', *J. Proc. Inst. Sew. Purif.* (1957) Pt 1, 16.
3. STANDER, G. J. and FUNKE, J. W. 'South Africa reclaims effluents as industrial water supply', *Wat. Wastes Engng* **6** (1969) D-20
4. BONDERSON, P. R. 'Quality Aspects of Waste Water Reclamation', *Proc. Am. Soc. civ. Engrs* **90** (SA5) (1964) 1; discussion, **91** (SA3) (1965) 124
5. TEBBUTT, T. H. Y. 'Sewage effluents as a source of water', *Effl. & Wat. Treat. Jnl* **5** (1965) 565
6. CILLIE, G. G., VAN VUUREN, L. R. J., STANDER, G. J. and KOLBE, F. F. 'The reclamation of Sewage Effluents for Domestic Use', *Proc. 3rd Int. Conf. Wat. Pollut. Res., Munich* **2** (1966) 1
7. STANDER, G. J. and VAN VUUREN, L. R. J. 'The Reclamation of Potable Water from Sewage', *Wat. Pollut. Control* **68** (1969) 513
8. PARKHURST, J. D. 'Progress in Waste Water re-use in Southern California', *Proc. Am. Soc. civ. Engrs* **91** (IR1) (1965) 79; discussion, **91** (IR4) 95
9. AMRAMY, A. 'Waste Treatment for Groundwater recharge', *Proc. 2nd Int. Conf. Wat. Pollut. Res., Tokyo* **2** (1964) 147

10. Taylor, L. E. 'A Report on Artificial Recharge of Aquifers for Hydrology', *J. Br. Wat. Wks Ass.* **45** (1963) Suppl. to No. 386
11. Stevens, D. B. and Peters, J. 'Long Island recharge studies', *J. Wat. Pollut. Control Fed.* **38** (1966) 2009
12. Jansa, V. 'Artificial Replenishment of Underground Water', *Proc. Congr. int. Wat. Supply Ass. Paris* (1952) 147
13. Evans, J. O. 'Ultimate sludge disposal and soil improvement', *Wat. Wastes Engng* **6** (1969) 45
14. Sopper, W. E. 'Wastewater renovation for re-use—key to optimum use of water resources', *Wat. Res.* **2** (1968) 471
15. Hershkovitz, S. Z. and Feinmesser, A. 'Sewage reclaimed for irrigation in Israel from oxidation ponds', *Wastes Engng* **33** (1962) 405
16. Liebmann, H. 'Parasites in sewage and the possibilities of their extinction', *Proc. 2nd Int. Conf. Wat. Pollut. Res., Tokyo* **2** (1964) 269
17. Murray, H. M. 'The incidence of *Ascaris* Ova in Pretoria Sludge and their Reduction by Storage and Maturation in Large Heaps', *J. Proc. Inst. Sew. Purif.* (1960) Pt 3, 337; also Silverman, P. H. and Guiver, K., *J. Proc. Inst. Sew. Purif.* (1960) Pt 3, 345
18. Oswald, W. J., Golueke, C. G., Cooper, R. C., Gee, H. K. and Bronson, J. C. 'Water Reclamation, Algal Production and Methane Fermentation in Waste Ponds', *Proc. 1st Int. Conf. Wat. Pollut. Res., London* **2** (1962) 119
19. Thorstensen, A. L. 'Refuse Composting with Sewage Sludge', *Wat. Pollut. Control* **66** (1967) 525
20. Jaag, O. 'Present-day Problems in Composting', *J. Proc. Inst. Sew. Purif.* (1958) Pt 4, 436
21. Imhoff, K. and Müller, E. 'Die heisse Vergärung von stichfestem Schlamm', *Gesundheitsingenieur* **61** (1938) 160
22. Müller, W. J. 'Schlammbehandlung und Krankheitserreger', *Gesundheitsingenieur* **75** (1954) 187
23. Keller, P. 'Sterilization of sewage sludges', *Publ. Hlth, Johannesburg* (Jan. 1951) 11; *J. Proc. Inst. Sew. Purif.* (1951) Pt 1, 92
24. Haury, M. and Eckert, J. 'Das neue Zentralklärwerk der Stadt Darmstadt', *Städtehygiene* **12** (1961) 185
25. *Rep. City of Bradford Sewage Cttee* (cf. *Wat. Pollut. Abstr.*, 1959, 1961, 1963, 1966)

PART C

TREATMENT OF SEWAGE AND OTHER WATER-BORNE WASTES

9

GENERAL CONSIDERATIONS

INTRODUCTION

IN THIS section, the term 'wastes', except where otherwise qualified, includes all ordinary water-borne wastes, including domestic or municipal sewage and industrial or trade wastes.

As a general rule, all waste waters must finally be disposed of into natural waters, save perhaps in some cases of low-rate disposal onto land. Such disposal could endanger health by infection or contamination of drinking and bathing waters, beach sand, foods such as oysters and other fish, or food plants. Odour or other aesthetic nuisance, damage or danger to fish and other animal life and interference with the normal use of surface or groundwaters might also be involved. Some treatment of waste waters, therefore, is commonly necessary before disposal. It is most important to realize that in such treatment, the emphasis must be laid on the effective removal of polluting matter—for example, as sludge-forming solids or by gasification—after which the waters, more or less purified, are disposed of by dilution and dispersal and, ultimately, by assimilation in natural environments, in rivers, lakes, groundwaters or the ocean.

The methods of treatment adopted must be sufficient to ensure the necessary degree of purification required to suit the means of disposal. The designer of treatment works has to consider both treatment and final disposal as one integral design, however complicated, in order to develop a sound and proper solution for each problem. Thus, all arrangements for disposal are involved, including any necessary preparation of receiving waters or lands, and the appropriate means of distribution and discharge.

Whereas the final disposal of the polluting matter of wastes such as sewage depends largely on material changes which are due mainly to biological activity, the techniques of waste treatment include also processes which are almost entirely physical or chemical. Accordingly, waste-treatment processes may be classified broadly as physical (or mechanical), chemical and biological.

Physical or mechanical processes chiefly directed towards the concentration and separation of suspended matter only, include screening, mechanical filtration, mechanical flocculation, sedimentation and flotation. Heating, evaporating and centrifuging are other means of less general application. These are discussed in subsequent sections. In addition, other processes of a novel nature, using sound, light or electrical effects, have been proposed or adapted for specific cases but are not considered in detail.

In addition to very finely divided particulate matter which settles only very slowly, colloidal matter, both organic and inorganic, is commonly present in wastes in significant amounts. Such solids may be aggregated by the addition of coagulants such as ferric chloride, and may thus become readily settleable or filtrable; correspondingly, dissolved matter may be precipitated by chemical reaction, such as proteins by chlorine solutions. However, only a limited number of practical techniques, based on simple chemical reactions, can be generally applied in waste treatment.

Some colloidal matter may be precipitated without coagulants by agitation alone (mechanical flocculation). On the other hand, coagulated proteins, such as egg albumin after cooking, are readily dissolved by 'ferments' or enzymes which may be secreted or excreted by bacteria or protozoa, just as by animals or in the human stomach. In these and similar ways, the solids content of wastes, both dissolved and suspended, may be altered during treatment, so that physical, chemical and biological processes may be involved together.

Distinction must be made at the outset between mineral (inorganic) and organic matter as regards waste treatment. Mineral matter is to be defined as that which is not normally susceptible to decomposition by biological agencies alone, save perhaps by highly specialized organisms in restricted environments. Thus black coal is mineral, not organic, in this sense. The mineral matters encountered in waste treatment are perhaps as many and various as the organic, but in general they are biologically inert and respond only to physical or chemical treatment.

In the main, the organic matter of sewage is the residue of the life processes of human beings, including food residues, which consist of complex animal and plant tissue made up principally of carbon, hydrogen and oxygen, commonly also with nitrogen and sometimes sulphur, phosphorus and other elements, usually with a large proportion of water. Proteins and related compounds containing nitrogen and appreciable amounts of sulphur and urea—an end-product of protein decomposition in the human body—are specially important because from their decomposition the most objectionable odours arise. Wastes from food and other industries may contain similar substances, but in many cases specific classes of organic matter occur almost exclusively; for instance in sugar wastes, although still a mixture of complex substances, they consist very largely of carbohydrates and contain little nitrogen or sulphur. From this viewpoint, a most important characteristic of organic wastes is the susceptibility of the organic constituents to rapid decomposition by biological activity. This lends itself to utilization in the processes of waste treatment.

In all living systems, there is a balance between processes of oxidation and reduction, and the organic residues occurring in wastes thus consist of substances in various stages of oxidation or reduction. In biological treatment, the enzyme systems of bacteria and other organisms are again effective in bringing about further biochemical changes. When sufficient free oxygen is available, aerobic conditions, favouring the development and active life and growth of aerobic organisms, result in a substantial oxidation of the organic matter, finally to carbon dioxide, water, nitrogen dioxide, sulphate and so on, involving many biochemical reactions. Different reactions follow if there is insufficient free oxygen supply and anaerobic or reducing conditions develop; oxidation of some substances results, with a corresponding reduction of others, so that, taking an apparently simple example, the oxygen of nitrates serves to oxidize some proteinaceous matter yielding ammonia or simply nitrogen.

MICRO-ORGANISMS AND WASTE TREATMENT

The use of biological treatment processes involves an understanding of the factors controlling the activity of the effective organisms.

Of these, bacteria are the principal agents.[1–3] They belong to the plant kingdom, being members of the group which includes fungi, moulds and yeasts. They are microscopically small, and most consist of minute individual cells which contain a living complex protoplasm, characteristic for each type of organism. This has a high moisture content and may be likened in appearance to egg albumin but is composed of many highly complex substances together with simpler inorganic and organic compounds. Most bacteria multiply (reproduce) simply by the splitting of one cell to make two approximately equivalent ones which develop independently until each again splits, yielding four cells and so on; in a suitable environment this may happen at intervals of one hour or even less. They become visible to the naked eye only as colonies, i.e. aggregations of very large numbers of individual cells which are usually suspended in a 'slimy' fluid medium spreading over surfaces, often of plants or stones. Together with other solid matter, such colonies form the flocculent aggregates which are characteristic of, and essential to, the activated-sludge process (see p. 217) where such flocs, suspended mechanically in the aerated sewage liquor, effect rapid biochemical decomposition of simple and complex organic substances. Bacterial slimes are here associated with other microscopic organisms such as protozoa; indeed, it has been shown that these alone can accomplish similar changes.

The bacterial cells of a particular species are generally similar in shape and size but different species may vary considerably. Three basic shapes are generally distinguished; the *coccus*, essentially spherical, mostly only one micron or less in diameter; the *bacillus*, generally 'rod-shaped', sizes varying from less than one micron to over six microns long (mostly < 3–4μ) and up to one and one-half microns diameter; and the *spirillum*, in which the cells are convoluted. The cells of the cocci occur characteristically in chains (streptococci), bunches (staphylococci), or more simply, in pairs or groups of four cells (tetrads, sarcinae). The bacilli exist mostly singly, but some species tend to form chains; the spirilla singly. Some bacterial cells are provided with one or more movable, extremely fine hair-like appendages (flagellae) by means of which they can move themselves, others are non-motile.

The bacterial cell is composed essentially of a membrane (the cell wall) enclosing a watery protoplasm which contains the cell nucleus (sometimes other specific bodies also). The cell wall is surrounded by an adherent 'slime' layer, in some organisms relatively very thick, in others variable according to the environmental circumstances. Overall, the bacterial unit is made up of about four-fifths water and one-tenth each of organic and inorganic solids, and most bacteria are killed by excessive evaporation from the outer slime layer. However, some are able to form structures called spores, and the spores of some organisms are specially resistant to outside influences such as chlorination, moderate heat or prolonged desiccation (similarly with encysted forms of protozoa, etc.). *Clostridium welchii* (p. 441), a comparatively large bacillus, strictly anaerobic, and *Bacillus anthracis*, a large aerobe, provide examples.

Most bacteria can grow only by utilizing organic substances such as certain carbohydrates, proteins or simpler compounds (particular species may use even hydrocarbons), i.e. they are *heterotrophic*. There are some types, however, the *autotrophic* bacteria, which can utilize fully oxidized carbon (carbon dioxide or carbonate) as a source of carbon for cell growth, their vital energy requirements being obtained by other oxidations. Thus some thiobacilli, in the presence of carbon dioxide, ammonia (or other nitrogen source), water (and phosphate, etc.) use the oxidation of sulphur compounds; others oxidize ferrous or manganous compounds, Of the heterotrophic bacteria, those called *saprophytic*, which specifically use organic debris ('dead' organic matter), effect the decomposition of the organic pollutants of sewage and other water-borne wastes.

Protozoa are single-celled organisms belonging to the animal kingdom, generally 10–20 times larger than the bacteria. As with these, there are many types, but most are predatory upon the bacteria or microscopic algae and do not feed directly on organic residues. There is a very large number of different species showing great variety of shape and size. Most are motile, either by amoeboid distortions of the whole cell or by the movement of cilia or flagellae.

Bacteria, microscopic algae, simple fungi, protozoa and many more types of living organisms develop naturally in the presence of decomposable (biodegradable) organic matter, their numbers and specific types depending on its composition. The excessive growth of these, the simpler organisms, is controlled in turn mainly by larger animals, accompanied in some cases by higher plants, progressively the worms, crustaceans, snails, etc., and finally (in the receiving waters) the fish. From this point of view, the process of biological treatment of wastes and their disposal may be seen as a series of bacterial decompositions leading to a complex food chain, with individual cycles of chemical changes.

Bacteria are sensitive to both acidity and alkalinity, and for most bacterial types a neutral medium is optimal. Large sudden changes in pH may be intolerable, but most species can readily adapt themselves within a range of about one or two pH units; thus pH 7–8 or even pH 6–8 may be tolerated, although the reduction in activity with changing pH can be appreciable. There is also an optimum temperature range for each type of bacteria, with limits of both active growth and toleration without injury. They die naturally if the medium in which they are suspended dries up; otherwise they may be killed or even disintegrated, or temporarily 'paralysed', by disinfection, by chlorine, ozone or permanganate, or other destructive or toxic substances. Their metabolism proceeds as a result of diffusion through the bacterial cell walls which allow transfer only of fluids such as water, certain gases and dissolved substances. Insoluble matter, such as the suspended-matter content of sewage (including colloidal solids) is decomposed *outside* the bacterial cell by the action of enzymes excreted by the bacteria (exoenzymes), soluble decomposition products being then further decomposed *inside* the cell by other enzyme systems (endo-enzymes). Accumulated end products are eventually excreted from the cells. Bacterial cells are also more or less affected by the concentration of the dissolved substances inside and outside the cell, in so far as this determines the solution (osmotic) pressures of the internal and external systems. When the external pressure is greater, water is withdrawn from within the cell which consequently shrinks, and this may affect or inhibit the activity of the organism; when it is less, the organism tends to absorb water from the environment and may become more efficient. As with other environmental factors, there are optimum conditions for particular types of organisms; however, the living matter of the cells has powers of compensation which partly offset the effects of changes in the environment within the limits of tolerance. It is for such reasons that bacteria inured to dilute aqueous systems such as domestic sewage are less active when dispersed in sea-water of moderate salinity.

Bacterial metabolism is continuous, provided that the food supply is maintained and undue accumulation of waste products prevented. Hence decomposition efficiency is greatest in continuous systems such as percolating filters (p. 197) and sludge-digestion tanks (p. 260) when properly designed and operated.

Almost all 'biological' sewage-treatment processes utilize aerobiosis, that is, biological activity in the presence of dissolved or free oxygen or both. It is to be noted that the separation of polluting organic matter from wastes cannot be attributed entirely or even largely to the direct metabolic activities of bacteria or other living organisms. The principal basic process rather is that of adsorption, a surface effect arising primarily from the complex of flocculent matter and slime associated with the living organisms. Not only the enzymes but other excretions from the living cells, including waste products of their metabolism together with residual humus-like matter, make up this complex active surface. Obviously, the living organisms are essential to this mechanism, not only by providing essential constituents of the active substrate but also by continuously removing part of the adsorbed matter, so that adsorption is continuous.

The processes of adsorption would soon come to a stop if the adsorbed matter were not broken down by living organisms as fast as it accumulates. End-products must likewise be removed continuously or before they become excessively concentrated. This is achieved partly by biochemical oxidation, so that oxygen is required continuously and must always be available, usually from atmospheric air, and dissolve into the water at a rate sufficient to maintain aerobic conditions. The main end-products of aerobic oxidation by bacteria are carbon dioxide and nitrate, from the carbon and nitrogen content of the organic matter and ammonium compounds, and sulphate from the sulphur (with water from the hydrogen).

Anaerobic biological systems are used for waste treatment too, particularly for more concentrated wastes such as sludges from sedimentation tanks or very strong raw wastes, e.g. from abattoirs or associated meat-packing industries. Before modern processes were developed, septic tank treatment of sewage was widely used; many small plants are still operating. For more concentrated suspensions, such as sewage sludge, in which the greater proportion of the solids are insoluble, anaerobic processes are well suited for preliminary treatment (see p. 315). The sludge-digestion tanks of sewage treatment works use anaerobiosis under conditions so controlled that 'alkaline fermentation' ensues. The several species of bacteria which are active under these conditions effect decomposition of nitrogenous organic matter, with the formation of ammonia, while fatty acids (from soaps, waste foodstuff and body wastes) and other carbonaceous matter are also decomposed, the system remaining essentially neutral; considerable volumes of gases are evolved, mainly methane and carbon dioxide. If the anaerobic process is uncontrolled, 'acid (anaerobic) fermentation' usually ensues naturally as a first stage. Under such conditions nitrogenous compounds are decomposed according to entirely different biochemical reactions. Carbonaceous matter is readily decomposed but yields mainly acidic products, e.g. acetic and butyric acids, produced in such amounts from fats and carbohydrates that the pH drops progressively and rapidly to values below 5 or 4·5. Highly obnoxious odours are developed. Such large pH changes result, of course, in the death of many types of bacteria, and the accumulation of end-products finally brings the fermentation almost to a standstill. Limited volumes of gases are evolved, mainly carbon dioxide and hydrogen from carbohydrate–protein decomposition by intestinal types of bacteria, with some methane and hydrogen sulphide and other odoriferous compounds. Subsequently, a much slower rate of decomposition by these and other bacterial species gradually decomposes the acid compounds and liberates basic substances until finally alkaline fermentation prevails. The natural sequence of 'acid' and 'alkaline' fermentation may require many months for digestion, whereas continuous control of the latter allows digestion time to be reduced to one or two months at most (see p. 261).

The control of conditions to ensure continuous biological activity at the highest practicable levels of treatment efficiency is thus paramount among the techniques of wastes engineering, and from this field of bio- or biochemical engineering must come much of the future development of treatment processes. To improve the practical efficiency of processes dependent on biological activity, whether aerobic or anaerobic or both, as is common, offers the greatest promise of improvement in all directions—which means finally an improved overall economy in constructions, operations and maintenance.

WASTE-TREATMENT PROCESSES

In Tables 3.2–3.4 the composition of sewage was set out in terms of mineral and organic matter, both dissolved and suspended. Part of the latter is either readily settleable or rises to the surface, but a substantial part is so highly dispersed that it is kept in suspension by the slightest motion of the sewage. This includes living organisms, e.g. the eggs or other reproductive parts of parasites such as liver and blood flukes, or of fungi, bacteria, etc., which may cause diseases of man or animals. Other water-borne wastes are likewise made up partly of dissolved and partly of suspended solids, but their relative proportions vary widely according to their origins. Wastes from certain chemical factories, for example, may be practically wholly soluble and at the same time wholly mineral; those from fat-rendering may comprise almost entirely insoluble organic solids.

Nevertheless, most of the techniques useful for treating domestic sewage are applicable to other water-borne wastes, whether from pastoral, mining or manufacturing industries.

The commonly applied processes may be listed[4] as follows:

For coarse suspended solids: screening
For heavy settleable solids: grit chamber or detritus tank
For floating solids including fats and oils: skimming, using grease-traps or skimming tanks
For fine suspended solids: flotation, skimming and sedimentation; chemical precipitation and fine screening, mechanical filtration
For highly dispersed, including dissolved, organic solids: biological treatment, including irrigation, intermittent sand filters, trickling or percolating filters, sewage (oxidation) ponds, septic tanks or anaerobic ponds, activated-sludge works
Deodorization: chemical treatment using chlorine, ferric chloride, etc., or biological treatment
Disinfection or sterilization: chemical treatment, using chlorine,

copper sulphate, etc.; heat treatment; biological treatment (as above).

The basic processes may conveniently be dealt with under three main headings.

Mechanical Treatment

1. Screening and mechanical filtration are determined by the particle size of the suspended solids.
2. Skimming: applicable only to suspended matter of lower density than the liquid. The effective particle densities may be modified by 'flotation' whereby gas bubbles adhere to solids and float them to the surface.
3. Sedimentation: applicable only to suspended matter of relative density greater than that of the liquid.

Chemical Treatment

The process of simple aeration is neither purely chemical nor purely physical (that is, 'mechanical') but in most cases is linked inevitably with biological factors; it is, therefore, not listed here.

1. Coagulation or chemical precipitation: using chemical reactions to modify the characteristics of suspended and dissolved solids.
2. Deodorization: specific chemical reactions of malodorous compounds are used.
3. Disinfection or sterilization: to destroy dangerous organisms or control the growth of specific pathogens, etc.

Biological Treatment

That is, the systematic use of the natural activities of living organisms:

1. In natural environments, either in the soil or in natural bodies of water; or on the land, e.g. using grass plots;
2. In artificial environments, such as intermittent sand filters, septic tanks, trickling filters, activated-sludge plants.

The various processes may be compared by the respective proportions of polluting matter removed or destroyed in the treatment of a particular waste. These can be measured in various ways: in sewage, for example, by the content of suspended matter or by that of bacteria, or by the B.O.D. test. For ordinary municipal sewage of average strength, the expected 'purification' efficiencies for various treatments are given in Table 9.1; it should be noted that screenings or sludges removed by treatment have to be disposed of also.

In estimating the purification to be expected, allowance must be made for any special local conditions, particularly for inevitable weaknesses in operation; for example, lower efficiencies will result if loadings on sewage treatment works are too high (indeed also if they are much too low); if structures are inadequate so that some sewage

Table 9.1

PERCENTAGE REMOVAL* OR 'PURIFICATION'

Treatment processes	*Measured by*		
	Suspended solids tests	*Bacteriological tests*	B.O.D. *tests*
Mechanical			
1 Fine screening only	5–20	10–20	5–10
2 Primary sedimentation only	40–70	25–75	25–40
Chemical			
3 Chlorination only of raw sewage	nil	90–95	15–30
4 Chemical precipitation and sedimentation	70–90	40–80	50–85
Biological			
5 Primary sedimentation, high-rate trickling filtration and secondary sedimentation	65–92	70–95	65–90
6 Primary sedimentation, low-rate trickling filtration and secondary sedimentation	70–92	90–95	80–95
7 High-rate activated-sludge treatment, including preliminary and final sedimentation	65–95	70–95	50–90
8 Conventional activated-sludge treatment, including preliminary and final sedimentation	85–95	90–98	80–95
9 Intermittent sand filtration	85–95	95–98	90–95
10 Nos 5–9 with chlorination of effluent	—	98–99	—

* Screenings, sludges, are reckoned as completely disposed of.

bypasses treatment; where sludge is disposed of with the effluent; or as a result of the return of poor-quality digestion tank (supernatant) liquor (p. 275). The efficiencies given above for sewage treatment processes will be attained only if adequate works are available and maintained and operated properly.

COST OF WASTE TREATMENT

Waste-treatment works can be regarded as well designed only if the costs of construction and operation are reasonable in relation to their utility.[5–9] The community, authority or other organization from whose domestic or industrial activities the polluting wastes are derived have generally the responsibility of preventing any undue nuisance, harm or expense to others arising as a result of the disposal of the wastes. This duty determines the degree of the waste treatment required.

Costs of construction of treatment works depend largely on the dimensions of the works required, for example, the volumetric capacities of sedimentation and sludge-digestion tanks, the area and volume of trickling filters, and so on. It is, therefore of the utmost importance for economy that volumes and areas should be adequate but not unnecessarily large. Estimates are best taken on the basis of cost per unit of contained volume—for example, cost per gal (or m^3) sedimentation tank or per yd^3 (or m^3) of percolating filter. For irrigation works, area is generally a better basis. Sedimentation and aeration tanks are the more expensive of the structures most commonly used, digestion tanks being about 40% and trickling filters about 50% cheaper per unit of constructed volume.

To compare estimates of costs of treatment by different techniques it is best to scale these against a unit related directly to the source of the wastes. For sewage treatment, the scale would be the number of persons contributing; slaughterhouse wastes might be related to that of animals slaughtered and may be extended to a population equivalent basis by suitable units of comparison, appropriate to the community, which might be 0·18 lb (82 g) of suspended solids, 0·12 lb (55 g) of B.O.D.$_5$ per population equivalent or other such values (see Table 14.3). However, the costs of sedimentation works depend not very much on the amount of solids but rather on the volume of waste flow and possibly on the nature of the solids, particle size and shape, etc. Equivalent figures may likewise be useful for the design of sludge-treatment units, but differences in the composition of the particular sludges must be considered; similarly costs for B.O.D. removal by percolating filtration or activated-sludge treatment cannot easily be related to simple population equivalents.

DISPOSAL OF SOLIDS IN RELATION TO TREATMENT OF WATER-BORNE WASTES

The disposal of solid refuse such as domestic garbage and street sweepings, from residences, industries and communal areas, is not commonly the concern of sewerage authorities. However, it will be realized that in most places some of this refuse finds its way into sewers, by accident or design, through private plumbing and drainage fixtures and with street washings. The varying proportions of such solid wastes occurring in sewage, even of separate sewerage systems, partly explain the wide variations found in the composition of sewages from place to place and from time to time, particularly the relative quantities of screenings and settleable (sludge-forming) solids. Development of disintegrating apparatus (see p. 144), designed to reduce the larger particles caught on screens to a size permitting their return to the sewage for treatment with the other fine suspended matter or else direct discharge to sludge-digestion tanks, has encouraged the use of water-carried waste systems for the disposal of solid wastes such as garbage; this is gradually growing in popularity. In some places the garbage is collected separately as a special solid waste and delivered by truck to the treatment works; more recently, it has been ground at the source, using small garbage grinders (such as 'disposal' units) mounted, for example, in the kitchen sink, so that the ground material passes into the house drainage system.[10–12]

The amount of garbage varies widely from place to place, but on average may amount to, say, 0·3–0·6 lb (0·14–0·27 kg)/hd daily with 60–85% moisture, and 65–85% organic matter in the dry solids. Investigations in a modern European community with high living standards showed that, if all households were equipped with garbage grinders, the pollutional load of sewage would be substantially increased: the raw sewage settleable solids nearly doubled and the settled sewage B.O.D.$_5$ increased by about 15–30%. Substantial enlargement of treatment works would be required.

The volume of total refuse, about 0·2–0·3 gal/hd d (0·9–1·4 l), is large compared to that of dried digested sludge (0·03–0·06 gal/hd d (0·14–0·27 l), see p. 290), closer to the volume of digested sludge as drawn from the digesters (0·09–0·17 gal/hd d (0·41–0·77 l), p. 257), and it may be treated by composting alone or along with digested sludge. The best process for the treatment of refuse before disposal is by rotting in compost heaps or beds, aiming to produce a fertilizer of high humus content. This process requires oxygen (from the atmosphere), water, sufficient nitrogen and traces of other essential elements. Municipal refuse normally contains enough nitrogen for bacterial metabolism but other refuse may not; thus composting of straw requires addition of up to 1·4% by weight of nitrogen, usually as liquid manure or ammonia. Coal ash may be harmful and should be separated. Composting with sewage sludge is dealt with elsewhere (see p. 115).[13]

The material is composted in heaps up to 17 ft (5·2 m) high which are sprayed with water from time to time. Drainage from the heaps should be returned with this water, thereby conserving not only water but also essential minerals and other constituents. The composting process may require about 2–6 months. The compost is screened to remove unsuitable solids, then ground. The final volume is about 40% of the original, amounting, therefore, to about 20 ft^3 (0·6 m^3)/1 000 persons daily. It may be spread for fertilizing at approximately 50 yd^3/acre (96 m^3/ha) yearly, equivalent to using refuse from a population of about 180 persons/acre (450/ha).

Heat is produced during composting. If the fresh material is screened and ground, a fertilizer with latent heat capacity is obtained, useful for hot beds in the same manner as horse manure.

WASTE TREATMENT AND ODOUR NUISANCE

Good treatment works are practically free of odours. To this end it is essential that the wastes be maintained in a fresh condition throughout the drainage system and pipelines and channels leading to the treatment works (see pp. 7–12). Where free-flowing pipes or open channels having good grades are used, the waste usually absorbs sufficient oxygen from the atmosphere to maintain freshness and remain free from foul odours, but in closed lines with long travel time, in pipes running full or where the grades, and hence the velocities in free-flowing pipes are too low, hydrogen sulphide and other such foul-smelling compounds are generated, the waste being seeded continuously with anaerobic organisms derived from slime coatings.[14–17] Any sludge deposits which may accumulate also contribute to rapid fouling. Pumping systems and their associated pressure mains require special consideration (see pp. 10–13). Where the wastes contain relatively high concentrations of sulphate, correspondingly high contents of sulphides are likely to build up unless overall detention times are restricted, particularly if the dissolved oxygen content of the flow reaching the pumping station is already low.

Ventilation of sewers should be given particular attention. Provision must be made not only for the flow of the aqueous wastes but, in addition, of sufficient ventilation air. Normally there is no need to pay particular attention to the ventilation for combined sewerage systems, allowances for stormwater flows in wet weather providing plentiful room—usually more than 50 times that occupied by sewage—for air flow under dry-weather conditions. However, the design of separate sewers calls for special attention in that respect, because the less favourable conditions might require some consideration of the interrelationships of oxygen consumption and reaeration in the flowing sewers, particularly of course where the time of travel is likely to be longer than usually allowed for. Normally, the provision for peak flows in wet weather and allowances for groundwater infiltration (pp. 13, 17) leave space for air in dry weather, but the designing engineer should always assess requirements and allowances for ventilation.[17] The larger the ratio of surface area to volume, the greater the aeration of the flowing wastes and the relative velocity of waste and air; therefore, increased turbulence also increases the rate of oxygen uptake but likewise the amount of odour emanating from the waste. The rate of oxygen usage in fresh wastes depends upon their oxygen demand, both chemical and biochemical, so that fresh conditions are the more easily maintained the lower the concentration of oxygen-consuming substances.

In practice, it is not always possible to maintain freshness or, in some cases, to prevent septicization (see pp. 8–12), unless special technical precautions are taken. In some cases rising mains may be aerated with success or wastes may be chlorinated[18,19] before discharge into them or at other critical places. Prior to discharge into treatment works, highly septic wastes should be pretreated to restore them to a state of freshness, both to facilitate treatment and to reduce odours. Pre-aeration applied in such a way as to avoid emanation of odours, addition of chlorine or other oxidizing agents or the mixing (recirculation) of oxidized effluent into the incoming wastes are the commonest techniques used for the freshening of raw wastes prior to treatment. By such means (which, however, may be relatively costly), septic wastes can be sufficiently refreshed, but it is generally preferable and nearly always more economical, to avoid septicization. According to Heukelekian,[19] chlorination of sewage to oxidize hydrogen sulphide requires about 10 times as much chlorine as would have been needed to prevent septicization. (Theoretically, 1 lb (0·45 kg) hydrogen sulphide requires only about 2 lb (0·9 kg) elemental chlorine or 0·47 lb (0·22 kg) oxygen (= 1 lb (0·45 kg) sodium nitrate), but in practice the required proportions may be higher.)

Care must also be taken to conserve the freshness of the wastes as far as possible during treatment. Pipelines and channels must provide suitable flow conditions or may be aerated continuously. Septic tanks should only be used for the special cases of isolated buildings or small unattended works. Detention times in sedimentation tanks should be restricted according to the dissolved oxygen content of the incoming sewage and its rate of oxygen consumption, preferably not exceeding a few hours, and the settled sludge should be removed frequently or continuously. Storage tanks must, of course, be treated like settling tanks and kept free of sludge and undesirable slimes, provision being made for some continuing aeration if necessary or desirable. If care is given to all parts of the works, no objectionable odours will arise from settling tanks, screen chambers, pumping wells or trickling filters.

So far as these parts of the works are concerned, odour difficulty is readily resolved, but this is not so easy with the handling and treatment of sludge. Fresh sludge derived from fresh wastes is practically odourless only as long as it is kept under water containing sufficient dissolved oxygen. Once separated from the wastes, objectionable odours may arise, unless the sludge is immediately and thoroughly dried and kept dry, or brought at once under the regime of alkaline fermentation. Raw sludge is susceptible to acid fermentation (p. 260) which tends to develop within a few hours, producing a strong and very objectionable sour type of odour similar to that from dirty pig-pens. Such odours may arise anywhere in the works where residues from screens, grit chambers, skimmings or sludge are exposed to the air, particularly where varying water-levels may occur, as in channels, for example of irrigation areas, sludge wells and drying beds carrying incompletely digested sludge. Such residues may be especially objectionable when wetted again after partial drying. Odours arising from organic sludges may be controlled by maintaining scrupulous cleanliness in and around all treatment works units. Screenings should be disposed of as soon as possible either by burying, burning, digesting or disintegration and return to the waste flow at least several times daily. Detritus from grit chambers should be washed and disposed of promptly, and floating solids and grease (scum) skimmed off frequently and disposed of at once. Sewage sludge should not be stored but transferred to the digestion tanks or treated otherwise without delay. Sludge-drying beds should be used only for digested sludge; raw sludge requires liming or artificial drying, using either mechanical filters with heat drying or incineration. When artificial drying is used, the hot moist gases must be burnt to avoid odours, or the sludge may be incinerated[20] at temperatures above 1 500°F (800°C).

With due regard to cleanliness and proper operation, a well-designed waste treatment works may be maintained efficiently without the occurrence of any odour nuisance either outside or inside the works. Careful planning of the area with the provision of suitable plant life, including shrubs and areas of well-kept lawn, and hence the encouragement also of birdlife, greatly assists cleanliness and appearances and provides suitable areas for the burying of residues and their proper absorption and final disposal. The falling of leaves from trees must be considered in relation to the treatment units.

Odours resulting from errors in design or operation may not be readily overcome. They may demand considerable attention, such as spraying surfaces or areas with chlorinated water or bleaching powder, or even drastic measures such as the enclosure of treatment units. The latter usually results in local concentration of odours, requiring provision for deodorization along with mechanical ventilation. Chlorine and activated carbon are possible deodorants, and combustion of the air is also possible. Other techniques for treating such air have been described, including diffusion into activated-sludge tanks and pumping through trickling-filter beds.[15, 21]

BIBLIOGRAPHY

BOLTON, R. L. and KLEIN, L. *Sewage Treatment—Basic Principles and Trends*, 2nd Edn (1971) Butterworths, London

ECKENFELDER, W. W. *Theory and Design of Biological Oxidation Systems for Organic Wastes* (1965) Univ. of Texas; *Industrial Water Pollution Control* (1966) McGraw-Hill, New York

ECKENFELDER, W. W., Jr and O'CONNOR, D. J. *Biological Waste Treatment* (1961) Pergamon, Oxford

FAIR, G. M., GEYER, J. C. and OKUN, D. A. *Water and Waste Water Engineering*, Vol. 2 (1968) Wiley, New York

HAWKES, H. A. *The Ecology of Wastewater Treatment* (1963) Pergamon, Oxford

ISAAC, P. C. G. (Ed.) *The Treatment of Trade Waste Waters and Prevention of River Pollution* (1957) University of Durham; *Contractors' Record and Municipal Engineering*, London

ISAAC, P. C. G. (Ed.) *Waste Treatment* (1960) Pergamon, Oxford

MCCABE, B. J. and ECKENFELDER, W. W., Jr (Eds) *Biological Treatment of Sewage and Industrial Wastes*, 2 Vols (1956, 1958) Reinhold, New York

MCKINNEY, R. E. *Microbiology for Sanitary Engineers* (1962) McGraw-Hill, New York

NEMEROW, N. L. *Industrial Waste Treatment* (1963) Addison-Wesley, Reading, Mass.

SAWYER, C. N. *Chemistry for Sanitary Engineers*, 2nd Edn (1967) McGraw-Hill, New York

SOUTHGATE, B. A. *Pollution and Conservation* (1969) Thunderbird, London

REFERENCES

1. BAARS, J. K. 'Bacterial Activity in Pollution Abatement', *J. Proc. Inst. Sew. Purif.* (1965) Pt 1, 36; HARKNESS, N. *J. Proc. Inst. Sew. Purif.* (1966) Pt 6, 542

REFERENCES

2. WILSON, I. S. 'Some Problems in the Treatment of Effluents from the Manufacture of Organic Chemicals', *J. Soc. chem. Ind. Lond.* (1967) 1278
3. BOARD, R. G. 'Biological Aspects of Organic Waste Treatment', *Wat. Pollut. Control* **67** (1968) 614
4. ISAAC, P. C. G. 'Principles of Waste Treatment', *Effl. & Wat. Treat. Jnl* **3** (1963) 480, 645; **4** (1964) 80, 174, 316, 516; **5** (1965) 138, 196
5. CALVERT, J. T. 'Costs of construction of sewage treatment works and their influence on design', *J. Proc. Inst. Sew. Purif.* (1962) Pt 2, 131
6. LOGAN, J. A., HATFIELD, W. D., RUSSELL, G. S. and LYNN, W. R. 'An analysis of the economics of wastewater treatment', *J. Wat. Pollut. Control Fed.* **34** (1962) 860
7. JAMES, R. P. B. 'Techniques and Recent Developments in Sewage Works Design', *Wat. Pollut. Control* **69** (1970) 62
8. CUPIT, J. V. 'Economic Aspects of Sewage Works Design', *Wat. Pollut. Control* **68** (1969) 166
9. BRADLEY, R. M. and ISAAC, P. C. G. 'The Cost of Sewage Treatment' *Wat. Pollut. Control* **68** (1969) 368
10. BURKE, C. E. and HILLIER, W. H. 'A Report on the Effect of Garbage Grinding on Sewers and Sewage Purification', *J. Proc. Inst. Sew. Purif.* (1956) Pt 1, 110
11. WATSON, K. S. and CLARK, C. M. 'How Food Waste Disposers Affect Plant Design Criteria', *Publ. Wks, N.Y.* **93** (1962) No. 6, 105
12. BOWERMAN, F. R. and DRYDEN, F. D. 'Garbage, detergents and sewers', *J. Wat. Pollut. Control Fed.* **34** (1962) 475
13. JAAG, O. 'Present-day Problems in Composting', *J. Proc. Inst. Sew. Purif.* (1958) Pt 4, 436
14. VIEHL, K. 'Über die Ursachen der Schwefelwasserstoffbildung im Abwasser', *Gesundheitsingenieur* **68** (1947) 41
15. POMEROY, H. 'Generation and control of sulfide in filled pipes', *Sewage ind. Wastes* **31** (1959) 1082
16. MÜLLER, W. J. 'Die Frischhaltung des Abwassers in Leitungen', *Gesundheitsingenieur* **73** (1952) 164
17. Technological Standing Cttee, *Control of Sulphides in Sewerage Systems* (1971) Butterworths, Sydney
18. CHANIN, G. 'Solving Odor Problems by prechlorination of flow, paced by automatic continuous monitoring of H_2S content of atmosphere', *Wat. Wks & Wastes Engng* **1** (1964) 42
19. HEUKELEKIAN, H. 'Utilization of chlorine during septicization of sewage', *Wat. Sewage Wks* **95** (1948) 179
20. SAWYER, C. N. and KAHN, P. A. 'Temperature Requirements for Odor Destruction in Sludge Incineration', *J. Wat. Pollut. Control Fed.* **32** (1960) 1274
21. CARLSON, D. A. and LEISER, C. P. 'Soil Beds for the Control of Sewage Odours', *J. Wat. Pollut. Control Fed.* **38** (1966) 829

10

MECHANICAL TREATMENT PROCESSES

SCREENING

SCREENING is commonly used as a preliminary treatment prior to other treatment processes, or as the only treatment of particular wastes prior to admixture with others or to discharge into a public sewerage system. As screening is effective only for the separation of relatively coarse suspended solids, the degree of purification is in most cases low. For ordinary municipal sewage, the efficiency for removal of solids using fine screens varies from about 5 to about 20%, compared with that from sedimentation tanks of between about 40 and 70% of suspended matter, actual figures being variable for different plants and sewages. The screening plant would normally have a volumetric capacity of about one-fifth that of a sedimentation plant. With most sewages, screening is the more costly process as to cost per unit quantity of solids removed; on the other hand, it may be relatively more efficient and economical for some industrial wastes, e.g. those from vegetable canning. In general, screening alone is sufficient only where the final disposal of the partially treated wastes is permissible on account of high dilution ratios in the receiving waters or other favourable circumstances. Although it has the advantage of relatively small area requirements, the screenings are often very unpleasant in character and in most cases should be disposed of as fast as they accumulate, for hygienic reasons as well as odour control, especially in tropical or subtropical countries. Accordingly, screens should be totally enclosed and the access of flies, etc., completely prevented in all plants where foul wastes are treated.

There are a number of different types of screens, falling into two distinct classes: first, bar screens or racks, and second, mesh screens or sieves.[1]

Bar screens or *racks* are usually constructed as a series of metal bars arranged in one plane across a slightly expanded channel, and inclined upwards in the downstream direction. Less frequently the bars are arranged vertically, spaced for example in a circle to form a basket grate. Bar screens fall into two groups, *coarse* having spacings wider than 1½ in (38 mm) and *fine*, with openings between ½ in (13 mm) and 1 in (25 mm) approximately. The bars are commonly circular, otherwise rectangular, in section, spaced to suit the wastes being treated. Coarse bar screens with openings between $1\frac{1}{2}$ in (38 mm) and 2 in (51 mm) are recommended for the preliminary treatment of raw municipal sewage ahead of sedimentation tanks, provided that the sewage solids are already partly disintegrated as a result of their travel down the sewers to the works. Broken-up faecal matter and paper, etc., pass through. These screens serve mainly to withhold larger pieces of extraneous solids which might otherwise cause blockages in sludge lines and other works structures. The slope is usually about 30° but for mechanical cleaning say 75°. The channel must not be unduly expanded, or sand and silt might separate. A bypass overflow channel, usually fitted with vertical bars spaced 4 in (102 mm) apart, is essential (Fig. 10.1).

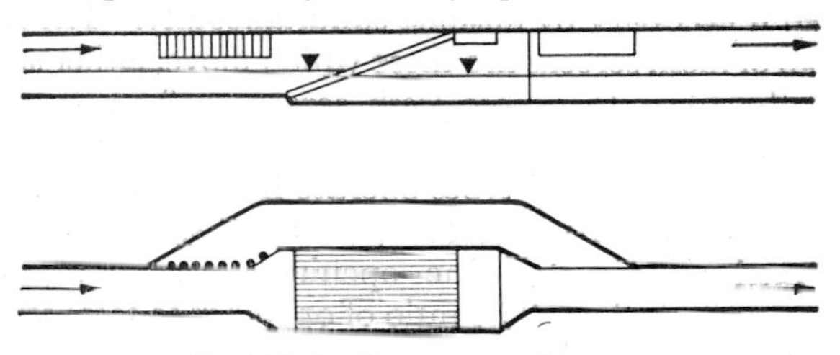

Fig. 10.1. *Bar screen and bypass*

Coarse bar screens normally remove less than 0·1 ft³ (3 l) *per capita* yearly of solids from municipal sewage, fine from 0·1 (3 l) to about 0·3 ft³ (8 l). The screenings may be buried, incinerated or digested. Incineration leads to odour nuisance, unless a temperature of about 1 500°F (816°C) or higher is reached. About 3 500 ft³ (100 m³) of digester gas is sufficient to burn 1 ton (1 000 kg) of screenings. It is becoming common practice to disintegrate screenings, after which they may be returned to the waste flow or added to digestion tanks.

Cleaning is performed by rakes the tines of which fit between the bars and allow the screenings to be drawn upwards out of the flowing waste. The material may then be transported by conveyors or wheeled buckets, usually after dewatering on simple perforated trays draining through small screens. Manual cleaning of coarse-bar screens is usually sufficient if regular attention is given, otherwise it may be done mechanically, usually automatically when the upstream level builds up to a predetermined height. This type of arrangement has become common in connection with sewage sedimentation plants with mechanical sludge-scraping machinery, fine-bar screens with bars spaced about $\frac{3}{4}$ in (19 mm) being used for protection of scrapers and to avoid blockages in the sludge pumps. Manual cleaning of these is impracticable and it is usual to place the bars in a much steeper plane, as is permissible with mechanical cleaning plants. Because much organic matter may then be screened out, treatment of screenings by disintegration has developed as common practice.

Mesh screens serve a somewhat different purpose. The openings are commonly formed by interweaving metal wire or textile fibres. More material is, of course, screened out by ¾ in (19 mm) mesh than by a ¾ in (19 mm) bar screen. The removal of screenings requires different techniques. Flushing with screened or works effluent or water, as most often used, is suitable for non-fibrous screenings. Fibrous or elongated light solids are removed more readily by brushing, assisted if necessary by flushing simultaneously. The most successful types of mesh screens are those which move, either steadily or intermittently, so as to expose fresh screen surfaces to the waste continuously. The clogged surfaces are cleaned before re-use. Trommel and 'endless-belt' types operate in this manner.

Screens consisting of perforated plates are similar in action to ordinary mesh screens but the openings are usually circular, elongated or rectangular. The ratio of openings to total screen area is generally lower than for mesh screens, but they are more readily cleansed. The Riensch–Wurl screen, which is well known in Europe, consists of a revolving disc plate perforated in this way and rotating in a slightly inclined plane so that the waste flow is screened continuously by some segment of the disc, screenings being carried out of the flow and removed by brushes and flushing jets. Ordinary sewage yields up to 1·0 ft^3 (30 l)/hd yearly. The moisture content after draining is 85–90% and the volatile (organic) solids content about 85% of dry solids. In small plants, the fine screenings are buried, in larger installations they may be composted with municipal (street) refuse (p. 115). They may also be digested in heated digestion tanks (p. 285), but coarse indigestible solids should first be separated.

Stationary-frame-type (fixed-mesh) screens cannot usually be cleaned in position but must be removed temporarily for cleansing. Fine mesh screens with mesh down to ¼ in (6·4 mm) openings or less are used after sedimentation tanks in some works, for example to reduce clogging of sprays of trickling filters or irrigation areas. All such installations must also be provided with bypass overflows. Much finer screens may be used for treatment of some industrial wastes, such as recovery of waste wool or cotton fibres from wet processing of textiles.

Table 10.1

SCREENING OF MUNICIPAL SEWAGE

Type of screen	*Openings*		*Cleaning*	*Volume of screenings*/hd year		*Approximate percentage removed*	
	in	mm		ft³	l	*Suspended solids*	B.O.D.
1.	2–3	50–75	manual	up to 0·02	up to 0·6	—	—
2.	$1\frac{1}{2}$–2	40–50	manual	0·02–0·1	0·6–3·0	up to 2	up to 1
3.	$\frac{1}{2}$–1	10–25	mechanical	0·1–0·3	3–8	2–5	1–3
4.	$\frac{3}{8}$	10					
5.	down to $2 \times \frac{1}{32}$	50×1	mechanical	0·2–1·0	6–30	5–20	5–10

1., 2. Coarse bar. 3. Fine bar. 4., 5. Fine (Woven mesh or plate screens, including Riensch–Wurl screens (5)).

The efficiency of screening units varies greatly, depending upon the composition of the wastes to be treated. For municipal sewage of normal composition, approximate figures are shown in Table 10.1.

Mechanical filters, including rapid sand and 'magnetite' types, may also be regarded as screens but usually remove much finer particles. For raw wastes, load allowances are very much less than for fine screens,[1] and filters are not normally used as a preliminary treatment but may serve for intermediate or final ('polishing') processes. They may be suitable for certain industrial wastes, particularly those consisting of fine slurries of inorganic solids or such as are readily flocculated. Mechanical filtration is discussed in greater detail later (see p. 171).

Micro-straining,[1] a recent technique using drum screens rotating on a horizontal axis and fitted with woven stainless-steel fabric having 10^5 openings about $50\mu \times 100\mu/m^2$ ($150/mm^2$) has also been applied successfully to the filtration of effluents.[2–4]

DISINTEGRATION[5]

As mentioned above, provision for the speedy and permanent disposal of screenings is very important. The maintenance of clean conditions within and around waste treatment works requires scrupulous attention to all details and on this odour control also depends. The time-honoured means of disposal of screenings is burial, involving removal and partial drainage; some uncleanliness, odour and unsightliness are inevitable. Incineration is less objectionable but plant is relatively costly. The modern practice of disintegration provides the most hygienic and, usually, the simplest means of disposal; the coarse solids separated as screenings are reduced in size to permit further treatment and disposal along with other suspended matter which passes through the screens. There are two distinct methods of handling the screenings of disintegration plants: first, by separating them either manually or mechanically and discharging them separately into the disintegrator; second, by taking these and some waste water directly to the disintegrator from which the disintegrated solids are returned with the waste water to the main flow. Some odour nuisance is inevitable with the first method because the screenings are exposed, but when they are kept under water, as with the second method, this is avoided.

Disintegrators vary in design. Three arrangements have been widely used:

1. The screenings are delivered substantially free of water to a

shredding mill consisting of a high-speed cylinder fitted with cutting knives which rotate on a horizontal axis so that they just clear the bars of a semi-circular screen fitted below. The screenings are chopped by the knives until the shredded material falls through the screen, directly into the raw waste flow or into a special container or conveyor for transport to the sludge tanks (Fig. 10.2).

2. A disintegrator pump, which is a specially designed centrifugal pump with replaceable cutters mounted on the impeller. The

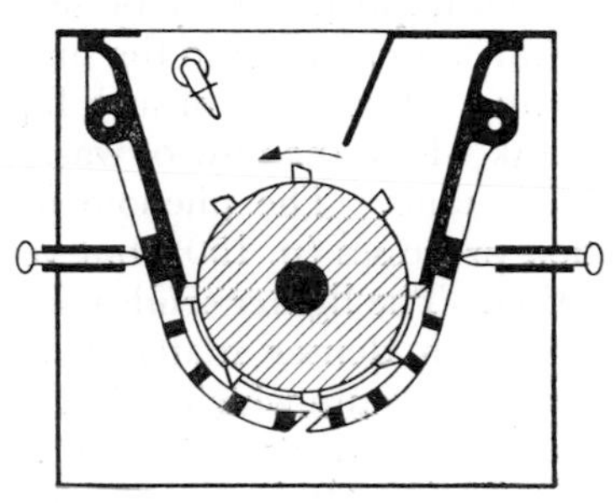

Fig. 10.2. *Shredding mill*

disintegrators may be arranged to pump the screenings, together with some screened waste for dilution, back into the waste channel upstream of the screens (Fig. 10.3). The operation is usually automatic, for example, simultaneous with the operation of mechanical screen-cleaning apparatus when the

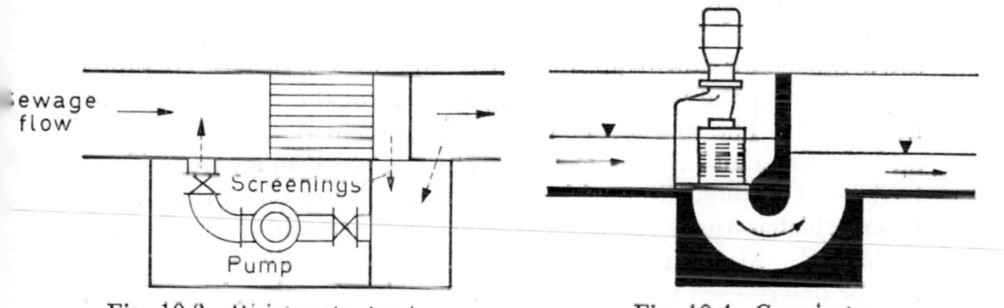

Fig. 10.3. *Disintegrator pump*

Fig. 10.4. *Comminutor*

level rises upstream of the screens. In small plants, use of the disintegrator only once or twice daily may be sufficient.

3. Another, widely adopted, arrangement takes the whole of the waste flow through the disintegrating machine or 'comminutor'[5] which consists of a vertical trommel or cylindrical screen rotating about a vertical axis (Fig. 10.4). A fixed cutter, or

series of cutters, disintegrates the material caught outside the screen so that it passes through the trommel to the downstream channel. Grit chambers or detritus tanks should be provided upstream of such apparatus. Another form has vertically mounted bars against which a travelling cutter moves up and down.

SKIMMING

Oils, fats and other such impurities may be separated by allowing the material lighter than water to rise to the surface, whence it may be removed by skimming. The rising of such liquids and solids can be observed in any tank where the flow of wastes is stilled, as in an ordinary sedimentation tank. The phenomena are analogous to those in sedimentation processes (p. 150), but vertical velocities are in the opposite direction. The lighter solids rising to the surface are usually contaminated with denser material and form a heterogeneous layer of scum. The scum which builds up on sedimentation tanks of municipal sewage works may contain relatively small proportions of fats and oils (20–30%), but well-designed skimming tanks will separate scum containing much higher proportions of fats, oils and greases (50%). Larger particles of fats, oils, etc., rise more rapidly than smaller ones; the latter may also aggregate to form larger particles which then separate much more quickly. Moreover, particles carried upward by adherent gas bubbles become more and more buoyant as they rise because of the natural expansion of the gas.

The design of skimming tanks is straightforward only where they are arranged for downward or horizontal flow, the surface area then depending on the minimum rising velocity of the impurities to be floated off. The simplest efficient arrangement is a relatively long and shallow tank with inlet and outlets designed to secure uniform horizontal through flow with a minimum disturbance of the scum layer, which is usually retained behind a baffle suitably located at the outlet end. Minimum rising velocity for each particular case must first be determined by experiment; the design can then be made on the basis of the required skimming-tank area for the estimated flow, and the main dimensions fixed by consideration of hydraulic factors and relative construction costs. The required surface area, A, the basic design figure, is given by the relationship

$$A = \frac{720 \times \text{waste flow, ft}^3/\text{s}}{\text{min. rising velocity, in/min}}\,\text{ft}^2 = \frac{\text{waste flow, l/s}}{\text{min. rising velocity, mm/s}}\,\text{m}^2$$

Thus a rising velocity of 10 in/min (4·2 mm/s) allows removal from below the surface—say, from the bottom of the skimming tank—of a volume of skimmed effluent equal to 10/720 ft³/s = (1 440/72) × 60 × 6·24 = 7 500 gal/ft² (363 × 10³ l/m²) tank surface daily.

Skimming may be used also for the separation of particles heavier than water by techniques of flotation, i.e. adhesion of gas bubbles to solids. This has been done successfully by either blowing compressed air into the waste or by dissolving air in it under pressure (for instance, in pumps), excess air then being released as numerous fine bubbles after discharge of the waste into skimming tanks or, similarly, by reducing the pressure above the tanks. Dissolved gases then are released, thereby floating solids to the surface. Flotation processes are used widely in the mining industry for separation of

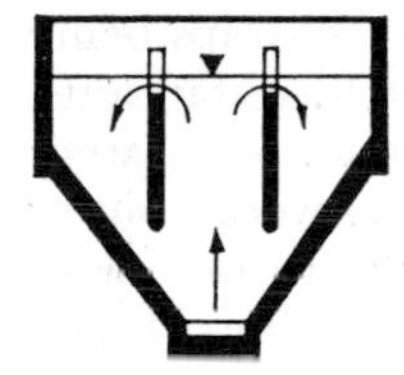

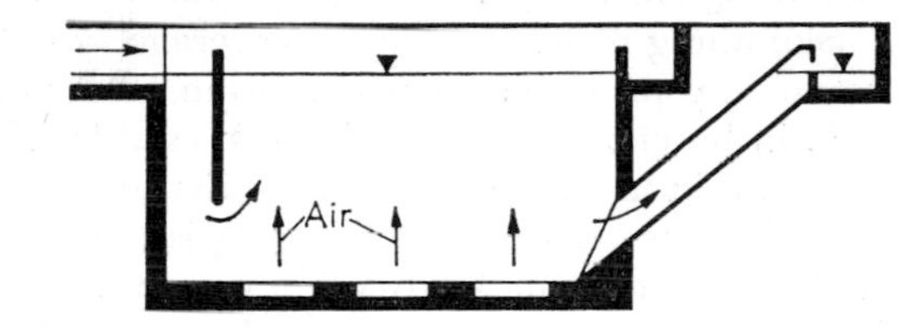

Fig. 10.5. *Aerated skimming tank*

ores or coal, using reagents to encourage adhesion of air bubbles to ore particles. These include 'collectors', usually organic nitrogen or sulphur compounds (such as alkyl xanthates), and frothing agents such as oils, fats, resins or glues. Flotation processes have been applied to the treatment of water-borne wastes, for example of wool scouring (no agents), paper mills with addition of alum and glue, meat packing where chlorine addition may be useful, and for the thickening of activated sludge.

In the treatment of domestic and municipal sewage, skimming or flotation tanks may be useful adjuncts to sedimentation plants, particularly where the proportions of floating grease are unusually high and the removal of most of it before sedimentation could simplify the arrangement for skimming the larger tanks. A suitable design is shown diagrammatically in Fig. 10.5. The aeration and skimming tank should be elongated rectangular, trough-shaped, with a relatively large surface area. Turbulence is confined to the inner aeration zone from which the floating foam and scum pass through adequate but not large openings to the stilled zones beyond, where heavier solids are freed and sink to the bottom. These leave the tank with the effluent of skimmed sewage which is taken off from the bottom at the far end in such a way that all sludge and silt are scoured through. Foam and scum are taken off manually or mechanically fromt he stilled zones to small decanting tanks and the skimmings added to the digestion tanks, except in special cases where they are burnt because of high mineral oil content. Detention is about 3 min for the average design of aeration tank, and the air consumption about 30 ft^3/1 000 gal sewage (180 l air/m^3 sewage). Since some freshening of the sewage also may be achieved by such pre-aeration, this may in some cases warrant detention up to a maximum of, say, $\frac{1}{2}$ h, using about 100 ft^3 air/1 000 gal sewage (615 l air/m^3 sewage).[6] However, most of the 'grease' (oil, fat, soaps, etc.) of municipal sewage may not be separated even then. The addition of chlorine to the compressed air at a rate of 1–$1\frac{1}{2}$ p.p.m. based on average sewage flow has been found to improve skimming efficiency in some particular cases.

Skimming is used widely in sewerage systems, grease-traps being commonly provided for wastes from kitchens, hostels, restaurants, etc., in butcheries and small-goods establishments and, on a larger scale, for industrial works such as abattoirs, rendering works, butter factories, etc. Regular desludging is required in most cases, with daily or more frequent attention to scum, and grit arrestors should be installed upstream to prevent the access of grit to the grease-tank. Under the name of 'Save-alls', grease-tanks are widely used in the food industry for recovery of fats, etc., for re-use. For large workshops and, particularly, oil depots, skimming to remove both light and heavy oils is essential, but usually settled sludges should also be separated because of oil contamination. Light-oil traps for garages are designed on a similar basis to prevent accumulation of significant quantities of explosive vapours in the sewerage system. In all these particular cases, relatively high efficiencies (say, $\sim$ 80% removal) are obtained with properly designed apparatus.

SEDIMENTATION

Most of the suspended particles in municipal sewage and many industrial wastes are too fine to be separated by mechanical screening. Since most of the particulate matter in municipal sewage is of greater density than water, substantial removal is possible by means of simple sedimentation.[6–14] Many industrial wastes may also be so treated efficiently. As noted above, the skimming may be combined with sedimentation, suspended matter being removed in settling tanks as sludge from the bottom and as scum from the surface.

It is important to distinguish between granular and flocculent suspended matter. The first includes solids such as sand, coal dust, sugar-beet waste solids, etc., the individual particles of which settle independently and at constant rates determined by relative densities and sizes. Flocculent suspended matter includes solids such as precipitated iron or aluminium hydroxides, activated-sewage sludge solids or those characteristic of trickling-filter effluents; such solids are loosely aggregated and do not settle freely. However, under favourable circumstances individual aggregates tend to combine into progressively larger and effectively denser ones, so that their rates of settling increase progressively, some of the minute non-settleable particles (colloidal and pseudo-colloidal matter) also becoming attached to them. It should be noted, however, that these aggregates are also readily broken down again, for example, as a result of excessive turbulence or other physical disturbances, such as ultrasonic vibrations. In such circumstances the rates of settling may decrease.

The suspended matter of fresh municipal sewage is partly granular and partly flocculent, and many other wastes also contain both types of solids. On the other hand, some wastes carry practically only granular, others only flocculent solids. When designing sedimentation plants these fundamental characteristics have to be considered.

It is also important to distinguish between sedimentation tanks of predominantly horizontal and predominantly vertical (upwards) direction of stream flow. Horizontal-flow tanks in their simplest efficient form are relatively elongated, rectangular in cross section. In vertical-flow tanks the influent is introduced below the surface, and the flow subsequently rises upwards steadily to overflow from the surface. They are generally either square or circular in plan section.

Rates of sedimentation may be studied simply by Imhoff cone tests (pp. 30, 31) from which approximate rates of settlement can be determined for given proportions of the total settleable matter. A sedimentation curve may be prepared in this way (cf. Fig. 10.8 below). This is directly applicable to the design of settling tanks with horizontal flow. Although sedimentation in an Imhoff cone is not in any way similar to that in vertical-flow tanks, the results of such tests may nevertheless be used to obtain comparative data suitable as a basis for design, in particular where samples of wastes to be treated are available beforehand. If this cannot be done comparative data must be sought or estimated from general information.

Sedimentation of Granular Solids

The factors governing the sedimentation of granular solids were studied originally by Stokes and later, with special reference to water and sewage treatment by Hazen and others.[14] As already stated, the settleable solids most frequently encountered in waste-treatment practice are of such particle sizes that they tend to settle through water at constant rates which depend on the effective sizes and relative densities of the individual particles. Thus it has been found that fine coal particles of about 4×10^{-4} in (0·01 mm) effective diameter settle at about 3 in/h (0·02 mm/s). Therefore, when waste water carrying fine coal in suspension is discharged into a sedimentation tank, it can be anticipated that all particles having effective diameters of about 4×10^{-4} in (0·01 mm) or greater will have settled out of the upper foot (metre) of water within 4 h (13 h). Such waste waters can, therefore, be purified to this extent by treatment in simple sedimentation tanks at a rate equal to $\frac{1}{4}$ ft^3 waste flow/ft^2 tank surface h ($\frac{1}{13}$ m^3 waste flow/m^2 tank surface h), equivalent to 37 gal/ft^2 tank surface daily (1 800 l/m^2). The required surface area for continuous treatment is given by the formula

$$\text{Required surface area} = \frac{\text{rate of flow of waste}}{\text{minimum settling velocity allowed for}} \qquad 10.1$$

The minimum settling velocity can be determined from the results of the Imhoff cone test if available. If the depth of sample in the test cone is measured, the observed time required for satisfactory sedimentation gives the minimum settling velocity directly:

$$\text{Minimum settling velocity} = \frac{\text{depth of test sample in cone}}{\text{efficient settling period}} \qquad 10.2$$

The loading rate of sedimentation tanks is best expressed in terms of volume of waste treated per unit area. A simple transformation of the former equation shows that the efficient loading rate is numerically equivalent to the settling velocity of the smallest particle:

Efficient loading rate, ft^3/ft^2s (or m^3/m^2s) = minimum settling velocity, ft/s (or m/s) of Eqn 10.2.

Then the direct equations in common units, respectively gal/ft^2d, and m^3/m^2d, are as follows:

$$\text{Loading rate} = 12{\cdot}5 \times (\text{in/h})\ \text{gal/ft}^2\text{d} = 24 \times (\text{m/h})\ \text{m}^3/\text{m}^2\text{d} \qquad 10.3$$

Fair gives the settling velocities in water for various particle sizes (Table 10.2), based on assumed specific gravities of 2·65 for quartz sand, 1·5 for coal, 1·2 for sewage solids, and a water temperature of 50°F (10°C).

The table shows, for example, that sand particles of 3·9 × 10^{-2} in (1 mm) diam. settle 8 000 times faster than those of 0·02 × 10^{-2} in (0·005 mm) diam. Sewage solids of 3·9 × 10^{-2} in (1 mm) diam. settle much slower than sand but 15 000 times faster than typical solids of 0·02 × 10^{-2} in (0·005 mm) diam.

In horizontal-flow tanks, settling velocities determine the time

Table 10.2

SETTLING VELOCITIES IN WATER AT 50°F (10°C) (after Fair)

Nature of particles	Diameter, in × 10^{-2} (mm)						
	3·9 (1·0)	2·0 (0·5)	0·79 (0·2)	0·39 (0·1)	0·20 (0·05)	0·04 (0·01)	0·02 (0·005)
	Settling velocities,* in/h (mm/s)						
Quartz sand (s.g. 2·65)	19 800 (140)	10 200 (70)	3 200 (22)	950 (67)	240 (1·7)	11·8 (0·08)	2·35 (0·016)
Coal (s.g. 1·5)	6 000 (40)	3 000 (20)	1 020 (7)	300 (2)	58 (0·4)	3·14 (0·02)	0·58 (0·004)
Suspended solids of domestic sewage (s.g. 1·2)	4 800 (30)	2 400 (17)	710 (5)	118 (1·3)	31·5 (0·3)	1·18 (0·008)	0·31 (0·002)

* The effects of flocculation are not allowed for (p. 149).

required for solids to reach the bottom. Obviously, this is the minimum efficient detention period. The horizontal stream velocity must not be such as would prevent settling or disturb the settled solids. In sedimentation tanks of usual design, horizontal stream velocities of about 10 ft/min (50 mm/s) tend to roll settleable sewage solids along the bottom; for activated sludge, this critical velocity may be as low as 4 ft/min (20 mm/s).[6] These velocities differ according to the dimensions of the structures but can be determined readily by model experiments. Although those generally reached in practice are much lower than the critical, the rates of travel in sedimentation tanks should be checked, particularly for shallow tanks such as settling beds (p. 168), taking any variations in section properly into account.

In vertical-flow tanks, the settling velocities of the particles determine the maximum permissible upward velocity. Thus, according to Table 10.2, coal particles of $0{\cdot}2 \times 10^{-2}$ in (0·05 mm) diam. will not be settled out unless the upward velocity of flow is less than 58 in/h (0·4 mm/s). Detention time in vertical-flow tanks has no significance in relation to the sedimentation of granular solids, whether coarse or fine.

Sedimentation of Flocculent Solids[6, 11–14]

Tabulations of settling velocities cannot be used for sedimentation-tank design when the suspended matter to be removed is flocculent, e.g. in chemical precipitation and settling activated sludge. The same laws govern the rates of sedimentation of individual particles but flocculent particles are continually changing their individuality and the settling velocities vary accordingly. It is not possible, therefore, to fix a settling velocity for particles of any given size, and the simple design basis of Imhoff cone test results set out above for granular solids does not apply.

The behaviour of the flocculent particles in sedimentation tanks depends on a number of factors, mainly those which influence their relative movements and, hence, opportunities for their aggregation. *Detention* or *flowing-through times* matter more than the surface area; depth is also important.

The coagulation of flocculent solids, especially in vertical-flow tanks, is assisted by 'sludge-blanket screening' effects: the passage of the waste upwards through the zone of settling sludge aggregates, increasing the chance of absorption of very fine particles.[12] The upward speed of the waste flow is not, therefore, limited by the settling velocity of the finer particles but by the settling velocity of the coagulated aggregates. Generally, this has to be determined by actual trials with the waste to be treated, but the usual limits—suitable, for example, for the settlement of activated or chemical precipitation sludge—are from about 1–2 in/min (0·42–0·85 mm/s). Such velocities are high enough to prevent the settlement of relatively coarse granular organic solids of low density, such as distinctly visible particles of coal dust or cellulose of about $0{\cdot}2 \times 10^{-2}$ in (0·05 mm) diam., but in the presence of flocculent suspended matter these also become entangled in the flocculent aggregates, adding to their effective size and density and increasing their settling velocity.

Because of this behaviour of flocculent suspended matter, sedimentation-tank design cannot be based on area alone, and the best basis then is detention period, provided that the flow velocity of the waste does not exceed about $\frac{3}{4}$–2 in/min (0·32–0·85 mm/s) upwards and about 4 ft/min (20 mm/s) horizontally.

Density Currents in Sedimentation Tanks

It is obvious that sedimentation tanks designed in accordance with the properties of the suspended matter will not attain their full efficiency unless the waste flow is distributed practically uniformly over their cross sections. Hydraulic conditions in the inlet and outlet sections require careful attention in all cases. The possibility of *density currents* must also be considered.[13,15] The incoming waste loaded with solids is usually denser than the settled upper layers and tends accordingly to stream towards the bottom. This may be sufficient to set up and maintain a counter-flow of clarified waste backwards from the effluent weirs, so that the basic tank design fails and sedimentation efficiency is significantly reduced.

Such currents may also occur if the density of the waste varies substantially, owing for example to changes in the content of salts or other dissolved matter or in temperature.

In ordinary sewerage systems, any changes in density are usually gradual enough to obviate such current effects. In special cases, however, as with intermittent trade waste discharges, sudden changes may be large enough to affect the sedimentation process and if these cannot be avoided, special designs may have to be adopted. Density-current effects may also be important for final sedimentation tanks of activated-sludge plants. They can be avoided if the tanks are made fairly long, with a ratio of depth to length of at least 1 : 20. In one case, where density currents in a final settling tank of an activated-sludge plant were spoiling its performance, Gould reversed the travel direction of sludge scrapers and the location of the sludge hoppers in relation to the effluent weirs to take advantage of the density-induced currents, with considerable improvement in sedimentation efficiency.[13] The entrainment of air in the entering wastes or the release of dissolved gases likewise gives rise to density-current effects.

GRIT CHAMBERS OR DETRITUS TANKS[6,16–19]

The type of sedimentation plant considered under this heading is designed to settle out granular solids of relative densities and particle sizes such that they separate readily from water and tend to segregate from other fine granular and flocculent solids even when dispersed in sludges.

Even in domestic sewerage systems, some extraneous solids, such as silt, sand and grit, enter the sewers fortuitously or may be introduced from bathrooms and laundries. Trade-waste discharges frequently include factory floor washings which also introduce 'grit', and processes such as wool scouring and hide cleansing carry large quantities of dirt into the drainage system. The flow velocity in sewers is usually high enough to ensure that all these solids are finally discharged into the treatment or disposal works; any deposits which may form are usually flushed through during peak discharges under wet-weather conditions. In combined sewerage systems, relatively large quantities of grit may enter the system and reach the outfalls during periods of rainfall. Even under these circumstances, much of this heavier solid matter travels along the bottom of the sewer; it may, therefore, in some cases be diverted from the main stream with only part of the flow, which is returned to the sewer after separation of the grit. Treatment in grit chambers, whether applied to the whole flow as usual, or to part only (the underflow), is directed at separating only non-decomposable, mostly inorganic solids. In most cases, an exact separation is not possible, even with re-washing of the grit (see below), but with careful design this will be clean enough to be odourless and to drain rapidly, and it may be used around the grounds or perhaps even for sludge-drying beds.

The proper basis of design is the settling velocity of the smallest and least dense of the particles to be removed. Taking, for example, sand particles of $0{\cdot}39 \times 10^{-2}$ in (0·1 mm) effective diameter, Table 10.2 gives the settling velocity as 950 in/h (67 mm/s), so that the proper surface-loading rate from Eqn 10.3 is $950 \times 12{\cdot}5 = 11\ 900$ gal/ft^2 d (582 000 l/m^2 d) of grit chamber surface. Grit chambers or detritus tanks may be designed for either horizontal or vertical flow. Theoretically, the same area is required in either case.

Horizontal-flow Grit Chambers

Experience has shown that a satisfactory separation of sand from sewage particles can usually be obtained by reducing the flow velocity to about 1 ft/s (0·3 m/s) during passage through an elongated channel. Technically, such a design may be arrived at by computing the required surface areas and determining the cross section corresponding to this velocity. Storage is then provided by additional capacity below this section to hold deposit equal to several days' peak discharge, extending the length of the chamber if necessary and thereby ensuring efficient grit removal even under unusual circumstances such as collapse of a sewer or excessive stormwater inflow (see *Examples*, pp. 157, 347, 349 and Fig. 10.6).

Unless provided for by special means, variations of waste flow result in significant deviation of the flow velocity through the grit chambers from the design, and special provision is necessary to maintain it practically constant. This may be attained, for example, by the provision of a Venturi flume downstream of each chamber, so proportioned that at all rates of flow through it the velocity in the chamber is maintained at about 1 ft/s (0·3 m/s). Another type of design provides for the construction of one wall (occasionally both walls) with a curved surface developed so that the cross section of the settling chamber varies directly according to the equation

$$\frac{\text{Waste flow}}{\text{Cross section}} = \text{constant}$$

Other designs provide for movable baffles (sometimes automatic) or for the use of several chambers which are brought into operation as

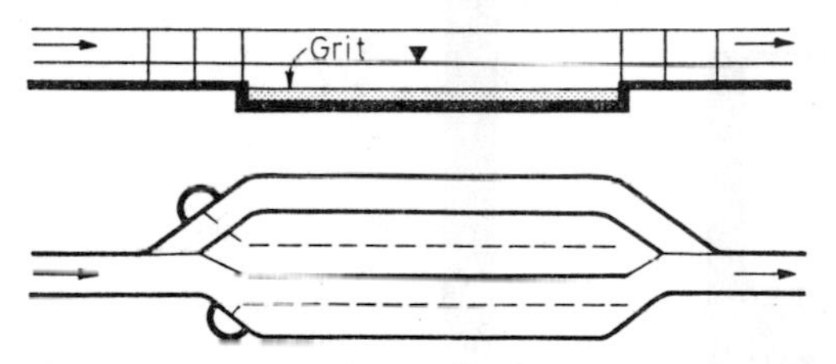

Fig. 10.6. *Horizontal-flow grit chamber (Essen type)*

required. Constant velocity is likewise obtained by the installation of a proportional-flow weir at the end of the grit chamber with the orifice so shaped (Fig. 10.7(*a*)) that the discharge is a linear function of the depth of flow.

Another device is the aerated grit chamber.[16–19] Its cross section must be large enough for the flow velocity along the chamber, at maximum sewage flow, not to exceed 1 ft/s (0·3 m/s). In order to prevent lighter (organic) solids from settling at lower sewage flows, compressed air is used in such a way that a spiral flow is generated and grit deposited on the floor of the chamber is swept into a collecting sump channel (Fig. 10.7(*b*)). The capacity of the aerated grit chamber usually provides for two minutes or more detention of maximum sewage flow. The grit is usually removed from the sump channel mechanically and as a rule is already clean enough, deposition being readily controlled by adjustment of the air flow to about 1–2 ft^3/h (1–2 m^3/h) per ft^3 (m^3) total grit chamber capacity.

Where no such provisions are made, the flow velocity through the chambers usually falls during off-peak periods so much that considerable proportions of organic matter contaminate the grit and sand deposited; then some washing of the deposit is required before disposal. Two methods in use are:

1. The grit is removed from the chamber and washed in suitable apparatus, providing a suitable flow of grit-free effluent to wash out the organic matter, which is then returned to the works influent channel;
2. A removable aerator is used to blow air through the deposit,

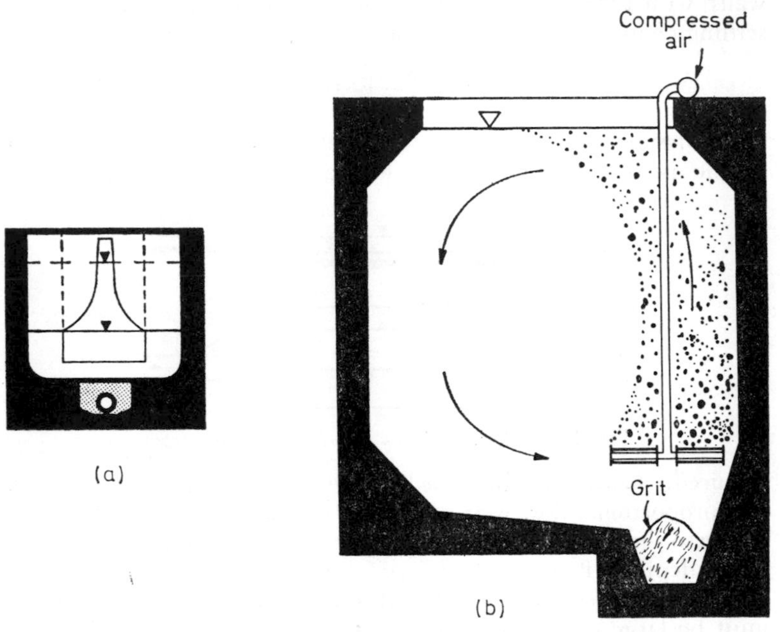

Fig. 10.7. *(a) Specially shaped outlet baffle to maintain constant velocity through horizontal grit chambers (b) Aerated grit chamber*

the separated organic matter being carried away by waste flow and the sand settling again.

The quantity of grit removed from combined sewerage systems usually amounts to 5–15 yd³ (3·8–11·4 m³)/1 000 persons yearly, the lower figure applying to densely built-up sewerage areas (about 100 or more persons/acre (250/ha)), the higher to urban areas with about 20–40 persons/acre (50–100/ha).

In separate systems the amount of grit entering the sewers is usually negligible; in some cases, however, such as sandy areas or seaside towns where sea-bathing is an important feature, significant amounts of grit may enter separate sewers, and about 5 yd³ (3·8 m³)/ 1 000 persons yearly may be anticipated.

The Essen grit chamber (Fig. 10.6) is typical for horizontal flow. It is provided with an under-drain so that it can be drained after diverting the waste through other chambers, the drainage being returned to the waste flow. See also Fig. 10.7(*a*).

Vertical-flow Grit Chambers

These have also been designed, and where adequate surface area is provided have operated successfully. Since the flow distribution across the tank section is uniform vertically, a relatively precise separation of grit, sand and other inorganic solids may be achieved. Moreover, the grit can be collected in a hopper at the tank bottom and is readily washed using compressed air before discharge. However, an appreciable tank depth is necessary to achieve uniform distribution of flow, and this considerably increases the cost of such structures, the surface area required being independent of the depth.

Chambers of special shapes or incorporating particular baffles have been designed to make full use of the available area without unduly increasing the volume, or to solve design problems for widely varying flows. Thus grit chambers may be funnel-shaped, with the waste flowing upwards along a spiral path, advantage being taken of the whole surface area without undue wastage of volume. In another design, vortex action is created by introducing the flow tangentially into the tank, the heavier solids settling in the central 'core'.

Disposal of grit varies according to relative quantities, nature and cleanliness. From small installations it is removed by hand. In larger units, mechanical grit collection by submerged scrapers, endless-chain mechanisms, etc., allows cleaning without interruption of operation. Similarly, travelling grit pumps may be used, and pumping can be combined with washing of the grit to reduce the proportion of organic matter. Disposal usually finally involves burial, but clean grit may be used as land filling, for paving or on sludge-drying beds.

Example—A city of 80 000 inhabitants is expected to contribute sewage at a rate reaching a peak of 4 000 gal/min (300 l/s) D.W.F. just before midday at the treatment works. Designing on a basis of sand particles of $0{\cdot}39 \times 10^{-2}$ in (0·1 mm) effective diam., Table 10.2 gives the minimum settling velocity of particles 0·1 mm diam., S.G. 2·65 at 950 in/h (67 mm/s). The flow rate of 1 000 gal/min equals, say, $5{\cdot}8 \times 10^6$ gal/d (26×10^6 l/d)) and hence (Eqn 10.1, 10.3)

$$\text{Required surface area} = \frac{5{\cdot}8 \times 10^6}{12{\cdot}5 \times 950} = \sim 500 \text{ ft}^2 \ (47 \text{ m}^2)$$

The peak D.W.F. corresponds to a rate of 10·7 ft^3 (0·3 m^3)/s. On this basis, a velocity of 1 ft (0·3 m)/s requires a cross section of 10·7 ft^2 (say 1 m^2), say 7 ft (2 m) wide with a depth of 18 in (0·5 m). The surface area required is 500 ft^2 (47 m^2), whence length should be 500/7 = say, 71 ft 6 in (23·5 m). The design would then provide three chambers each 3 ft 6 in (say 1 m) wide, 71 ft 6 in (23·5 m) long, one chamber being a stand-by unit, used during cleaning of the others in turn. If the storm flow also passes through these units, the surface loading rates will be higher. However, the third unit may be also used during storm flows; thus for a flow of 32 ft^3 (0·9 m^3)/s (about 6 × D.W.F. average in this case) the surface loading rate would be twice as great, reckoning on 750 ft^2 (70 m^2) in all, the corresponding minimum settling velocity would be about 1 900 in/h (13 mm/s), and sand down to about 0·006 in (0·15 mm) particle size would be separated, which may be considered satisfactory. The chamber is accordingly designed with inlet and outlet channels sized and graded for a discharge of 4 000 gal/min (300 l/s) when running 18 in (0·46 m) deep; it would be constructed with a floor 9 in (0·23 m) below the invert, allowing for about one week's storage of grit. Total volume up to 30 in (0·76 m) above the invert $= 3 \times 250 \times (2{\cdot}5 + 0{\cdot}75) \simeq$ 2 500 ft^3 (70 m^3).

The alternative design of a 500 ft^2 (46·5 m^2), say 23 ft (7 m) square-section detritus tank for vertical flow would require a minimum of, say, 7 ft (2·1 m) depth above inlet or about 3 500 ft^3 (99 m^3) plus grit storage hopper, or say, 4 000 ft^3 (112 m^3) in all. Grit removed would be approximately of the same size as above (minimum $0{\cdot}39 \times 10^{-2}$ in (0·1 mm) at peak dry weather) but presumably much cleaner, and the volume would be about 20% less. The tank would be subdivided into two 16 ft (4·9 m) square, with a third as a stand-by; the total volume then required would be about 6 000 ft^3 (170 m^3).

SEDIMENTATION TANKS FOR MUNICIPAL SEWAGE TREATMENT

Normally, municipal sewage contains significant amounts of granular and flocculent solids, so the calculation of sedimentation tank dimensions is not straightforward. Either of two methods may be used:

1. An appropriate detention period is first determined, and details are then arranged according to the rule that the larger the surface the greater the efficiency;
2. An appropriate surface loading rate is first determined, as for granular solids, and details are arranged by considering that the greater the detention period the greater the efficiency.

The first method using detention period as the basis, is generally satisfactory, most commonly used and preferable for treatments using chemical precipitation and for final settling tanks. By 'detention period' is meant the average, or 'theoretical', detention time calculated according to the formula

$$\text{Detention time} = \frac{\text{tank volume}}{\text{rate of flow}}$$

In practical designs the effective detention time may be appreciably

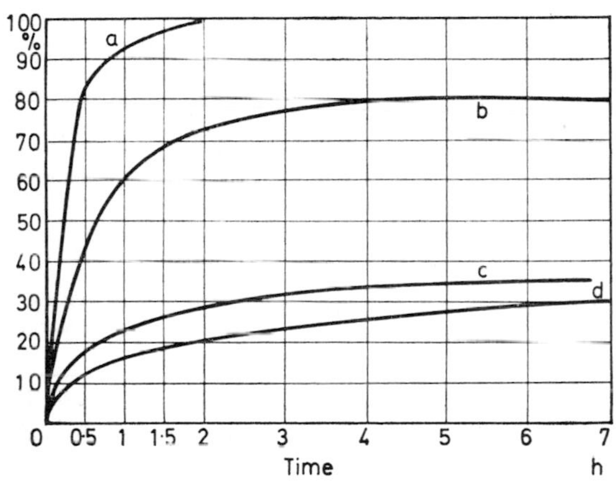

Fig. 10.8. *Relative removal of B.O.D. and suspended matter by sedimentation of municipal sewage, according to Sierp: (a) Settleable Solids (b) Total Suspended Solids (c) Biochemical Oxygen Demand (d) Oxygen consumed from permanganate*

less than the theoretical, determined by special tests.[8–11] Determination of the detention period appropriate for treatment of some particular sewage is, therefore, of great importance. When samples can be obtained, Imhoff cone tests may be used to establish sedimentation curves. Sierp prepared the data of Fig. 10.8. Such curves give an approximate indication of the efficiency to be expected from treatment by sedimentation using different detention periods.

As mentioned before, Imhoff cone tests results do not provide a basis exact enough for calculation of surface loading rates, since the suspended solids of municipal sewage include flocculent solids. The reason is that the particles do not behave independently and settling velocities are not constant, so that sedimentation in Imhoff cones (about 16 in (0·4 m) deep) is not directly comparable to that in tanks of greater depth; in practical tank design, this may be 7 ft (2·1 m) or more, as in the example given below. These tests are useful, however, in comparing different sewages and recording efficiencies of sedimentation by testing influent and effluent.

Other trials using apparatus similar to the prototype sedimentation tanks can be used to test proposed designs experimentally. Van der Zee, for example, used cylinders 4 ft (1·2 m) high to examine the effect of various velocities in vertical-flow tanks and concluded that, with mixtures of raw sewage and activated sludge, improved sedimentation resulted from the use of upward vertical flow, provided that velocities did not exceed about 90 in/h (0·6 mm/s), which corresponds to an upper limit of about 1 100 gal/ft^2 d (54 000 l/m^2 d) of tank surface.

Experience has shown that a suitable average detention period for municipal sewage is $1\frac{1}{2}$ h in tanks about 7 ft (2·1 m) deep, using as a basis the average daytime flow rate (12 h, 8 a.m. to 8 p.m.). If the average 24 hour flow rate is used, as is common in some countries, the corresponding detention time is about 2 h. For a 7 ft (2·1 m) deep tank, the equivalent surface loading rate is 84/2 or, say, 42 in/h (0·3 mm/s) = 525, say 500 gal/ft^2 d (24 400 l/m^2 d). For municipal sewage a design basis of $1\frac{1}{2}$ h daytime flow, equal to 2 h of average flow, gives in most cases the optimum size of sedimentation tank units for practical economy under dry-weather conditions. Generally it would prove incorrect to use either longer or shorter detention periods, shorter ones resulting in the carry-over into the effluent of solids which cannot be treated so economically in secondary treatment units, while larger tanks, though more efficient, are not economical—the additional costs are better spent on improving secondary treatment—and, moreover, favour the septicization of the sewage, which may offset the slight gain in sedimentation efficiency. This is especially true in subtropical or tropical climates where higher temperatures may improve sedimentation efficiencies if the sewage is kept fresh but septic conditions develop relatively quickly. Experience shows clearly that tanks of such size, that is, about 2 h detention, equipped for most cases with mechanical sludge-handling equipment, are more economical than large ones. Two hours detention time is also about the right size for two-storey tanks, such as Imhoff tanks, in relation to that required for sludge digestion.

The above figures provide economically satisfactory designs only for normal sewerage conditions where fresh municipal sewage is to be treated, excluding stormwater or excessive infiltration. In special cases, shorter detention periods or higher surface loading rates may be applicable, e.g. where the degree of purification necessary before disposal requires only the removal of the coarser particles from the waste. On the other hand, there are circumstances which may justify the use of much larger tanks, including the following cases.

1. If the wastes or sludge contain substances which inhibit or reduce the rate of septicization, possibly as a result of substantial amounts of certain trade-waste discharges, longer detention periods may be permissible. The sludge may then be retained in the tank for relatively long periods (even weeks or more), thereby obviating provision for frequent removal and with the additional advantage that it becomes denser. The tanks must then be enlarged accordingly.
2. If toxic, acid or other harmful industrial wastes are discharged irregularly in widely varying proportions to the total sewage flow, the possible harmful effects are less the longer the period of detention in the sedimentation tanks. These effects may be minimized by providing for mixing and balancing the wastes, both in composition and rates of discharge, in large tanks using many hours detention.
3. If under wet-weather conditions an increased flow exceeding twice the normal dry-weather flow is to be treated, additional capacity is required. This may occur not only in combined or partly separate but also in separate systems, owing to relatively large volumes of storm- or groundwater infiltration. This additional capacity may be provided by separate stormwater tanks or by increasing the capacity of the ordinary (dry-weather) settling tanks. In England, regulations require provision of adequate sedimentation for all flows up to six times average dry-weather flow, and it is common practice to enlarge dry-weather sedimentation tanks accordingly, thus averaging out variations in flow and strength, and detention periods of 6–12 h, even up to 24 h, are in use.

In such cases two-stage sedimentation may be advisable, with continuous sludge removal using mechanical or two-storey tanks for the first stage, and secondary tanks of simpler design.

The designer must take local conditions and factors into account in relation to the whole treatment works and final disposal of effluents. The more carefully the units of the sedimentation plant are designed, having regard to local circumstances, the better and more economical will be the secondary treatment and disposal.

The excess kinetic energy content of the inflowing waste must be dissipated so that the flow may be distributed across the tank section in accordance with the design.[8–11, 20, 21] Economy demands that this be achieved in a minimum of tank volume, because zones of turbulence are not effective for sedimentation and, hence, detract from the volume available for settling. Forms of baffling are commonly provided, in some cases with carefully shaped and spaced openings. Different arrangements have been reported as being applied successfully, involving diversion of the flow upwards, downwards, transversely or even backwards.[20] Such baffles tend to retain some floating matter, especially oils and fats and arrangements for skimming are usually necessary.

Outlet arrangements also require careful attention, because they likewise influence the pattern of flow through the tank and thus affect the sedimentation processes. Horizontal overflow weirs are usually provided so that the effluent stream is spread over the full tank width thereby avoiding high-velocity currents which interfere with sedimentation and tend to carry settleable solids away. Experience has shown that it is desirable to provide not less than 3 ft (1 m) length of weir for every 5 000–8 000 gal (75–120 m^3) waste flow/d. With large sedimentation tanks, this requirement necessitates multiple effluent weirs; recent designs have used several weirs in parallel. Very long weirs cannot be maintained truly level over their full length, except perhaps at considerable expense, and satisfactory distribution of the flow is more readily obtained by forming indentations at regular intervals. A suitable arrangement might be, for example, the provision of shallow V-notch weirs, say 2 in (50 mm) deep, spaced 6–12 in (0·15–0·3 m) apart, along the full length of the weir or double weir. A shallow surface baffle, at a suitable distance back from the weir, prevents escape of matter floating towards it.

The provision of relatively long weirs ensures that the water level in the sedimentation tank does not vary much over a wide range of flows. This is especially important in two-storey (Imhoff type) tanks (p. 166) because variations in the water level result in corresponding exchanges of the contents of the sedimentation and sludge compartments. A constant water level also simplifies the installation and use of mechanical sludge- and scum-removal equipment.

Additional capacity for storm-flow conditions (p. 161) can be provided by raising the water level temporarily to store an additional volume (p. 339). This may be done by the use of a secondary weir or of a series of circular or rectangular openings at different levels, or by means of floating or other weirs moving between upper and lower limits. Automatic adjustment of weir heights is sometimes provided.

SEDIMENTATION TANKS FOR WATER-BORNE WASTES OTHER THAN SEWAGE

The design of sedimentation plants for wastes other than municipal sewage normally involves the application of the same principles as have been outlined above, although some trade wastes are markedly different from sewage. The suspended solids may differ in effective specific gravity; some are readily fermented, some practically inert; the wastes may be much more concentrated or diluted than ordinary sewage. The economy of treatment by various processes may vary considerably from that for sewage, and the curves of Fig. 10.8 cannot be used for sedimentation-tank design except for wastes sufficiently similar to domestic sewage.

In many cases, particularly with factories of moderate size, it is possible to use batch-wise treatment; the waste is run into an empty tank until it is full or while manufacture is in progress, after which it is held quiescent until the suspended matter has settled at the tank bottom, the clarified supernatant liquor being then decanted from the sludge (by means of a trunnion, for example) or simply drained out of the tank by an opening above the level of the settled sludge. The settling period may be extended over a whole working shift or longer, especially if the wastes are not readily fermentable. If factory work is continuous, at least three batch tanks are usually required. In most cases, however, it is sufficient to provide for continuous flow through the sedimentation plant, even when the factory processes are intermittent. The types of plant used are generally similar to those used for sedimentation of municipal sewage, except for cases where the daily flow is small (say, less than 20 000 gal (90 000 l)). Smaller plants may take many forms to suit the special nature of the wastes but commonly consist of shallow tanks of rectangular plan and section, with provision of simple inlet and outlet baffles. Sludge is usually removed manually after draining the tank. The types of tank generally suitable for larger plants are those described below. Where the effluent is to be discharged into public sewers, relatively small tanks may be sufficient.

TYPES OF SEDIMENTATION TANKS

If the sludge formed as a result of sedimentation is fermentable, it must never remain in the tank long enough to be decomposed and lifted into the effluent by the gases formed. Not only is sedimentation efficiency reduced accordingly, but septicization of the waste may also interfere with secondary treatment.

The various types of sedimentation tank may be classified according to the means employed for sludge collection and removal.

1. *Tanks which Must be Emptied for Cleaning* (Fig. 10.9)

This type of tank is operated until gasification or, simply, accumulation of settled sludge begins to disturb sedimentation significantly. It is then laid off for cleaning. The supernatant waste is pumped or drained off; workmen enter the tank with rubber squeegees and hoses and the sludge is pushed to the lowest section, aided by flushing water, and drained or pumped away. The floor of such tanks can be allowed only a gentle slope, otherwise men cannot work safely and

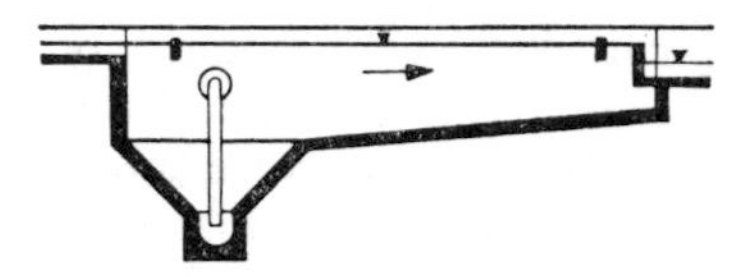

Fig. 10.9. *Manually cleaned sedimentation tank*

economically, except that a rectangular or hoppered sump may be built into the floor to facilitate the final removal of the sludge. At least one extra tank should be provided to allow for cleaning, as this may occupy several days. Such designs are practicable if the allowable operating period is long enough, and for municipal sewage treatment operating periods of 4–14 d are common. However, ordinary municipal sewage in subtropical countries cannot be treated efficiently in such sedimentation tanks because active decomposition of the sludge ensues within 2–3 d or even less, and it has to be removed at least daily to control septicization.

2. *Hopper-bottomed Tanks from which the Sludge may be Removed without Interrupting Operation and Collecting Mechanisms*

These tanks are provided with one or more conical or pyramidal hoppers expanding upwards to their full area (Figs 10.10 and

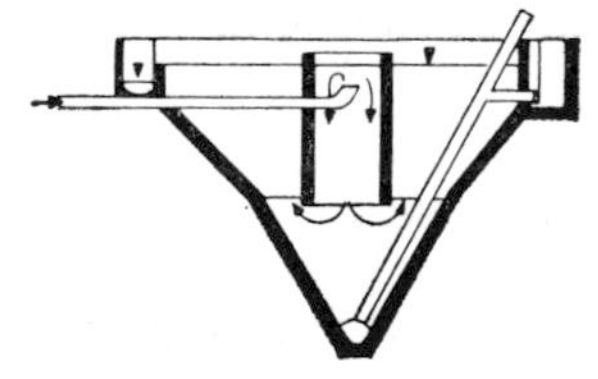

Fig. 10.10. *Vertical flow (Dortmund) tank*

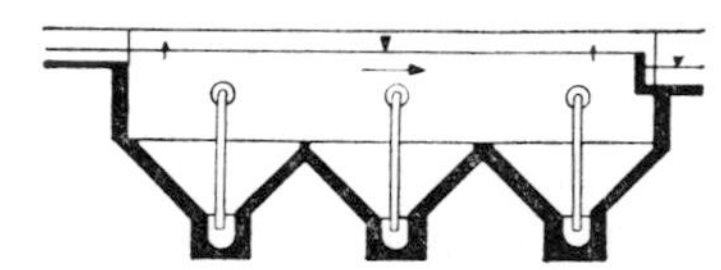

Fig. 10.11. *Multiple-hoppered horizontal-flow tank*

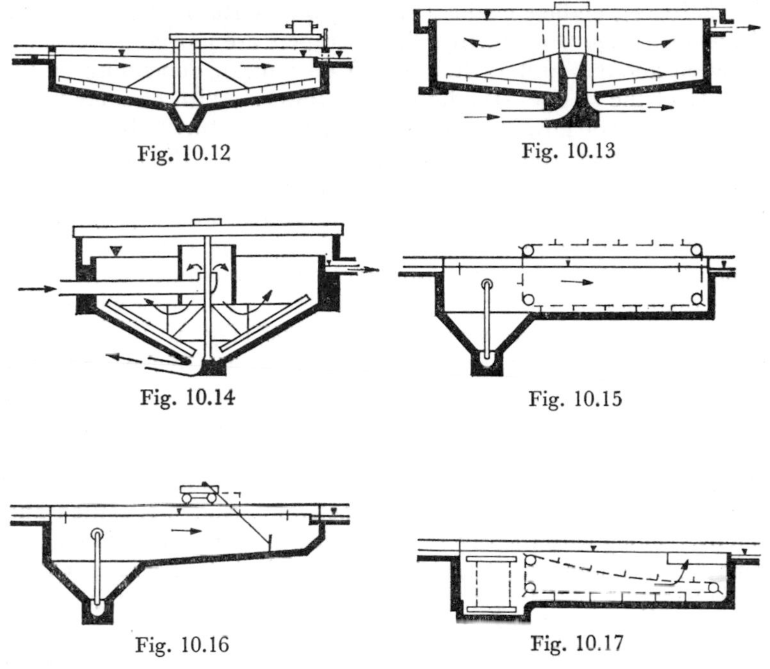

Fig. 10.12 Fig. 10.13

Fig. 10.14 Fig. 10.15

Fig. 10.16 Fig. 10.17

Figs 10.12–10.14 *show circular tanks with rotating sludge collection equipment;* 10.12 *and* 10.13 *with horizontal and* 10.14 *with vertical flow patterns. Figs* 10.15–10.17 *show rectangular tanks with horizontal flow;* 10.16 *with travelling* (*Mieder*) *scraper and* 10.15, 10.17 *with endless chain arrangements.*

10.11) so that the sludge is automatically collected at the bottom of the hoppers and can be withdrawn progressively through a sludge pipe with a mouth there, discharging through a valve or gate outside the tank. The types used include both vertical-flow (called Dortmund or Kniebühler tanks) and horizontal-flow tanks, but with such arrangements for sludge removal the first is generally preferred. Sludge is normally removed daily but may be more frequently.

3. *Tanks with Mechanical Sludge-Removal Equipment*

In this category are placed the many types of tank in which the sludge which settles to the bottom is moved underwater by mechanically operated scrapers to a sludge-collecting hopper or drain from which it is removed either intermittently or continuously. The tank may be so arranged that the scrapers move over the bottom along either circular or straight paths, using rotating arms (Figs

10.12–10.14), endless chains (Figs 10.15 and 10.17) or an external driving carriage (Fig. 10.16). The scrapers are moved at velocities between about 2 and 12 ft/min (10–60 mm/s), depending on the density of the sludge—the lighter the lower. With circular tanks, the direction of flow may be transverse (Fig. 10.12) but is more commonly either from the centre radially towards the tank circumference (Fig. 10.13) or using one similar to the Dortmund type (Fig. 10.10) but fitted with sludge-removal mechanism (Fig. 10.14).

Such tanks may include a small section (sludge compartment) to which the sludge is scraped, usually of a size calculated to hold one day's quantity; in consequence, the sludge may become denser.

Scraper mechanisms of the Mieder type (Fig. 10.16) normally operated intermittently with relatively long inactive periods, are more suitable for denser sludge solids such as the suspended matter removed in raw sewage sedimentation tanks. For lighter or more flocculent suspended matter, apparatus which operates continuously is preferable, especially in secondary activated-sludge tanks and for the primary of plants where excess activated sludge is settled and removed along with the raw sewage solids. Scum-removal mechanisms are commonly combined with mechanical sludge scrapers, the scum being moved along the surface of the waste to a suitable scum trough from which it may be drained off.

When tanks using sludge-removal mechanisms are designed, additional volume must be allowed for the mechanical apparatus and the collecting sludge, also for the zone of disturbance due to the operation of the mechanism.

4. *Tanks with Movable Sludge Pipes*

In place of, or together with, scrapers, it is also possible to provide pipes moved across the bottom of the tank and removing the sludge by suction. In this case, it usually contains more water than that collected after thickening in sludge compartments.

5. *Two-storey Sedimentation Tanks*

In these a sludge compartment is located directly below that for sedimentation, the two being usually connected by an elongated slot through which the settled sludge solids slide down freely and continuously. As shown in Fig. 10.18, the slot is usually so constructed that any gas bubbles or rising sludge solids are deflected upwards away from it. Such bubbles are vented by means of a separate opening to the atmosphere or into a gas-collecting compartment from which sludge gas may be collected (see p. 268).

Such a design can be advantageous for the rapid and continuous removal of a readily fermentable sludge without use of mechanical apparatus, thereby minimizing septicization of the sewage during sedimentation. It is then essential that only a negligible interchange of fluid between the sedimentation and sludge compartments should occur. Such a design is aimed at in the Imhoff two-storey tank (p. 265) which is dimensioned accordingly. These tanks (see Fig. 13.4, p. 266) are designed to provide for the continuous transfer of the sludge as it settles, through the slots into the sludge compartment, usually large enough for storage and digestion of two months' sludge, the sedimentation compartment providing for a settling period as short as possible consistent with adequate removal of settleable solids ($1\frac{1}{2}$–2 h usually). The whole of this compartment is available for sedimentation if inlet and outlet designs and other proportions are suitable.[22]

Other types of two-storey tanks used include Imhoff-type tanks in which only a small sludge compartment (1–7 d storage) is provided, without sludge-digestion space (Figs 10.18 and 10.19) and Travis tanks, not nowadays used but popular early in the century, especially in England. The Travis tank, though superficially similar in construction, is quite different in principle from the Imhoff type; the

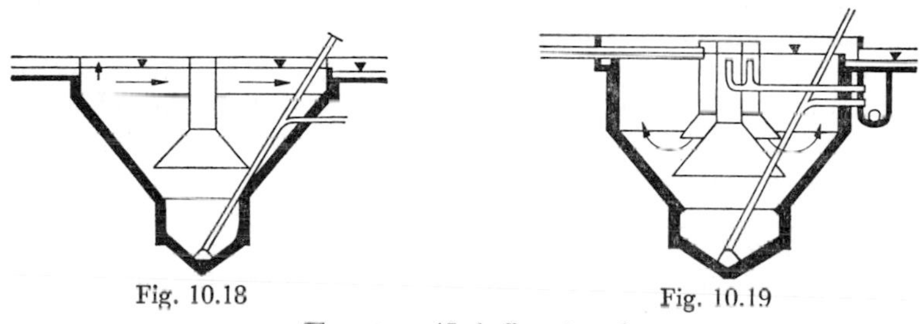

Fig. 10.18 Fig. 10.19

Two-storey (Imhoff-type) tanks

design provides for a positive flow of sewage through the sludge compartment, amounting usually to about 15–25% of the total flow through the sedimentation compartment. This arrangement was thought to favour digestion in the sludge compartment, but septicization of the settled sewage effluent, which is an undesirable feature, becomes appreciable. Imhoff-type tanks with small sludge-storage compartments, combined with separate digestion tanks, have an obvious advantage in separating the sludge automatically without use of machinery. However, unless the sludge compartment is well balanced, as in the case of properly operated Imhoff tanks, greater care is required in all details of design to achieve real advantage from the additional costs of these as compared with simpler tanks.

Imhoff-type tanks are usually designed for horizontal flow through the sedimentation compartments (Fig. 10.18), but it is also feasible to design upward-flow two-storey tanks (Fig. 10.19).

6. *Other Types of Sedimentation Basins*

Although these five types of tank are to be found generally in use at sewage treatment works, many other structures, including basins, beds, ponds and impounding reservoirs, have been constructed or adapted for the sedimentation of waste waters; the following are examples of other types of tanks which have been used successfully in certain cases.

The treatment of wastes in septic tanks is not strictly a sedimentation process but involves biological treatment under anaerobiosis and is, accordingly, dealt with in Chapter 12.

(*a*) *Settling Basins with Manual Cleaning*

These may be constructed using earthen dams to form a basin, usually with an average depth of about 3 ft (1 m). The bed is prepared with a positive fall leading to drains or a central drain to which the sludge may be removed. It is usually allowed to accumulate for about 6–20 d, after which draining of the basin and manual cleaning are necessary. Wastes producing inert sludges may be treated in this manner. The bed and walls of such basins may be dressed with concrete surfaces to facilitate cleansing and prolong their useful life.

(*b*) *Settling Beds*

These are flat basins (Fig. 10.20) similar to sludge-drying beds (see p. 288), and used alternately as settling basins (with under-drains closed) and as drying beds.[23] The waste liquor is drained away by opening the under-drain system, and the settled sludge allowed to

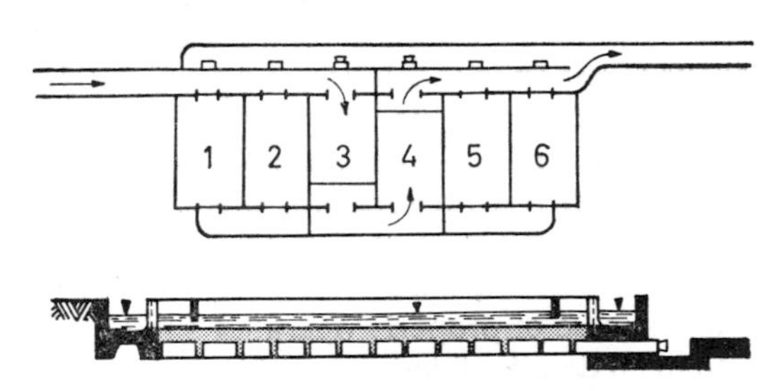

Fig. 10.20. *Settling-drying beds*

drain and dry *in situ* as with sludge-drying beds. A minimum of three pairs of beds is necessary, arranged for operation of each separately in series, using only one at a time.

The velocity through the beds should not exceed about 10 ft/min (50 mm/s). For municipal sewage, their depth is taken to about 8–16 in (0·2–0·4 m), allowing for satisfactory drying of the settled sludge after draining. Greater depths may be used for other types of wastes (for example, up to 12 ft (3·7 m) for those from coal mining). Although suitable for mineral wastes, settling beds can only be used for temporary sewage works because in most cases odour nuisance is inevitable.

(*c*) *Settling Ponds*

These are usually basins of simple design and construction formed in the earth, in which sludge is allowed to accumulate until the pond is so full of it as to be practically useless as a settling basin. The sludge is not removed but remains in the pond while another is brought into use. Natural ponds may be used in this way if protected from flooding by surface waters, and in the same way artificial pits such as clay-pit excavations of abandoned brickworks (or other ceramic works) may provide a means of waste disposal for many years. Their relationship to surface or groundwaters must always be considered. In some cases, a large proportion of the waste water may be re-used, or used for some other process, after treatment in such natural or artificial ponds. Artificial ponds may be equipped with drainage arrangements which can be controlled, so as to provide for alternative settling or draining, as with drying beds.

Ponds of such types have been operated successfully in the treatment of industrial wastes such as from coal mines, smelting works and chemical factories where the waste is practically inert, requiring then a purely physical treatment, in which case sludge accumulation during several years or more is provided for. Some of the sludge solids are then suitable for building up the walls of the ponds so that the level may be raised according to their rate of accumulation; when the mound has reached a critical depth, however, another disposal area must be found.

These types of ponds cannot be operated independently of biological agents unless the waste itself is inert or toxic; with municipal sewage or organic wastes such as from abattoirs and other food industries, septic decompositon usually results and an unsatisfactory effluent is produced.

However, large-capacity settling ponds may be useful for municipal sewerage systems under wet-weather conditions when their additional storage provides for the excess volume of flow, so that even such flows may be well settled. Batch discharges of trade wastes can also be 'averaged' by such means. 'Buffer' ponds like these can be an important means of protecting rivers, especially when used for water supply.

7. *Sewage Ponds and Impounding Reservoirs*

These also act as sedimentation basins. In most cases, however, such structures are used for biological purification, and even when raw sewage is discharged into them, this is more important than simple mechanical sedimentation.

PURIFICATION BY SEDIMENTATION

The results and efficiency of treatment by mechanical sedimentation are usually measured in terms of the suspended matter content of raw waste and settled effluent. The content of settleable solids, as determined by the Imhoff cone test, can be used also to measure sedimentation efficiency and does in practice provide figures which give better estimates of efficiency than the commonly used suspended solids contents. Practical Imhoff cone tests show that sedimentation tanks used for the treatment of domestic sewage, with about $1\frac{1}{2}$ h theoretical detention, remove sludge equivalent to about 80–98%, usually 90–95% of the settleable solids. Efficiencies measured by suspended solids tests largely depend on the waste being treated; usually, the fresher and stronger it is, the better the removal of solids for any given detention period. Of an average of about 500 mg/l of suspended solids (Table 3.3, Gooch test pp. 30, 31) present in ordinary domestic sewage (average standard), about 320 mg/l are settleable solids (2 h Imhoff cone test), and sedimentation-tank removals of about 250–300 $\times$ 10^{-5} lb/gal (mg/l) of suspended solids are common. However, if the concentrations in raw sewage are very low, the percentage removal may be lower than for sewage of normal strength and composition. B.O.D. tests may also be used as a measure of sedimentation efficiency, but with normal municipal sewages only about 35–40% of the B.O.D. is settleable, with weak sewages as little as 25%. Thus B.O.D. tests are not suitable for measuring sedimentation-tank efficiency, except where the settleable solids exert a relatively very high B.O.D. Fig. 10.8 illustrates generally the variation of sedimentation efficiencies for normal municipal sewages according to detention periods. The nature of industrial wastes, and hence of sewages containing relatively high proportions of these, may differ considerably from normal municipal sewage.

Scum separates from most wastes as a floating layer on sedimentation tanks and will usually give rise to odours if not removed continuously or at frequent intervals (maximum 3–4 h). Not only oils, greases or fats separate in this way but other solids of low effective density may enter into such scum layers, and it is found that with ordinary municipal sewage only about 20–30% of the scum consists of substances which would normally be classed as oils, greases or fats. If screenings are disintegrated and returned to the waste water to be settled, a larger quantity of skimmings will separate on the sedimentation-tank surfaces; it will be much less if flotation tanks are installed ahead of the settling tanks. In any case, however, provision for skimming of all such tanks is essential. Where mechanical sludge-removal equipment is installed, skimming devices should be included; otherwise manually operated tools may be used, consisting of flat 'spoons' of sieve wire mounted on wooden handles.

Skimmings can be disposed of by mixing with sedimentation-tank sludges with which the scum solids may be digested, thereby yielding an increased volume of digester gas (p. 286). If necessary, or where the quantity of skimmings is large, the scum may first be collected in a special well, the excess water being drained away and returned to the settled sewage before the solids are disposed of. Sedimentation-tank scum is quite distinct from that of sludge-digestion tanks (see p. 273).

Mechanical Flocculation

This can be used as a preliminary treatment before sedimentation to increase the efficiency of this process,[24, 25] especially in industrial districts where the sewage may be of such a composition as to favour some precipitation and flocculation of highly dispersed matter. A certain degree of flocculation can usually be effected by gentle agitation of the sewage for 10–20 min or more, using slowly rotating paddles. It is important that the flow velocity in the channels between flocculation and sedimentation tanks be kept below about 1 ft/s (0·3 m/s) so that flocculated particles are not again broken up. Sand and silt must be removed before treatment, or accumulation of deposits will occur. The apparatus and hydraulic principles used in chemical precipitation plants (p. 177) are also applicable to mechanical flocculation. It should be noted that aeration (p. 216) may likewise result in some flocculation.

MECHANICAL SAND FILTRATION

Mechanical filters must be distinguished clearly from biological filters such as intermittent sand filters (p. 192) and trickling (percolating) filters p. (196).

Mechanical filtration is common practice in water-supply treatment works (rapid sand filters), but its application to waste-water treatment normally meets with difficulties, mainly because with wastes such as municipal sewage the particles of filter sand stick together and the filter bed becomes completely blocked owing to the suspended and colloidal matter present. Simple backwashing, as used in water-treatment practice, does not suffice to maintain a satisfactory filtration medium; additional mechanical means such as stirring mechanisms or compressed-air agitation are, however, also necessary to 'clean' the filter during backwashing so that normal filtration can be continued. Pretreatment by chemical coagulation and sedimentation or by this and biological treatment tends to make the suspended solids of the wastes flocculent so that backwashing is facilitated.

The ordinary arrangement of filter beds using downward flow through a quartz sand bed for filtration, with upward (reversed) flow for backwashing, can serve for the treatment of waste waters or effluents; sand sizes must be significantly coarser than used for water supply (for settled sewage, say, sand retained on 16 mesh), and especially with a sand-bed depth of more than 18 in (0·4 m) requires more frequent and vigorous backwashing. The backwashing flow, of course, requires treatment, the solids flushed from the filter usually being removed by return to the preliminary sedimentation plant, the backwash water of about 5–10% of sewage flow thus recirculating with the waste.

Installations at sewage-treatment works have included Magnetite (or Laughlin) filters. This type of mechanical filtration system consists of a relatively shallow bed of graded magnetite sand carried on a non-magnetic wire screen. Flow is normally upwards but may also be downwards. Backwashing is carried out by temporarily loosening the magnetite under a travelling magnet.

Sand filtration of humus-tank effluents[26] has also been achieved by suspending beds of sand below the top water level of the tanks such that the effluent must pass upwards through these before overflowing from the tank.[10, 27]

Mechanical sand filters serve to improve sedimentation-tank effluents, especially from chemical or biological treatment plants, and help to maintain uniformity of quality, particularly in respect to appearances. Micro-straining[1-4] (cf. p. 144) is also applied to biologically purified effluents from filter beds or activated-sludge plants, and the improvement obtained has been reported to be similar to that achieved by rapid sand filtration.[28]

REFERENCES

1. Bowen, L. D., Lovell, J. W., Jr and Thistlethwayte, D. K. B. 'Screening and Microstraining of Water: Studies and Experiences', *Proc. 3rd Int. Conf. Wat. Pollut. Res., Munich* **1** (1966) 1
2. Truesdale, G. A. and Birkbeck, A. E. 'Tertiary Treatment Processes for Sewage Works Effluents', *Wat. Pollut. Control* **66** (1967) 371
3. Evans, S. C. and Roberts, F. W. 'Twelve months' operation of sand filtration and micro-straining plant', *Wat. sanit. Engr* **3** (1952) 286
4. Cassidy, J. E. 'The operation of the Hazelwood Lane Works of the Bracknell Development Corporation', *J. Proc. Inst. Sew. Purif.* (1960) Pt 3, 276
5. Campbell, L. J. 'Preliminary treatment processes in sewage purification', *Survr munic. Cty Engr* **111** (1952) 55: see also, *Mechanical Equipment for Sewage Purification Works*, Symposium Inst. Mech. Engrs, London (1962)
6. Seidel, H. F. and Baumann, E. R. 'Pre-aeration and the primary treatment of sewage' *J. Proc. Inst. Sew. Purif.* (1960) Pt 4, 444
7. Miller, D. G. 'Sedimentation: a Review of Published Work', *Wat. & Wat. Engng* **68** (1964) 52; see also Hanson, S. P., Culp, G. L. and Stickenberg, J. R., *J. Wat. Pollut. Control Fed.* **41** (1969) 1421
8. March, R. P. and Hamlin, M. J. 'An Investigation into the Performance of a full-scale Sedimentation Tank', *J. Proc. Inst. Sew. Purif.* (1966) Pt 2, 118
9. Villemonte, J. R., Rohlich, G. A. and Wallace, A. T. 'Hydraulic and Removal Efficiencies in Sedimentation Basins', *Proc. 3rd Int. Conf. Wat. Pollut. Res., Munich* **2** (1966) 381; see also Clements, M. S., *Survey L.G. Technology* **132** (No. 3985) (1969) 4
10. Bayley, R. W., Adams, R. J. and Melbourne, K. V. 'The Removal of Suspended Solids by Sedimentation from Filter Effluent', *J. Proc. Inst. Sew. Purif.* (1964) Pt 5, 405
11. Rebhun, M. and Argaman, Y. 'Evaluation of Hydraulic Efficiency of Sedimentation Basins', *Proc. Am. Soc. civ. Engrs* **91** (SA5) (1965) 37
12. Barnard, J. J. 'Clarification of Sewage using the Sludge Blanket Principle', *J. Proc. Inst. Sew. Purif.* (1963) Pt 6, 517
13. Gould, R. H. 'Improved Final Settling Tanks at Bowery Bay', *Civ. Engng, Easton, Pa* **13** (1943) 279; also Anderson, N. E., *Sew. Wks J.* **17** (1945) 50
14. Hazen, A. 'On Sedimentation', *Trans. Am. Soc. civ. Engrs* **53** (1904) 45 (repr. *J. Proc. Inst. Sew. Purif.* (1961) Pt 6, 521); see also Camp, T. R., *J. Proc. Inst. Sew. Purif.* **111** (1946) 895

15. ANDERSON, N. E. 'The design of final settling tanks for activated sludge', *Sewage Wks J.* **17** (1945) 50 and discussion by R. H. Gould, 63
16. NEIGHBOR, J. B. and COOPER, T. W. 'Design and Operation Criteria for Aerated Grit Chambers', *Wat. Sewage Wks* **112** (1965) 448
17. JOHNSON, D. 'General developments in sewage works design', *Survr munic. Cty Engr* **112** (1953) 347
18. CHASICK, A. H. and BURGER, T. B. 'Using graded sand to test Grit removal Apparatus', *J. Wat. Pollut. Control Fed.* **36** (1964) 884
19. 'Advances in Primary Treatment' Sub-Committee Report, *Proc. ASCE* **88** SA2 (1962) 105
20. CRANSTONE, J. A. 'Circular radial flow sedimentation—a new design proposal', *Effl. & Wat. Treat. Jnl* **5** (1965) 412
21. GILES, J. H. L. 'Inlet and outlet design for sedimentation tanks', *Sewage Wks J.* **15** (1943) 609
22. IMHOFF, K. 'A new method of treating sewage', *Survr munic. Cty Engr* (1909) May 21
23. IMHOFF, K. 'Sickerbecken und sparsame Ortsentwässerung', *Tech. Gembl.* **28** (1925)
24. City of Johannesburg, S. Africa, Annual Report of City Engineer for 1959/60
25. HURLEY, J. and LESTER, W. F. 'Mechanical flocculation in sewage purification', *J. Proc. Inst. Sew. Purif.* (1954) Pt 1, 61
26. NAYLOR, A. E., EVANS, S. C. and DUNSCOMBE, K. M. 'Recent Developments on the Rapid Sand Filters at Luton', *Wat. Pollut. Control* **66** (1967) Pt 4, 309
27. BANKS, D. H. 'An upward flow clarifier', *Survr munic. Cty Engr* **73** (1964) 21
28. *Rep. of Wat. Pollut. Res. Bd Lond.* (1952) 19, see also *Annual Reports* (1962–66).

11

CHEMICAL TREATMENT PROCESSES

AMONG the essentially chemical treatment processes of value in the treatment of wastes are precipitation and chlorination. Others include neutralization and chemical oxidation, which may be specially important for a variety of cases, as well as specific chemical treatments for particular cases only. Chemical processes may come more to the fore with increasing industrialization resulting in greater volumes of trade wastes being discharged into municipal sewerage systems or directly into natural waters; and with growing demands for control of nutrients.

CHEMICAL PRECIPITATION

In this process chemical coagulants are added to waste waters to yield flocculent precipitates (usually metallic hydroxides) which encourage the coagulation of the very fine suspended matter, including colloidal and some even more highly dispersed matter which would not otherwise be removed by ordinary sedimentation tanks. The efficiency and economy of such treatment varies widely—according to the nature and composition of the wastes; the suitability of a particular chemical as a coagulant also depends upon the character of the waste.[1–5]

Experience has shown that the best chemicals for the treatment of ordinary municipal sewage are ferric sulphate and ferric chloride. For coal-washing water, calcium hydroxide is specially suitable or, alternatively, a combination of potato starch and caustic soda. Iron hydroxide formed from iron filings and water has been used successfully for textile wastes.[6] Aluminium sulphate ('alum'), sometimes mixed with small amounts of ferric sulphate ('alumino-ferric'), applied at rates of 50–150 mg/l may produce good clarification, the best results of precipitation being obtained in the pH range 5·3–6·5. Apart from its normal usage as an alkaline material for pH control (p. 179), lime (calcium hydroxide) may also be useful for coagulation where the sewage contains wastes such as iron-pickle liquor, brewery effluents, etc., doses varying up to 250 mg/l and more.

Ferric sulphate is available as an industrial chemical; otherwise an effective coagulant may be produced by mixing chlorine water and ferrous sulphate solution, and this may be cheaper. Ferric chloride is best purchased also as an industrial chemical, alternatively as a solution containing about 40–50% $FeCl_3$, or as crystals hydrated with 60% or unhydrated with 98% $FeCl_3$ contents. Ferric chloride may also be prepared as required from chlorine water and scrap iron, using 2 lb (kg) chlorine/1 lb (kg) iron. Such a preparation is also suitable for oxidation and precipitation of sulphide in foul sewage.

In the U.S.A., chemical precipitation of weak municipal sewage is reckoned to require at least 35 p.p.m. of ferric chloride or about 53 p.p.m. of ferrous sulphate plus 8 p.p.m. of chlorine, strong sewage requires about half as much again. In Halle, Germany, which has

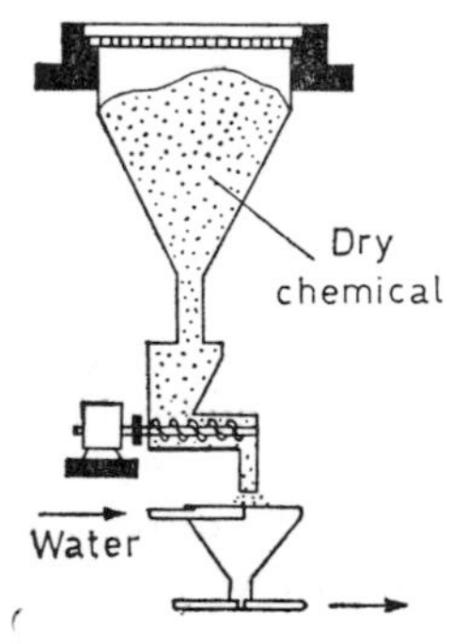

Fig. 11.1. *Feeding apparatus for powdered chemicals*

a very hard water supply, 20 p.p.m. of ferric chloride was found to be sufficient.[7] These figures may illustrate the significance of variations in waste composition.

The dosing of coagulant may be regulated by dry-feeding machines (Fig. 11.1) or by solution-feeding apparatus selected to suit the particular nature of the chemical to be used. Usually the coagulant is added finally to the waste as a solution (or suspension, e.g. when lime is used) containing up to 5 or 10% of the chemical; thus when using powdered crystalline iron or aluminium sulphate, the rate of feed of the coagulant might be regulated by a dry-feeder machine before being dissolved. Ferric chloride is best handled and regulated in solution because it is strongly hygroscopic. The rate of feed should be varied according to variations in the rate of waste flow, preferably automatically. The best coagulant and most economical dose should be found by experiment in each case, the choice depending on many factors. Often there is an optimum pH for best precipitation, requiring the addition of acid or alkali also, and automatic pH mechanisms may prove useful for plant control.

The coagulant must be mixed rapidly say for about 1 min throughout the incoming waste so that it may be uniformly effective, whereupon the floc starts to form and build up, usually without any lag period. After mixing, the waste enters a flocculation tank where a slight but definite turbulence throughout the mixture favours the gentle collision and aggregation of the floc particles. Usually there is a particular regime of turbulence best suited to the character of the waste to be treated and the plant and coagulants available; ideally each case requires experimentation to determine the optimum power input and flocculation time. In practice, a flocculation period of 10–20 min usually assures optimum coagulation, after which the waste is carried without additional turbulence to the sedimentation plant. Once a good floc has been built up, hydraulic conditions downstream must not be so turbulent as to redisperse it, and the channels and other structures must be designed to obviate this, so that the floc is still readily settleable. After coagulation the suspended solids are largely flocculent in character, so suitable types of tanks are either Dortmund or other such tanks with central inlet and mechanical scrapers (p. 165). The detention periods required for satisfactory sedimentation vary according to the nature of the coagulant and the waste. Periods of 1–4 h are usually sufficient, most municipal sewages requiring about 2 h.

The formation of a satisfactory floc may be assisted by special additives.[3, 5] For example, the Schmidt–Degener process developed in Germany uses waste brown-coal pulp added to sewage to increase flocculation and sedimentation efficiencies, similarly, in the U.S.A., waste-paper pulp has been employed; such processes generally may be used only in special circumstances. A further advantage of such additives may be in simplifying sludge disposal by drying and incineration.

The quantity of sludge resulting from chemical precipitation and sedimentation is usually much greater than that produced by plain sedimentation, not only because the moisture content is usually larger but also because the total amount of solids in the sludge is much greater, including the chemical precipitate with the additional solids coagulated from the waste being treated. Commonly the sludge volume is more than twice as great, especially if lime is added to another coagulant or additives such as brown coal or paper pulp are used. The sludge-handling and -disposal facilities must be enlarged accordingly. For disposal by sludge digestion, the digestion-tank capacity required for municipal sewage using chemical precipitation and sedimentation may be up to twice as large as for treatment by plain sedimentation and trickling filtration. At

Halle in Germany full-scale experiments using small coagulant doses show that the sludge resulting from ferric chloride treatment digests much more quickly and produces a greater volume of digester gas than ordinary sludge.[8] However, where higher coagulant doses are used, the proportion of added chemicals may so reduce the digestibility of the sludge that digestion is no longer efficient or economical. Chemical-precipitation sludge is generally quite suitable for dewatering by vacuum filtration (p. 291). It may be advantageous to employ sludge thickening before digestion or before mechanical dewatering (see p. 259).

Treatment by chemical precipitation is generally more efficient than unaided sedimentation, but less so than a complete biological treatment process. Its efficiency and economy may be improved by recirculating precipitated sludge from the settling tank to the mixing chamber, thus increasing the concentration of suspended matter in the flocculation tank to a level of about 2 000–3 000 p.p.m. Such recirculation may also reduce the amounts of chemicals required for precipitation. By using additives such as brown-coal pulp or the polishing processes of sand filtration and final chlorination, a stable effluent can be obtained and the overall efficiency may then approximate to complete biological treatment. Chemical treatment may be applied with success, especially to many trade wastes, and in some cases treatments with iron compounds combined with activated-sludge treatment (see p. 236) gives excellent results. Chemical treatment may also be more successful for municipal sewages containing a high proportion of industrial wastes, especially if these interfere with biological treatment processes so that they are less efficient than usual.

Continuously operated chemical precipitation costs about as much as complete biological treatment, although its efficiency is usually lower. In some cases it may be sufficient to use plain sedimentation for part of the year and chemical precipitation to increase efficiency when the capacities of the final disposal works or of the receiving waters are reduced, e.g. during dry seasons. The efficiency of treatment may then be greatly enhanced by using a coagulant without costly and large-scale construction being required. Similarly, chemical treatment may be used as a temporary measure to improve waste treatment in overloaded plants until additional works are available. It may also be suitable for pretreatment of trade wastes before admixture with normal sewage. Supernatant liquor (p. 275) from sludge-digestion tanks can be treated successfully by chemical precipitation.

pH CONTROL

Many industrial wastes are inherently strongly acid,[9–11] such as those from metal pickling, fruit canneries, cotton 'carbonizing' plants, wool dyeing, acid manufacture; or strongly alkaline, e.g. those from tanneries, textile scouring, vat and sulphur dyeing, laundries, acetylene manufacture. Some pH control may then be necessary for the protection of the sewerage system, to prevent disturbance of treatment works processes or to protect receiving waters or land-disposal areas. In general, these necessities arise from the susceptibility of the common construction materials to corrosion by acids or alkalis or from the toxicity of acids and alkalis to animal and plant life. Because of the concentration of certain industries in suitable areas, their sewage may contain an excess of acid or alkali such that special pretreatment of the whole flow may be more economical than separate processing of individual factory discharges, but in normal communities the neutralizing and buffering capacity of the domestic sewage exceeds the amount of extra acidity or alkalinity from the communal industries, and the average sewage is practically neutral. In this, the normal case, it is necessary, however, to protect the community's sewerage system from the injurious effects of acid or alkaline discharges, though maybe only locally.

Acid discharges are more or less neutralized by means of a commercially available alkaline material, the choice being usually determined by cost and suitability. Thus it may be more practicable to use caustic soda solutions to neutralize acids, with a simple drip-feed system or batchwise additions to acid tanks before draining, than to provide feeding apparatus and reaction tanks for hydrated lime which is not readily soluble and hence reacts more slowly. The materials used commonly include soda ash, lime, granular and powdered chalk, marble, limestone, dolomite and magnesite. The reactions of acids with caustic soda are virtually instantaneous. Soda ash also reacts rapidly and the simultaneous release of carbon dioxide assists mixing. Lime, which is used in suspension, normally requires considerable turbulence for continued reaction, especially in cases where relatively insoluble calcium salts are formed during neutralization. Powdered carbonates of lime and magnesium must also be suspended during reaction, longer reaction periods being necessary; allowance for the dissipation of the carbon dioxide released must be made by proper ventilation, necessary also where soda ash is used. Arrangements usually provide for reaction periods of at least 20–30 min after thorough mixing of the wastes with lime slurries, followed by sedimentation for removal of the insoluble matter, present in the acid wastes and the neutralizing agent or produced during neutralization. Granular materials such as marble are used sometimes simply by passing the wastes through stationary beds of the reagent—using gradings of 1–2 in (25–50 mm) or smaller—but such arrangements are seldom effective, even when long detention periods are available, usually on account of fouling, clogging and short-circuiting.

Alkaline wastes are usually neutralized only when the quantities are large enough to affect biological treatment units, or in special cases where the alkalinity may be injurious or must be removed before chlorination or disposal. Acid coagulants such as ferric salts or alum may be used to remove the alkalinity and produce coagulation (chemical precipitation), with beneficial results; otherwise sulphuric acid is used, normally the cheapest acid available. Carbon dioxide can serve in special cases.

Where wastes are discharged into municipal sewers, it is usually sufficient to ensure that the pH is brought within the range of about 5–9 or, where the volume of the waste is relatively large, about 6–8.

CHLORINATION[11–15]

Chlorine has numerous applications in the treatment of water-borne wastes: for disinfection, odour control, improved clarification and filtration, filter fly control, thickening of activated sludge and protection of beaches, streams and water supplies against infection. Its efficacy depends mainly on oxidation and toxicity to plant and animal life.[15]

Chlorine may be obtained either as liquid in cylinders or tank cars of various sizes or combined in solid form as chloride of lime (about 35% by weight of free chlorine when fresh), or as hypochlorite compounds containing varying proportions as high as 70% of free chlorine equivalent. The rate of addition for a particular purpose depends upon the nature and composition of the waste to be treated. In most cases, some of the added chlorine combines immediately with oxidizable compounds and so may not be available for the intended purpose. This *immediate chlorine demand* must in most cases be exceeded for efficient chlorination. Whereas in fresh municipal sewage it ranges from about 10 to 40 mg/l, that of septic sewage may be much greater. The reaction of chlorine with the sewage may effect a measurable reduction in its B.O.D., say up to about 2 mg/l of B.O.D.$_5$ reduction per mg/l of chlorine added.

Average dosing rates for disinfection are within the ranges given in Table 11.1.

Although complete disinfection usually requires more extensive treatment, satisfactory bactericidal effects are as a rule assured if a positive residual of chlorine is maintained for at least 15 min. Proper mixing with sewage ahead of a contact tank is required for good efficiency, the tank capacity providing for at least 15 min

Table 11.1

DISINFECTION BY MEANS OF CHLORINE

Water or waste to be disinfected	*Required chlorine dose rates*, mg/l	
	from	sometimes up to
Good quality drinking water	0·1	0·8
River water (untreated)	1	5
Effluents from biological treatment works treating municipal sewage	2	10
Raw municipal sewage	20–30	100
Pre-chlorination of settled municipal sewage for odour control	4	20

detention of the sewage after mixing. Some of the other applications of chlorine are based on the following:

1. When applied before sedimentation, the settled sludge remains fresh and is more readily thickened. Doses of the order of 15 mg/l may be effective. Fine fatty particles from industrial wastes such as wool-scouring, tannery and abattoir wastes are modified by suitable chlorination so that better separation is obtained. Effective doses amount to about 100 mg/l calculated on the industrial wastes, but in any case not less than about 10 mg/l of the mixed sewage.
2. Chlorine helps to control sewage septicity. When it is used to control sulphide conditions, ferrous sulphate or scrap iron may be used with it.
3. Gaseous chlorine added to the compressed air of aerated skimming tanks (p. 148) can improve the overall efficiency.
4. Where cheap ferrous sulphate is available, chlorine may also be used to convert the ferrous to the ferric state, which is often more effective as a coagulant.
5. Chlorine can be applied to the sewage before trickling filtration

to control clogging and ponding of filter beds (p. 211). Shock doses are used for short periods only, say 2–3 h. In the usual case of the filtration of settled sewage, up to about 50 mg/l of chlorine may be required. Several treatments may be necessary.

6. Where foaming of activated sludge is due to incoming septic sewage, pre-chlorination using about 5 mg/l may be effective for control. Thickening of this sludge is usually assisted by small doses of chlorine (about 0·2–0·5 g ($4{\cdot}4$–11×10^{-4} lb) *per capita* daily), reducing gasification and allowing moisture content to be lowered to around 96% in some cases (p. 239).
7. Chlorine may be useful to destroy the sulphides in supernatant liquors (see p. 236) or otherwise to improve their redox potential (p. 42), requiring perhaps 20–80 mg/l of chlorine, so that when returned to the raw or settled sewage they are readily treated without causing any significant reduction in treatment efficiencies which otherwise may result from direct supernatant return.

Where it is necessary to protect streams, beaches, waterworks, oyster farms, etc., from bacterial contamination, in some cases sedimentation followed by efficient chlorination may be as satisfactory as sedimentation followed by biological treatment. The first combination may offer a satisfactory treatment in cases where the waste is unsuitable for biological treatment. Installation of chlorination apparatus and plant is relatively simple, so that this may offer ready and economical means where treatment is required only temporarily, e.g. during periods of low river flow.

Chlorine may be used also to postpone the decomposition of wastes after discharge into natural waters. Thus discharge into a small river which joins a relatively large stream after a short course downstream may become feasible if the decomposition is delayed until reaching the large stream where sufficient dilution may be available. In this way, excessive fungal growth and local septicization of the stream may be avoided. However, it must be realized that purification and disposal of the wastes in natural waters is only possible because of biological activity there, and that in general chlorination only modifies the polluting compounds in the wastes but effects little real purification. Hence it cannot prevent decomposition and achieve purification over a long stretch of natural water, either in river, lake or ocean. Where it is used merely to delay decomposition in a short stretch of river or for local disinfection, the free chlorine is gradually absorbed by the mixed river water, followed by re-establishment of biological activity therein, with eventual absorption of the polluting load. It must be applied with great care because fish life in particular is sensitive to very low concentrations of free chlorine[16] and some chlorinated compounds.

The addition of chlorine to waters or sewages results in a reduction of alkalinity and an increase in the free carbon dioxide, so that the chlorinated waters or wastes become more aggressive. Generally the use of hypochlorite does not increase corrosiveness in this way and in some cases may be preferable to that of chlorine gas. Chlorine dioxide[17] is also a powerful disinfectant.

Chlorination may be a useful process for the treatment of industrial wastes (p. 315). In most cases it serves as a cheap but powerful oxidizing agent, in acid, neutral or alkaline conditions according to the chemical reactions involved, e.g. in the oxidation of toxic cyanides to cyanates, sulphides and sulphites, phenols, ferrous compounds and the bleaching of dyehouse wastes. It may also be of help as a coagulant in the precipitation of wastes with a high protein content, e.g. meat wastes; the required amount of chlorine may then amount to 100–1 000 mg/l or more.

OTHER CHEMICAL TREATMENT PROCESSES

Chemical processes are generally more readily applied to industrial wastes treatment than to municipal sewage flows because concentrations of reactive constituents are correspondingly greater in the wastes of factories than after dilution with domestic, or other industrial wastes, and the reclamation of wastes is better achieved before contamination with other wastes. Examples of various methods of treatment include such special techniques as

1. Ion-exchange treatment, useful for the removal of specific toxic substances or for the recovery of expensive compounds,[18] e.g. recovery of chromates from plating wastes may satisfy both purposes.
2. Removal of sulphides from tannery wastes by the use of flue gases[19] (containing CO_2), with possibilities of sulphide recycling.
3. Complexing of cyanides[20] with ferrous compounds, thereby rendering the cyanides insoluble and non-volatile.
4. Solvent extraction (i.e. removal and recovery) of phenols[21] (Phenosolvan-type processes).
5. Liming and distillation of ammoniacal liquors.

6. Chemical oxidation,[22] e.g. in the case of organic wastes by evaporation and burning in air, or using air or oxygen under pressure at high temperature, etc.
7. Nutrient removal processes or 'tertiary' processes for the 'advanced treatment' of effluents from biological treatment processes also may depend on chemical treatment; including, for example, removal of suspended solids and phosphates by coagulation with lime or alum; removal of organic matter by adsorption, electro-dialysis, ion exchange, reverse osmosis, nitrogen removal by air stripping of ammonia, denitrification, etc., and chemical precipitation of metallic compounds.[23, 24]

REFERENCES

1. GOCKEL, H. and ASENDORF, E. 'Theorie und Praxis der Schlammstofffällung', *Vom Wass.* **28** (1961) 94, 98
2. VAN KLEECK, L. W. 'Methods and Results of Operating Intermediate and Secondary Plants', *Wastes Engng* **28** (1957) 398
3. BURKE, J. T. and DAJANI, M. T. 'Organic polymers in the treatment of industrial wastes', *Proc. 21st Ind. Waste Conf. Purdue Univ.* (1966) 103; see also DAJANI, M. T. and GEARING, M. A., *Repr. Am. chem. Soc. Wat. Waste Chem.* **4** (1964) 91 and *Wat. Pollut. Abstr.* (1966) 1737
4. LUMB, C. 'Chemical precipitation of sewage', *Wat. sanit. Engr* **2** (1951) 209
5. SLEETHE, R. E. 'An Assessment of Polyelectrolytes for Sludge Conditioning at Worthing', *Wat. Pollut. Control* **69** (1970) 31: see also GARWOOD, J., *Effl. & Wat. Treat. Jnl* **7** (1967) 380
6. JUNG, H. 'Praktische Erfahrungen mit dem Niersverfahren bei der Reinigung gewerblich verschmutzter, besonders farbstoffhaltiger Abwässer', *Vom Wass.* **14** (1939–40) 216
7. MÜLLER, W. J. 'Chemisch-mechanische Abwasserreinigung in Emscherbrunnen', *Gesundheitsingenieur* **62** (1939) 659
8. MÜLLER, W. J. 'Die Ausfaulung des Schlammes der chemisch-mechanischen Abwasserreinigung', *Gesundheitsingenieur* **63** (1940) 405
9. HOAK, R. D. 'Acid iron wastes neutralization', *Sewage ind. Wastes* **22** (1950) 212: see also *Wat. Pollut. Res. Bd Ann. Rept* (1961) 31
10. ELLIS, H. M. 'Treatment of polluted water and effluents', *Effl. & Wat. Treat. Jnl* **4** (1964) 284
11. BESSELIEVRE, E. B. 'The Economical and Practical Use and Handling of Chemicals used in Industrial Waste Treatment', *Proc 12th Ind. Waste Conf., Purdue Univ.* (1957) Lafayette, 342
12. VON AMMON, F. 'Grundlagen der Abwasserdisinfektion mit Chlor und die Folgerungen sowie Richtlinien für Krankenhausabwässer', *Münch. Beitr. Abwass.- Fisch- Flussbiol.* **8** (1961) 174

13. SHUVAL, H. I., CYMBALISTA, S., WACHS, A., ZOHAR, Y. and GOLDBLUM, N. 'The Inactivation of Enteroviruses in Sewage by Chlorination', *Proc. 3rd Int. Conf. Wat. Pollut. Res., Munich* **2** (1966) 37
14. PITTS, H. W. 'Sub-residual chlorination', *J. Wat. Pollut. Control Fed.* **38** (1966) 1363
15. MOORE, E. W. 'Fundamentals of chlorination of sewage and waste', *Wat. Sewage Wks* **98** (1951) 130
16. WAUGH, G. D. 'Observations on the effects of chlorine on the larvae of oysters (*Ostrea edulis* (L.)) and barnacles (*Elminius modestus* (Darwin))', *Ann. appl. Biol.* **54** (1954) 423
17. BENARDE, M. A., ISRAEL, M. I., OLIVIERI, V. P. and GRANDSTROM, M. L. 'Efficiency of ClO_2 as a bactericide', *Appl. Microbiol.* **13** (1965) 776
18. MCCARVEY, F. X. 'Application of ion-exchange resins in the treatment of metal wastes', *Effl. & Wat. Treat. Conv.*, London 1962
19. RIFFENBURG, H. B. and ALLISON, W. W. 'Treatment of tannery wastes with flue gas and lime', *Ind. Engng Chem.* **33** (1941) 801
20. TARMAN, J. E. and PRIESTER, M. U. 'Treatment and disposal of cyanide-bearing wastes', *Wat. Sewage Wks* **97** (1950) 385
21. HUSMANN, W. 'Der heutige Stand der Abwasserreinigung der Kohleindustrie', *Münch. Beitr. Abwass.- Fisch- Flussbiol.* **11** (1964) 67
22. TAYLOR, C. 'Chemical oxidation—some laboratory experiments in the treatment of chemical trade wastes', *J. Proc. Inst. Sew. Purif.* (1952) Pt 1, 42; see also HUMPHREYS, F. E. and BAILEY, D. A., *Wat. Pollut. Control* **68** (1969) 93
23. STANDER, G. J. and VAN VUUREN, L. R. J. 'The Reclamation of Potable Water from Sewage', *Wat. Pollut. Control* **68** (1969) 513
24. Fed. Wat. Pollut. Control Administration, Summary Rept for 1964–67, *Advanced Waste Treatment* (1968) U.S. Dept of Interior Publication 10P-20-AWTR-19, Cincinnati, Ohio

12

BIOLOGICAL TREATMENT PROCESSES

GENERAL INTRODUCTION

WHEN mechanical and chemical treatment methods cannot ensure an adequate degree of purification before natural disposal, assistance is sought from micro-organisms such as those effective in self-purification of waters (p. 57 et seq.).[1,2] Methods using such agencies are referred to as biological. Either aerobic or anaerobic processes may be utilized, but the latter are specifically known as 'septic treatment', and biological treatment methods are usually taken to include only certain standard techniques such as activated sludge, percolating filter, intermittent sand filter, sewage irrigation and lagooning methods in which the processes involved are essentially aerobic. Anaerobic treatment methods include treatment in septic tanks and anaerobic lagoons and anaerobic digestion of wastes and sludges.

The principal feature of aerobic treatment is the utilization of natural oxygen resulting from artificial or natural aeration, using special means to intensify and maintain the activity of aerobic micro-organisms. This results in the production of flocculent particulate matter which is retained either as a slime carried on the structures used (e.g. the stones of trickling filters or the material of contact aerators) or as freely suspended particles in activated-sludge tanks. These flocculent aggregates absorb fine suspended, colloidal and indeed, dissolved matter from the waste. These solids are subsequently more or less broken down by living organisms or are changed in nature, so as to be readily separated from the purified water by simple mechanical means. The biological activity essential to these processes can be maintained only in the presence of dissolved oxygen, because oxygen is continuously consumed by the feeding organisms and must be replaced as fast as it is used up. Some oxygen is made available by the activity of photosynthetic organisms such as green algae, which are very important for the dissolved oxygen economy in the cases of oxidation ponds, but generally the uptake of atmospheric oxygen (pp. 58–62) is of paramount importance. The rate of uptake depends upon the interface between the water and the air, and the apparatus used for biological treatment must aim to extend this surface as much as possible, either by forcing air through the water in the form of fine bubbles, as in aeration tanks, or by spraying or spreading the water widely through the air or over large exposed surfaces, allowing sufficient time for oxygen absorption from the air. The flocculation of the solids provides greater surfaces and, hence, the availability of the dissolved oxygen for biological activity is improved accordingly.

Ten cubic feet (0·28 m³) of dry air measured at average atmospheric pressure and temperature contains about 0·175 lb (80 g) of

Table 12.1

AVERAGE LOADING RATES FOR BIOLOGICAL UNITS TREATING DOMESTIC SEWAGE

Biological treatment process	*Loading rates* (B.O.D.$_5$/d)			
	lb/yd³	kg/m³	lb/acre	kg/ha
Land treatment growing crops (p. 191)	—	—	8	9
Broad irrigation on grasslands (p. 191)	—		30	34
Intermittent sand filtration (p. 192)	—		60 150	70 170
Trickling filtration (p. 196)				
low-rate	0·2–0·4	0·12–0·24	3 000	3 400
high-rate	1·0–2·0	0·6–1·2	20 000	22 000
Activated-sludge treatment (p. 217)				
standard-rate plants	0·5–1·0	0·3–0·6	—	—
high-rate plants	2·0–6·0	1·2–3·6	—	—

1 acre = 43 560 ft² = 4 840 yd²; 1 acre × ft = 1 613 yd³

oxygen. The volumes of treatment-works structures must be designed so that a considerable excess of oxygen is always available where required, allowing for practical inefficiencies in its uptake and consumption. In activated-sludge plants, for example, only about 5–11% of the available supply of atmospheric oxygen is utilized in an operating plant, in a trickling filter only about 5% or less. However, intermittent sand filters may use much more of the available oxygen.

The dimensions of such biological treatment works must be such as to provide space sufficient for the activity of the zoogloeal slimes; consequently, a certain loading rate requires a corresponding minimum volume. In most cases, the amount of polluting matter provides a better basis for design, using the B.O.D. as a measure and taking the volume of the waste into account. Thus a volume loading of 100 000 gal (455 000 l) of waste with an average B.O.D.$_5$ of 500 mg/l would be reckoned as a total B.O.D. loading of 500 lb (227 kg). Table 12.1 sets out comparable loading rates which have been applied in practice to units treating normal domestic sewage.

Where estimates of the B.O.D. are not otherwise available, population contributing to the waste flow can be used as a basis. Thus, under average conditions, the B.O.D.$_5$ of domestic sewage after sedimentation varies from about 0·08–0·10 lb/hd d (say 35–45 g/hd d). The population equivalent of industrial wastes may also be used to calculate loading rates. Reckoning an average of 0·09 lb (41 g) of B.O.D.$_5$/hd remaining in settled sewage, sullage contains only one-third as much, that is, about one-third of the B.O.D.$_5$ of settled domestic sewage comes from kitchen and bathroom drainage, amounting to about 0·03 lb/hd d (14 g/hd d); the other two-thirds (about 0·06 lb (27 g)) results from water-closet drainage. The population equivalent for partial sewerage (sullage only) is then one-third person *per capita*.

In any case, it is necessary to consider whether special or local circumstances require modification of the average values. Thus loading rates should be reduced where:

1. An unusually high degree of purification has to be achieved;
2. A relatively high degree of nitrification is desired;
3. The waste water is relatively weak, so that additional volumetric capacity is required;
4. Additional provision for stormwater is required during rainstorms;
5. Preliminary treatment by sedimentation is relatively inefficient, inadequate or not provided;
6. The temperature of the waste is abnormally low;
7. The type of construction or the method of operation is likely to prove less efficient than usual;
8. The works are relatively small, so that normal attention and operation cannot be provided economically.

In the same way, other circumstances may justify higher rates of loading than usual.

The efficiency of such works is indicated by the difference in composition of the effluent by comparison with the influent. The degree of B.O.D. reduction is the best simple index of purification efficiency, a reduction of 90–95% of the B.O.D. of raw municipal sewage being commonly attained in treatment plants designed for complete biological treatment. In such plants, about one-third of the overall purification of sewage is achieved by preliminary treatments, including sedimentation, whereby the settleable suspended solids are separated and handled by special means such as sludge digestion before disposal. It is to be noted here that further quantities of solids result during biological treatment, and these are usually also separated in sedimentation tanks, called secondary or humus tanks, yielding additional decomposable sludges which require further treatment before disposal. The purification effected by this separation of solids is also credited to the efficiency of biological treatment, although actually the sludge solids when removed are not fully stabilized and have not used oxygen equivalent to the amount of B.O.D. removed.

Even during so-called 'complete biological treatment', only part of the organic matter present is actually broken down to the end products of complete oxidation, water, carbon dioxide, nitrate, sulphate, etc. The actual proportion so decomposed varies according to the processes used for treatment but is estimated to be roughly as follows, expressed as the relative amounts of B.O.D. actually satisfied:

intermittent sand filtration	nearly 100%
trickling filtration, low-rate	80%
trickling filtration, high-rate	55%
standard activated-sludge treatment	45%

These figures, however, are not directly related to the overall purification effected by treatment works incorporating these processes. As noted (see p. 131), it is the total amount of polluting matter effectively removed by treatment which provides the real measure of efficiency of treatment processes, not necessarily the relative degree of stabilization. The end products of decomposition include large amounts of carbon dioxide and perhaps some nitrate. Carbon dioxide makes effluents more aggressive to construction materials. Nitrates discharged into natural streams may promote excessive growth of algae, ultimately to the detriment of the stream, although nitrates are oxygen carriers and may be useful for the oxidation of sulphide or to freshen septic wastes. Moreover, although nitrate can be utilized for plant growth if carried into the soil by light, rain-like spray irrigation, it has no great effect in broad irrigation at normal rates of drainage. Nitrate is also of very little real value in natural waters because it does not become effective as an oxygen carrier until the free dissolved oxygen has already disappeared, being therefore of no use as a means to obviate deoxygenation or to prevent the destruction of fish life.[3] The operation of biological plants is usually arranged therefore to hold nitrate production within narrow limits. To achieve this, as much sludge as possible is separated by preliminary treatment and so kept out of biological oxidation units. Also, the shorter the effective detention periods of the solids in the biological treatment units, the less the extent of nitrification. Thus in activated-sludge plants aeration times and activated-sludge quantities should be kept as low as practicable for efficient operation.

On the other hand, it may be desirable in some cases to produce as much nitrate as possible—for example, if it is desired to recirculate effluent to the incoming raw waste in order to combat septic conditions and foul odours. This might be achieved by reducing flushing in trickling filters, using lower rates of loading, or, in activated-sludge plants, by holding the sludge quantity as high as possible with larger volumes of air and longer aeration periods.

LAND TREATMENT BY BROAD IRRIGATION

Land may be used not only for the final disposal of waste waters (p. 102) with or without preliminary treatment, but also as a means of treatment prior to final disposal into natural waters. Originally, however, waste-water treatment and final disposal were combined in the one process of broad irrigation or land filtration using areas of grassland. Pretreatment of the wastes was not provided or was limited to simple screening in most cases; sometimes partial sedimentation in earthen basins was provided. It has been realized only in the last few decades that treatment on the one hand and disposal on the other present two distinct aspects of land-treatment processes. Thus low-rate irrigation (p. 105) of waste waters after preliminary treatment, with or without fresh-water dilution, is to be regarded as a method of disposal rather than one of treatment. The waste water is prepared for irrigation by trickling filtration or activated-sludge treatment; such low-rate irrigation has been reviewed in Chapter 7 as a method of disposal. Broad, or high-rate, irrigation, as it may be called more properly, is not a method of disposal but one of biological treatment of waste waters.

The principal idea of high-rate irrigation with crop or grass filtration is to treat the waste water satisfactorily to such a degree as to permit its disposal into some particular body of receiving waters, either groundwaters or surface waters. The agricultural use of such areas is purely secondary, and re-use of water and nutritious constituents is likewise only a secondary consideration. Accordingly, the capacity required for broad irrigation is determined by the minimum area of the available land necessary to

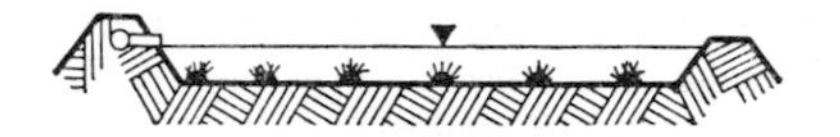

Fig. 12.1. *Flood irrigation*

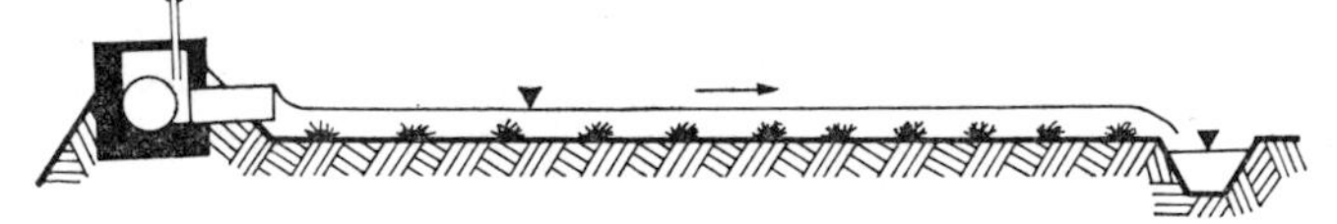

Fig. 12.2. *Surface irrigation*

Fig. 12.3. *Furrow irrigation*

produce the desired purification of the waste waters, not by that required to utilize best the water or fertilizing constituents, as in the case of waste disposal by low-rate irrigation. Of course, as a matter of economy, agriculture is used as far as is consistent with the main purpose of waste-water treatment.

For this technique of broad or grass irrigation, the area is subdivided into a number of relatively small fields, say 1–10 acres (0·4–4 ha) each in extent, according to the average gradients. Each area is under-drained by agricultural drain pipelines laid at a depth of about 3 ft (0·9 m) below the surface and 30–40 ft (9–12 m) apart, arranged to collect the effluent for discharge into the receiving waters. The waste is run on to the areas either by simple flood irrigation (Fig. 12.1), surface irrigation (Fig. 12.2), land filtration using shallow trenches (Fig. 12.3), by distribution from a relatively simple system of open ditches, or by more elaborate and costly systems such as spray irrigation in cases where suitable gradients are not available.

In the case of municipal sewage, land treatment is usually provided by a sewage farm owned by the sewerage authority. Treatment costs are in almost all cases considerably greater than the revenue derived from agricultural utilization of the farm, the difference being met by ordinary charges on the contributing population and industries. Hygienic problems arise from any such agricultural usage of sewage farms, and health authorities in various countries have prescribed methods of agricultural usage to guard against health hazards in the communities affected (pp. 106, 115). It is considered, for example, that at least 14 d should elapse before irrigated lands are used for grazing and that crops and vegetables for human consumption should be planted after irrigation and not beforehand. However, a number of health problems are still controversial and subject to considerable public opposition. These include principally the viability of intestinal parasitic ova and, to a lesser extent, other toxic organisms and compounds. In addition, odour nuisance is likely to arise and must be taken into account. However, the soil is effective for the treatment of many toxic and noxious wastes, organic or inorganic, and can be used for treatment of wastes which may be harmful to the biological activity of natural waters because of their toxicity. Land treatment, therefore, may become a very important method for the treatment of some industrial wastes in order to avoid their direct discharge into natural waters; care must then be taken to prevent harmful contamination of the groundwaters.

Load allowances for land treatment vary within a large range, depending upon the location and type of soil and the method of irrigation and drainage arrangements, to a lesser extent on climatic conditions. Agricultural areas growing crops or vegetables may treat waste water equivalent to ordinary settled municipal wastes from about 100 persons/acre (250/ha); for grasslands the equivalent of settled wastes from about 400 persons/acre (1 000/ha) of irrigation area may be assumed, taking average unspoiled natural soils as a basis. At these rates of loading, difficulties are likely to occur during periods of prolonged wet weather when the whole area becomes saturated with rainwater; then it may be helpful to provide emergency basins for storage and partial treatment by other means before direct disposal into suitable natural waters. Aerobic (p. 242) or anaerobic (p. 243) ponds offer other means for emergency treatment under such adverse conditions.

LAND FILTRATION OR INTERMITTENT SAND FILTRATION

Following the original system devised by Frankland, the technique of land filtration was developed in the U.S.A. more than 70 years ago and has not been altered since. It resulted from the application of land treatment without any agricultural use, aiming thereby at reducing the required area to the lowest practicable limit.[4] Usually some pretreatment by sedimentation is necessary before filtration.

Land filtration is possible only on suitable sandy soil and consists basically of cyclical processes including intermittent flooding of the soil or sand-bed, followed each time by a period long enough to permit complete drainage and natural aeration and recovery of the bed. Structurally the filtration area must, therefore, be arranged as a number of separate beds, 4–8 usually being satisfactory, none larger than about 1 acre (0·4 ha) in area. The beds are constructed by removing the top soil, from which shallow banks confining each area are constructed, the underlying sandy soil being loosened and

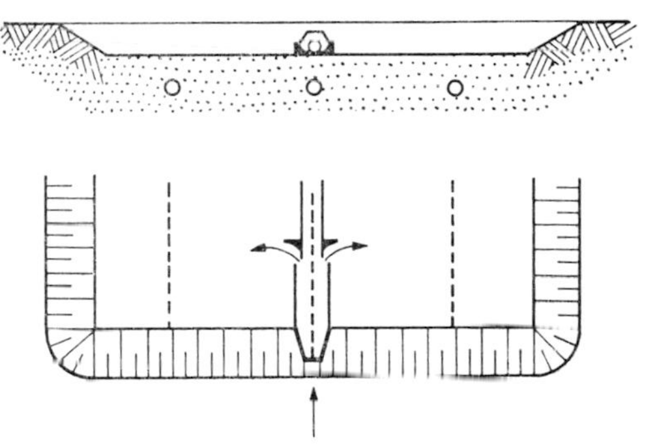

Fig. 12.4. *Arrangement of intermittent sand filtration bed*

under-drained. The surface must be laid exactly horizontally and the draining is effected by means of agricultural piping, 4 in (100 mm) or more in diameter placed about 30 ft (9 m) apart in a horizontal plane about 3 ft (0·9 m) below the bed surface (Fig. 12.4). Influent and drainage effluent pipelines are laid under the surrounding banks. Dosing tanks designed to ensure rapid and uniform flooding of the filter beds, and of volume equal to that of a single charge for any one bed, are usually provided, fitted with automatic valve or siphon systems to distribute the settled waste to the various beds in turn. Loading rates depend upon the actual character of the soil and the nature of the influent waste.

Soil Testing for Filter Beds

Particle sizes are determined using sieves to grade and separate particles upwards from about 0·1 mm size (150 mesh) and sedimentation methods for smaller particles, from which particle-size curves of composition are constructed. The usual sample size taken is

0·22 lb (100 g) (dry weight). From the curves, in which the particle sizes are plotted on a logarithmic scale against corresponding weight percentages of the sample, the particle sizes corresponding to the 10% and 60% weight percentages are read off. The rate of drainage of clean water through sand has been found to depend more or less on the particle size characteristic of the finest 10% of the sand. The characteristic (effective particle size) is, therefore, defined so that 10% by weight of sand particles are less than that size. This figure for suitable soils is 0·0080–0·020 in (0·2–0·5 mm). Soils for which figures below about 0·0080 in (0·2 mm) are characteristic cannot be used successfully because they are too tight; those characterized by figures greater than 0·02 in (0·5 mm) are not practicable because of uneven distribution during flooding, unless a finer

Table 12.2

RATES OF DRAINAGE OF WATER THROUGH SAND OF VARIOUS GRADES

Effective size of sand		*Rate of drainage of clean water*	
in × 10^{-1}	mm	ft^3/ft^2 h = ft/h	mm^3/mm^2 s = mm/s
0·079	0·2	2·6–7	2–6
0·118	0·3	7–14	6–12
0·157	0·4	14–28	12–24
0·197	0·5	28–42	24–36

covering bed is overlaid. In addition to this specification, the 'uniformity coefficient' is also specified, defined as the ratio of the particle sizes corresponding to the 60% and 10% weight percentages, respectively. Where the two figures, obtained as above, are say, 0·105 in (2·6 mm) and 0·015 in (0·38 mm), the uniformity coefficient is $\frac{0{\cdot}105}{0{\cdot}015} = 7$. Coefficients between 5 and 10 are common for soils; lower ones are found for water-graded sands. Uniformity coefficients below 5 are recommended for municipal sewage. Free volumes or void capacities are commonly 25–40% in the dry-filter medium, giving void capacities for air uptake in the wet drained soils of about 20–25%. These figures are not normally specified.

On the above basis the rates of drainage of clean water are approximately as given in Table 12.2.

The beds are operated by filling with 2–4 in (50–100 mm) of waste, equal to between 45 000 and 90 000 gal/acre (50 000 l/ha and 1 000 000 l/ha) or about 1–2 gal/ft^2 (50–100 l/m^2), allowing not more than 5–15 min for filling. The waste floods the bed and should then drain uniformly away, driving the air downwards through the soil and drawing fresh air from above, the water usually disappearing within 1–2 h. The bed is normally charged only once daily, but with strong wastes may require several days for recovery, but weak wastes may allow 2–4 fillings per day. When drainage requires 4 h for disappearance, clogging may be overcome by allowing continuous aeration (resting), without any further charges at least until a layer of dried sludge can be removed from the surface when, after a light disturbance by raking, it may again be flooded and the bed returned into the normal cycle. However, the permeability of the bed diminishes gradually until replacement of the top layer eventually becomes necessary. After several decades, the whole bed must be replaced if efficient treatment is to be maintained.

For effective winter operation in colder climates, up to 1 ft (0·3 m) of flooding is necessary to maintain suitable conditions within the bed, the filtration area then operating under an ice cover through which vents must be maintained, a sufficient number of loose sand heaps reaching above the ice cover being adequate. Cleaning and resettling for the next winter should be completed during spring and summer.

The basis of loading rates may be established as follows: assuming a daily flooding, the air supply to a bed 3 ft (0·9 m) deep containing 20% voids is replaced once daily, amounting to about 26 000 ft^3 (740 m^3) containing about 460 lb of oxygen/acre (520 kg O_2/ha) which, on the basis of 0·08 lb (36 g) $B.O.D._5 = 0·08 \times 1·46 = 0·117$ lb (53 g) $B.O.D._{20}$/hd d gives a theoretical allowance of about 4 000 persons/acre (10 000/ha). In practice, a loading of about one-half may be reckoned, say, 2 000 persons/acre (5 000/ha). Of course, if the applied waste is weak and allows several charges per day, the aeration of the bed is increased accordingly. On the contrary, very strong wastes with a high B.O.D. may require dilution with recirculated effluent or other clean water in order to maintain the oxygen balance inside the bed. In practice, loading rates for settled sewage corresponding to 800–2 000 persons/acre (2 000–5 000/ha) have been applied successfully, the higher loadings being allowable with the coarser grades of sand. This might be taken to be about half for raw sewage, say 400–1 000 persons or equivalent per acre. Intermittent sand filtration is also very suitable for 'polishing' treatment of effluents from trickling-filter installations, for which loadings of 4 000–10 000 persons/acre (i.e. 10 000–25 000/ha, about 5 times settled sewage figures) are practicable.

The purification efficiency of treatment by land filtration (intermittent sand filtration) is relatively very high, the reduction of B.O.D. usually exceeding 90% and bacterial reductions 95%. The effluent is normally very low in turbidity and colour, showing a sparkling and clear appearance by comparison with trickling-filter effluents. This method of treatment is especially suitable for small plants because of its inherent simplicity and is applicable not only for municipal sewage but also many types of industrial wastes, including those from breweries, distilleries, paper mills and wool-scouring plants.

TRICKLING FILTRATION

Trickling or percolating filters are a natural development from intermittent sand filters in which the effective agent was recognized as biological activity, dependent upon proper ventilation to ensure aerobic decomposition and stabilization of the pollutional content of water-borne wastes. In order to reduce the land area required, it was necessary to develop a deeper filter bed with better ventilation, such beds requiring a coarser filter medium which allows better ventilation at much higher surface-loading rates.

The original form was that of the so-called *Contact Beds* which were operated in a manner basically similar to intermittent sand filters. They consisted of watertight basins filled with coarse inert particles such as coke or stone of about 3 in (76 mm) size, with carefully laid checker brickwork, or in other such manner. Sewage was run into each basin in turn, the filling medium being completely immersed while the sewage remained quiescent for a period of, say, 2 h, during which suspended and dissolved matter was absorbed on the active slimes covering the medium. Thereupon it was rapidly drained away and the bed allowed an empty resting period of, say, 10 h or more for oxidation of the organic matter which had been absorbed onto the media.

Contact beds are no longer used because loading and efficiency are relatively low and they clog very quickly; they were soon superseded by trickling filters which, by contrast, are operated continuously and freely ventilated at all times.

Principles of Trickling Filtration

Purification of waste waters containing organic substances by treatment on trickling (percolating) filters[5–24] is a biological process depending primarily on absorption and adsorption of both soluble and suspended matter from the waste water into and onto zoogloeal slimes which develop and proliferate on the surfaces of the filtering medium over which the water trickles or flows, followed by further complicated processes of decomposition and synthesis of both the soluble substances and sludge-forming solids. Some of the solids are thereby removed from the wastes. These decomposition and oxidation processes result from the activity of bacteria, various forms of fungi and protozoa, together with other living organisms which feed upon them preventing unrestrained accumulation and clogging.

When wastes are fed continuously or with only short interruptions onto such filter beds, so that they trickle or percolate over the stones or other material filling the beds, a covering of zoogloeal slime (composed basically of fungal organisms including bacteria, which are normally present naturally in such wastes) develops gradually during a ripening period of from several weeks to a month or more. After this, full purification of the wastes can proceed continuously, provided that aerobic conditions are maintained within the bed, the degree of purification being dependent upon the efficient surface area of filter media available for direct contact with the trickling wastes. The effective zone for aerobic biological activity consists only of the slimes down to about 2 or 3 mm in depth below the slime surface, because deeper zones become septic.[25]

The time required for 50% of a salt, added to the influent distributed uniformly over a filter, to pass through the filter (the half-displacement period) provides a measure of the rate of percolation. Percolation at low rates (see below) normally corresponds to 20–60 min for half-displacement of soluble mineral salts through a 6 ft (1·83 m) deep filter, varying from time to time for any one filter according to the respective amount and character of the film. Recently, suitable radioactive isotopes have been used in place of salt.[11, 19, 26]

Percolation of water down through the bed also results in some flushing effects. Portions of active as well as decomposed or dead and inactive slime material are progressively washed down through the bed with the percolating water and finally appear as a flocculent suspension known universally (but unfortunately) as humus. Discharge of humus solids from filters goes on all the time, but the amounts discharged may vary widely from time to time and filters which are in good condition remain active despite this continuous loss of slime. Naturally, this flushing effect is greater with greater rates of application of water, and those commonly used today vary so widely that it has become necessary to distinguish between 'low-rate' and 'high-rate' filters. Oxidation of the organic matter within the beds proceeds further when the amount of solids in these is higher (as in low-rate filters) than where the accumulation of solids is largely prevented by increased flushing effects, as in high-rate filters.

In low-rate, so-called standard filters, the rate of application of waste waters is low enough to ensure a nearly complete decomposition of the organic matter within the filter, and these may be described as sludge-containing filters. In consequence of such decomposition, the amount of solids finally discharged is relatively low, and very little slime is discharged continuously. However, larger amounts of accumulated slime are unloaded from the filter at intervals of many months (perhaps 2–4 times yearly) during unloading periods lasting for some weeks. The humus solids from such filters tend to contain high proportions of inorganic or earthy matter and to be rather granular than flocculent, forming denser sludges which are not readily decomposable. In low-rate filters, ammoniacal and nitrogenous organic compounds are usually well enough oxidized to yield high proportions of 'oxidized nitrogen', principally in the form of nitrates with some nitrite.

On the other hand, high-rate filters operate under conditions permitting only a limited amount of slime to accumulate on the surface of the filter medium. Excessive slime is readily flushed away into the effluent, and the slime differs considerably from that of low-rate filters, although it develops from the same type of organisms, because it is of more recent growth and the organic matter is decomposed only to a relatively small extent. Similarly, the effluents from high-rate filters contain relatively little oxidized nitrogen. Thus high-rate filters achieve the first and essential part of the purification process, the absorption and adsorption of organic and other substances on to zoogloeal slimes but not much oxidation or stabilization of the polluting matter, which must therefore be flushed out as slime and sludge, removed by sedimentation and stabilized by sludge digestion or some other process. High-rate filters may, therefore, be described as 'flushing trickling filters'. The process is not merely mechanical, as such a name implies, but depends primarily on the maintenance of living slimes continuously active on the filter media.

Other successful processes using trickling filtration have been developed.[9, 11, 14, 15, 20, 27, 28] With the arrangement of two filter beds in series, in the process known as *two-stage filtration*, the effect is much the same as for a filter with a depth equal to the sum of those of the primary and secondary filter, but there can be some advantage with two relatively shallow filters because they are more readily ventilated, also because additional aeration may result from the secondary distribution of the primary effluent. This advantage may be comparatively insignificant, however, in relation to construction costs, which are much lower for a single deep filter. A settling tank may be provided to clarify the primary effluent before distribution onto the secondary filter, which may then be designed with a lower volume and smaller-sized filling medium.

Alternating Double Filtration[29] is another type of two-stage filtration which has been developed in England. Each filter may be provided with its own settling (humus) tanks and piping and channels are connected with pumps and so arranged that one of each of a pair of filters may be used alternately as primary and as secondary filter, reversing the sequence regularly from day to day or otherwise as found expedient or desirable. This technique can be applied to low-rate filtration and permits appreciable increases (say about twice normal rates) in overall loading rates with the same degree of purification including nitrification. Alternating double filtration has been successfully used especially in the treatment of dairy wastes where single filters soon become choked with fatty material and fungal slime.

Recirculation of effluent may be helpful in the filtration process. Treatment of strong wastes or relatively strong sewages may be more successful if the strength is lowered by dilution. Normally this can only be achieved by recirculating the filter effluent, also the most effective method, since some biological activity, including enzymes with oxygen and nitrates, is then introduced into the influent waste. This refreshes the waste and acts as a buffer against acidity, reduces odours and generally supports subsequent aerobic purification. Recirculation may also be required for some high-rate filtration plants to maintain the higher surface-loading rates essential. Normally a portion of the effluent is returned after sedimentation and mixed with the settled waste before it is distributed on the filter bed.[30]

Generally, wastes which are to be treated by trickling filtration should first be given partial treatment such as sedimentation. Primary sedimentation tanks provided for this purpose should be well designed, allowing a detention period of about 1½ h. Strong sewage, or sewage dominated by industrial or other strong wastes, may possibly require more effective pretreatment,[30] e.g. chemical precipitation (p. 175) or bio-aeration (p. 234). Better pretreatment has a similar effect to dilution and permits higher rates of application to filter beds.

Secondary sedimentation tanks for trickling filters are called *humus tanks*. Generally, it is necessary or desirable to remove sludge from filter-bed effluents before their final disposal. Such humus sludge flushed from filter beds, especially from low-rate filters, may be relatively harmless in rapid streams but tend to be deposited on stream beds and there to undergo digestion (see pp. 65–74). Therefore, except in favourable circumstances, filter effluents are treated finally by sedimentation in humus tanks. This is practically essential for high-rate filter plants and should allow for additional loadings due to recirculation. The solids are usually mostly flocculent in character (see p. 152). Hopper-bottomed (vertical-flow) and flat-bottomed sedimentation tanks are in use, detention periods of $1\frac{1}{2}$–2 h reckoned on peak dry-weather flow being common, with surface loading rates up to about 8 ft/h (0·7 mm/s), say 1 200 gal/ft^2 d (60 000 l/m^2 d).

Efficient separation of solids from filter effluents in humus tanks requires frequent or continuous removal of settled sludge. This is best disposed of by digestion after admixture with raw sludge from the pretreatment tanks. The moisture content of such sludges may be as low as about 92% from low-rate and about 95% from high-rate filters, the solids being similar to activated-sludge solids though somewhat more decomposed (but see p. 189). The solids contents of humus sludges from municipal sewage treatment works correspond to an additional final output from the sludge digestion tanks of about half as much as is derived from the raw sludges from pretreatment plants and must, therefore, be taken into account for digestion plant designs and the digestion tanks, etc., enlarged accordingly. (For average sludge data see Table 13.1, p. 257.)

Constituent Details of Trickling Filters

The many materials which are in use for filter media,[9, 10, 21, 22] i.e. the filling material of filters, include water-resistant rock fragments such as crushed basalt or volcanic lava, slag, refractories such as brickbats, ceramics, coke, timber, plastics, etc. Crushed rock is most commonly used, lavas having an advantage in lower effective density. Stones with a rough surface tend to ripen more quickly and are generally more suitable than smooth-surfaced stones as media for low-rate filters but show no advantage in high-rate plants.

The total slime surface and hence the active surface per volume unit of filter, depend on the shape and size of the filling material. Theoretically, smaller stones should be superior, because surface area is greater the smaller the size. Thus broken rock of size 1½–3 in (38–76 mm) effective diameter exposes about 800 ft² of surface per filled yd³ of rock (100 m²/m³ rock). A size range of 1–1½ in (25–38 mm) about 1 700 ft²/yd³ (200 m²/m³ rock) and filters filled with 1–1½ in rock might be expected to purify about twice as much sewage as those with 1½–3 in material. However, although the total space for water, air and slime is about the same in each case, the dimensions of the interstices are very much smaller for the finer rock (only about one-eighth as large) and this effectively limits rates of percolation and ventilation and the relative quantity of slime growing. Consequently, it is found in practice that the use of very fine material does not result in very greatly increased efficiency by comparison with relatively coarse, actual improvements of up to perhaps 50% at most resulting from sizes corresponding to more than tenfold increases in surface area. In England, it is customary to fill beds with rock of about 1½–2 in (38–51 mm) but laying 4–6 in (102–152 mm) rock at the bottom for about 1 ft (0·3 m), sometimes using a special topping layer of ½–1 in (13–25 mm) at the surface. The recommendation in Germany is for filling of about 1½–3 in (38–76 mm) uniformly throughout the bed, and similarly in the U.S.A. a maximum of 2½–4 in (60–100 mm) size has been found most satisfactory. Thus a uniformly distributed filling of 1½–3 in (38 76 mm) size, with possibly a coarser layer at the bottom, appears to be the best for most cases; for shallow filters, a maximum size of 2½ in (64 mm) is suggested.

The depth of the filter bed must be limited to suit the size of medium, having regard to natural ventilation, unless mechanical (forced) ventilation is provided. In England, beds of 4½–9 ft (1·37–2·75 m) in depth are common, a typical arrangement being 6 ft (1·83 m) depth of rock of about 1½ in (38 mm) size. In Germany, depths of 6–14 ft (1·83–4·27 m) with media of 2–3 in (51–76 mm) size give satisfactory results, and in the U.S.A., with 2½ in (64 mm) rock, beds 3–8 ft (1·0–2·5 m) in depth are used. The stronger the waste the deeper should be the filter bed, within these limits deeper beds yielding more highly purified effluents; but, as stated above, adequate natural ventilation becomes more difficult in deeper beds. If satisfactory ventilation can be ensured, greater depths may be used; filter depths of up to 26 ft (8 m) have been used successfully for filtration at very high rates.[31]

The filter-bed floor must be so designed and constructed that the percolating waste is run off freely and without forming sludge deposits. Free flow of air from the atmosphere must be ensured both upwards and downwards between filter floor and surface. These requirements are best ensured by constructing the filters with double floors; the upper one supports the filter medium and is provided with generously designed openings for ventilation and drainage, leaving ample space free for ventilation between upper and lower floors, and suitable drainage channels are built into the lower one for collecting the percolated waste and discharging it freely to the outside of the filter. There should be adequate openings in the side walls between floors for inspection and flushing at regular intervals. Designs should aim at the least possible restrictions of flow, both of water and of ventilating air, because it is only in this way that an adequate balance can be assured throughout all sections of the filter, thus attaining maximum efficiency under all conditions.

Proper design of filters requires actual calculations of the desired rates of ventilation so that ample provision is made for air movement as required throughout the filter and the channel openings. A high-rate filter may have been calculated for a desired air flow amounting to 1 ft/min (5 mm/s) through the full section of the bed.

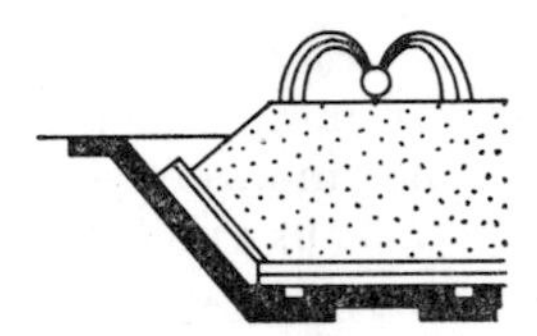

Fig. 12.5. *Fixed sprays*

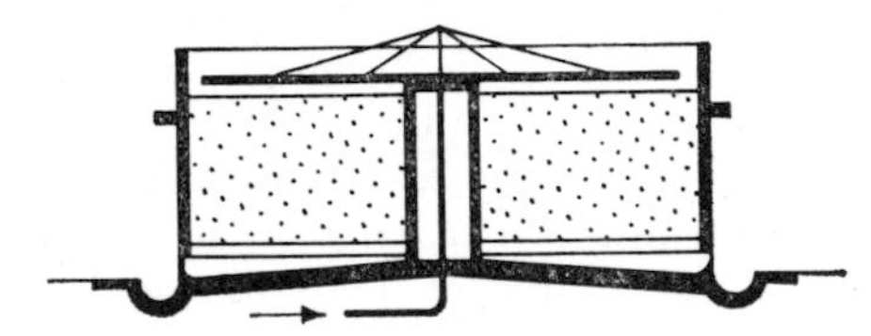

Fig. 12.6. *Rotary distributor*

Allowing, for example, 3 ft/s (0·92 m/s) for the velocity of air flow into the openings around the base of the filter, this necessitates a total area of openings equal to 1/180, or about 0·5% of the surface area of the filter; the provision of about 1% is better. Otherwise the sides of the filter beds should be enclosed in such a way as to avoid marked changes in temperature within the active zones of the bed due to atmospheric changes, particularly colder conditions. It should also be noted that airtight side walls prevent the free escape of flies, except upwards against the percolating waste, and favour positive natural ventilation.

Distribution of percolating water over the surface of the trickling filter requires special equipment, both stationary and moving apparatus being used for this purpose. *Fixed-spray distribution* (Fig. 12.5) is achieved by discharging the water through spray jets arranged at suitable intervals over the surface of the media. The waste is sprayed upwards into the air and falls back onto the surface over an annular circular or rectangular area, depending on the design of the nozzle. Such systems are usually operated intermittently from dosing tanks, and the variation in depth in the tanks during discharge causes the wetted annular zone to contract progressively, so that the area covered by each spray nozzle is approximately uniformly dosed during each cycle. A 3–7 ft (0·9–2·1 m) head of water is required for operation, including dosing-tank provision. Fixed sprays are not used for high-rate filter installations.

Except for fixed sprays of very special design, *rotary distributor* (Fig. 12.6) installations provide better distribution, and for the simple types generally used, only about 2–3 ft (0·6–0·9 m) of head is required. They have the disadvantage that circular filter-bed areas are necessary, involving a loss of area between each filter. Either spray jets or simple nozzles are arranged at suitable intervals on radial arms,[13, 20, 32] and the distributor is driven simply by the reaction of the water, by water-wheels or electric power. With low-rate filters it is usual to control the discharge onto the filter beds automatically from dosing tanks. These should have a capacity of at least 1 min storage at peak flow and can be operated automatically by means of a siphon.

Rectangular areas may be dosed using travelling distributors, which are in some respects similar to the rotary but travel backwards and forwards along the beds, drawing the waste to be treated by siphonage from lateral channels. Normally only one travelling-distributor arm is fitted to each bed, so that dosing is much more intermittent than with rotary distributors. High rates are scarcely practicable. Tipping trays which fill and spill alternately one side and the other do not distribute the waste uniformly but may be cheap and efficient enough for small installations (p. 375).

Ventilation of Trickling Filters

The volume of air required[33] for the biological activity of trickling filters may be determined from the oxygen demand of the percolating wastes. Taking an average raw sewage of 375×10^{-5} lb/gal (mg/l) B.O.D.$_5$ content, it may be assumed that about 130×10^{-5} lb/gal (mg/l) will be removed by primary sedimentation, and allowing a total purification of 90%, we may expect an effluent containing 38×10^{-5} lb/gal (mg/l) of B.O.D.$_5$ If that of the humus sludge flushed from the filter amounts to an equivalent 40×10^{-5} lb/gal (mg/l), then the filter bed removes $375 - 130 - 40 - 38 = 167 \times 10^{-5}$ lb/gal (mg/l) B.O.D.$_5$, equal to about 250 B.O.D.$_{20}$ ($167 \times 1{\cdot}46$). Now 1 ft^3 of air contains 0·017 lb (1 m^3 contains 0·27 kg) of oxygen, hence an oxygen demand of 250×10^{-5} lb/gal (mg/l) corresponds to one of 0·016 lb/ft^3 (0·25 kg/m^3) of sewage, so that $\frac{16}{17}$ ft^3 ($\frac{25}{27}$ m^3) of air is sufficient for each ft^3 (m^3) of sewage. The volume supplied must be much larger, however, because the utilization of atmospheric oxygen in operating plants usually is relatively inefficient, for trickling filters less than 5%.

Natural ventilation of shallow (up to 3 ft (0·9 m) deep) filters depends mainly on air interchange through the filter surface effected by the uptake of oxygen within the filter and a resultant air movement into and out of the filter medium, but in deeper filters ventilation air must pass also through the bottom. Therefore, in most filters, natural ventilation follows from differences in density between the atmosphere and the air inside, which may result in the movement of air either upwards or downwards through the filter. Moisture and carbon dioxide content may significantly affect the relative density of the air, but temperature is the principal factor. That of

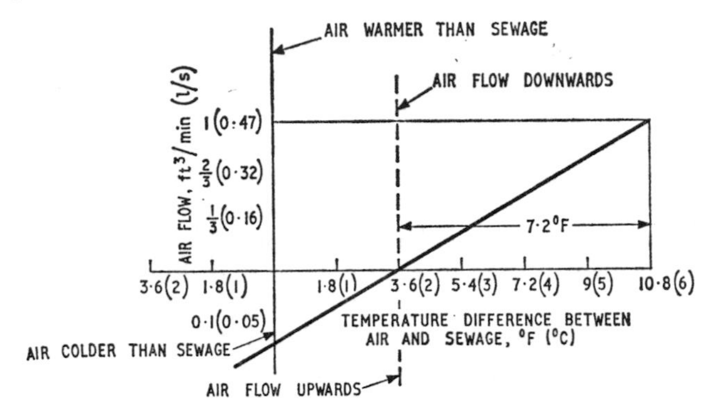

Fig. 12.7. *Air flow in relation to temperatures in trickling filters (after Halvorson)*

the filter air approximates that of the percolating water, and in temperate climates is usually colder in summer and warmer in winter relative to the atmosphere so that, correspondingly, the direction of ventilation is mainly downwards in summer and upwards in winter. However, especially in summer periods, the direction of ventilation may change twice daily, the waste-water and filter-bed temperatures remaining practically constant, while the atmospheric temperature commonly varies as much as 10–20°F (6–11°C) or more above and below the water temperature. Periods during which ventilation ceases, because of balance between the air of filter bed and atmosphere, normally last for only a few hours at most each day. According to Halvorson,[34] experimental data show that air–temperature differences of 10°F (6°C) (see Fig. 12.7) cause natural ventilation at the rate of 1 ft^3/min per ft^2 of filter-bed area, i.e. 60 ft^3/ft^2 h (18 m^3/m^2 h). Taking the high-rate filter dosed with sewage at the loading rate of 20×10^6 gal/acre d ($2{\cdot}27 \times 10^8$ l/ha d) $=$ 3 ft^3/ft^2 h $=$ 3 ft/h (0·9 m/h), a temperature difference of 10°F (6°C) would provide ventilation at the rate of $60/3 = 20$ ft^3 of air/ft^3 of sewage (20 m^3 air/m^3 sewage).

Since somewhat less than 1 ft^3 air/ft^3 sewage is sufficient in oxygen content, an oxygen-absorption efficiency of only 5% is adequate for filtration. This indicates that even relatively small temperature differences, say only 2°F (1°C), may generally be sufficient for adequate ventilation. Furthermore it has been found that wind velocities exceeding 2 mile/h (0·9 m/s) cause significant ventilation. Consistent with these findings, practical experience shows that if filters are properly designed and constructed, then natural ventilation is usually sufficient.

Artificial ventilation[35] is sometimes used for special cases. Filters which are totally enclosed, because of cold climates or to combat odour or fly nuisance, must be continuously ventilated mechanically, in the latter case from the surface downwards, otherwise alternatively upwards or downwards. Natural ventilation should always be available as an alternative. Halvorson recommends a minimum of 1 ft^3/ft^2 min (18 m^3/m^2 h) of filter surface. Rates of air flow about 60 times the sewage flow give satisfactory results. This relatively large air flow ensures adequate distribution of the air throughout the filter bed and may result in improved performance, especially in deeper filters.

Filter-bed Design Data[8–10, 18]

Both volume and surface area of filter medium must be taken into account for calculations of loading rates for trickling filters.

Volumetric loading rates are commonly expressed in terms of the total volume of waste applied daily per unit volume of filter medium over which the applied waste is percolated, using various units. The following average rates are used for design in the different countries:

Great Britain

Loadings of 100 gal/d per yd^3 for low-rate filters (based on 40 gal/hd d) (i.e. 600 l/d per m^3 based on 180 l/hd d).

With recirculation of 1 part effluent to 1 part settled sewage, up to about 200 gal/yd^3 d (1 200 l/m^3 d).

With alternating double filtration 200 gal/yd^3 d (1 200 l/m^3 d) (total volume of both filters) or more.

Germany

Usually 125 gal/d per yd^3 for low-rate filters (based on 35 gal/hd d) (i.e. 750 l/d per m^3 based on 160 l/hd d).
For high-rate filters, about 500–700 gal/yd^3 d (3 000–4 000 l/m^3 d).

U.S.A.

For low-rate filters 100–500 gal/yd^3 d (600–3 000 l/m^3 d).
For high-rate filters about 800 gal/yd^3 d (5 000 l/m^3 d) or more.

These are daily average dry-weather loading rates which distinguish between low-rate (sludge-containing) and high-rate (flushing) filters. These purely volumetric rates of loading had much more significance for the formerly widely used contact beds. With percolating filters, however, it is readily shown that within certain limits the filter-bed performance is not directly related to flow but depends on the content of polluting substances in the applied waste. For example, the normal low-rate loading of 100 gal/yd^3 (600 l/m^3) given above for Great Britain corresponds to the daily application of about 50 lb of B.O.D.$_5$/1 000 ft^2 (0·24 kg/m^2) of filter-bed surface, or about 8 lb/1 000 ft^3 d for a 6 ft depth (0·12 kg/m^3 d for 2 m depth). It is this total content of organic pollution (estimated by B.O.D. tests) which is important in determining what is required to ensure a stable (non-putrescible) effluent, and the following loading rates can be used safely if filter designs are good in all details.

Low-rate Filters

Great Britain

8 lb B.O.D.$_5$/1 000 ft^3 d, equal to, say 0·2 lb B.O.D.$_5$/yd^3 d (0·12 kg/m^3 d).
Using 1 : 1 recirculation or alternating double filtration up to 15 lb/1 000 ft^3 d (0·4 lb/yd^3 or 0·24 kg/m^3 d).

Germany

11 lb B.O.D.$_5$/1 000 ft^3 d, say 0·3 lb/yd^3 d (0·18 kg/m^3 d).

U.S.A.

5–25 lb B.O.D.$_5$/1 000 ft^3 d, say 0·14–0·70 lb/yd^3 d (0·08–0·40 kg/m^3 d).

High-rate Filters

Germany

Up to 70 lb B.O.D.$_5$/1 000 ft^3 d, say 1·9 lb/yd^3 (1·1 kg/m^3 d).

U.S.A.

Commonly 40–70 lb B.O.D.$_5$/1 000 ft^3 d, say 1–2 lb/yd^3 d (0·6–1·1 kg/m^3 d), although 3 lb/yd^3 d (1·8 kg/m^3 d) or more has been used.

The quoted quantities of B.O.D.$_5$ are based on laboratory determinations of the B.O.D.$_5$ of the settled wastes applied to the filters. Actual values may not be available for design but can be estimated from the number of population contributing sewage and according to the living standard, or by population equivalents of industrial process wastes. Average domestic sewage after preliminary treatment in sedimentation plants may be taken to contain about 0·08 lb (37 g) B.O.D.$_5$/hd d and, hence, for low-rate filters loadings range (8–15)/0·08 = 100–190 persons per 1 000 ft^3 (2·5–5 persons/yd^3; 3·5–7/m^3) and for high-rate filters, 500–900 persons/1 000 ft^3 (12–25 persons/yd^3; 18–32/m^3). As always, local or special conditions must be investigated and loading rates taken for lower or higher values accordingly.

Borrowed also from the design of contact beds, a design basis referred to the surface loading of trickling filters is in common use, though not suitable for low-rate trickling filters. With contact beds, ventilation is available from the top only and controls the activity, so that efficiency depends on the surface loading rate, but with trickling filters, where ventilation is provided through the top and bottom openings, performance depends less on sewage load to surface ratio than on the loading rate per unit volume of filter-bed medium.

For high-rate filters, however, the surface loading rate is important because on it depends the flushing effect characteristic of them. According to Halvorson, the surface loading for high-rate filters should normally exceed 10 × 10^6 gal/acre d (110 × 10^6 l/ha d) and never be less than 7 × 10^6 gal/acre d (80 × 10^6 l/ha d) (low-rate filter loadings are usually 1–3 × 10^6 gal/acre d (11–34 × 10^6 l/ha d) only). Standard designs for H.R. filters in the U.S.A. provide surface loading rates between about 160 and 850 imp. gal/ft^2 d (say 7–38 × 10^6 gal/acre d or 110–340 × 10^6 l/ha d) but usually not less than 12 × 10^6 gal/acre d (140 × 10^6 l/ha d). Higher rates are also used, efficiencies being reduced accordingly, up to a limit of about 100–125 × 10^6 gal/acre d (1 100–1 400 × 10^6 l/ha d) above which flooding conditions are approached.

In the U.S.A., sewage is commonly relatively weak and loading rates of 10–20 × 10^6 gal/acre d (110–230 × 10^6 l/ha d) may be easily reached, using ordinary types of filters without special provision. For stronger wastes, such as European sewages of normal strength, it is necessary to construct deeper filters with provision for recirculation, to provide extensive pretreatment or considerable dilution. Dilution with fresh water is seldom feasible, and dilution by recirculation of effluent has become standard practice, pumping from the overflow of the final sedimentation tank back to the filter-dosing tanks. Preferably the mixture of settled waste and recirculated effluent should have a B.O.D.$_5$ not higher than 100–150 p.p.m. and the proportion of effluent recirculated be at least high enough

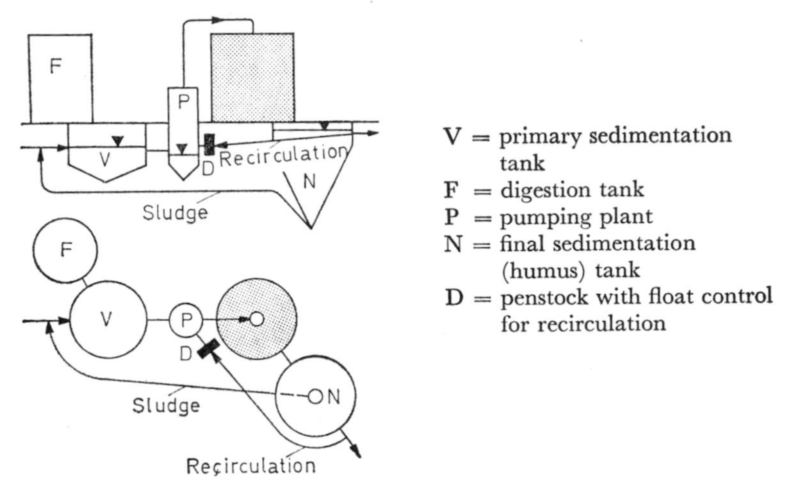

Fig. 12.8. *High-rate filtration with recirculation of humus tank effluent*

accordingly. This proportion is commonly referred to as the *recirculation ratio*, expressed as the ratio of the combined volumes of settled waste plus recirculated effluent to settled waste volume. A recirculation ratio of 3 (1 + 2) is that of a case where the mixture applied to the filter consists of two parts of recirculated effluent to 1 part of settled waste.

The effective degree of purification achieved by recirculation is increased if the primary sedimentation tanks are made larger and included in the recirculation cycle. The raw waste is thereby freshened and some biological purification is initiated.

Figs 12.8 and 12.9 show diagrammatically the two basic design arrangements for high-rate filters using recirculation, one showing also, as a variant, the combination of primary sedimentation with sludge digestion in an Imhoff two-storey tank. Recirculation is obtained by means of the pump, *P*, with a capacity greater than the incoming sewage flow; the proportion recirculated can be maintained by means of a valve controlled by the level of the final sedimentation tank, except that during the lower night flows a smaller pump may be used. When the inflowing waste is diluted by relatively clean water, e.g. in sewerage systems by stormwater, the amount of effluent recirculated is reduced in proportion.

The first design (Fig. 12.8) shows that only the final or humus tank is included in the recirculation arrangement. The actual case illustrated shows the return of the sludge from the humus tank into

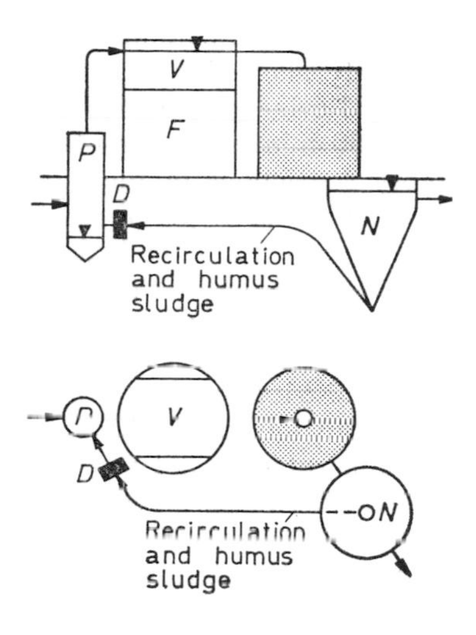

Fig. 12.9. *High-rate filtration with recirculation of filter effluent with humus sludge through primary sedimentation tank (for key to letters see Fig.* 12.8)

the primary sedimentation tank, either by pumping or by gravity according to the respective levels, which is an arrangement sometimes adopted.

In the second design (Fig. 12.9) the recirculation arrangements include the primary sedimentation plant in the cycle. The actual case illustrated shows the recirculated effluent drawn from the bottom of the humus tank, whereby any sludge solids which are separated from the final effluent are also returned automatically to the primary sedimentation tank. The arrangement for recirculation shown minimizes the loading of the humus tanks, which are kept small accordingly, but requires an enlargement of the primary sedimentation tank to provide for the recirculated flow. It is to be noted that, if the filter effluent is high in nitrate content—say, containing about 10 p.p.m. nitrate nitrogen or more—the recirculation of the effluent may cause lifting of sludge in primary sedimentation tanks because of production of gaseous nitrogen during denitrification.

Recirculation including the filter bed only, without using primary or humus tanks for recirculated effluent, is also possible, of course, but yields a lesser degree of purification. In such a case a grit chamber should be installed to separate particles of broken-down filter material from the recirculation flow before a pumping.

The problem of rest periods in the dosing of filter beds is of considerable importance. With low-rate filters, it is usually not practicable to maintain continuous operation of the distributors because of the wide variation in the rates of flow of the influent raw wastes and short rest periods are commonly allowed for, at least during times of lower flow. Studies by Levine and Halvorson showed that regular rest periods (say 10–30 min or perhaps less) may be harmful because the temporary restrictions of the food supply reduce bacterial activity. Rest periods also give *Psychoda* and other filter flies opportunity to escape from the filter beds in larger numbers, increasing local nuisance. With high-rate filters, however, the rates of dosing of the beds are so high that continuous operation is readily maintained and rest periods are not usually provided. Ordinary rotary and travelling distributors do not dose the whole area of the filter continuously but any particular area intermittently. Thus a four-arm distributor rotating at 1 rev/min may splash a portion of the waste to be treated on each corresponding area of the filter four times in each minute—with wide spacings of openings, perhaps only once—whereas a distributor with eight branched arms may increase the frequency to 16 times/min. The numerous experimental observations of dosing frequencies and rest periods reported in the literature give ample indication that no general specification for an optimum sequence of dosing is feasible but make it quite clear nevertheless that, at least in some cases, a control of the time schedule of dosing leads to improved overall purification, apparently as a consequence of a corresponding control of the growth of filter slimes.[32, 36, 37] Adjustments of the distribution system can be used at the same time to balance effects of variations in waste flow and, when arranged properly, can assist in reaching rates of dosing large enough for improved flushing effects. More radical changes in dosing cycles are used not uncommonly to help balance growths in the filter medium, e.g. temporary shut-downs of a particular filter bed for periods of hours or days. The disadvantages of such long rest periods include the escape of larger numbers of filter flies and the production of odour which have been avoided by enclosing the filters and ventilating them by mechanical means.

Maintenance of Trickling Filtration

It is necessary to prevent clogging of trickling filters in order to maintain operation. Clogging is indicated by a sudden increase in the B.O.D. of the settled (humus tank) effluent and by ponding of the applied waste on the surface of the bed. Ponding is caused by an overgrowth of algae and fungi on surfaces in the uppermost layer, but does not indicate a complete clogging of the bed. This algal or fungal surface film seriously interferes with normal ventilation of the bed so that proper operation is impossible. Ponding can be overcome in various ways, including chlorination (p. 182) of the applied waste for a short period (using about 20–50 p.p.m. of chlorine) or resting the filter for several days or recirculating effluent, so that the algae or fungus dies and may be readily flushed away. A similar result may usually be achieved by forking the surface of the bed so as to loosen and invert the upper 6 in (0·15 m), say, so that the surface overgrowths are dislodged and flushed away. High-rate filters, or those with sufficiently high dosing rates, operate without any ponding or clogging, at any rate if the design is correct and the medium not too fine, and old clogged filters have been put into order simply by recirculating and thereby increasing the surface loading rate. Adjustment of the dosing cycle to obtain increased flushing effects was referred to above.

Odour arising from trickling filters is minimized by maintaining the wastes in a fresh state during pretreatment. Septic wastes may be freshened before distribution onto filters by pre-aeration, by chlorination or by using recirculation of effluent. If filters are totally enclosed, using artificial ventilation as a means of odour control, it may be necessary to provide for deodorization of the waste air from the filters, e.g. by recirculating the air.

Annoyance from the escaping filter flies, such as *Psychoda* spp., may be controlled by the application of toxic chemicals onto the filter from time to time in carefully measured amounts. Chlorine solutions are commonly used, also suspensions of insecticides such as D.D.T. or Gammexane.[38–40] Their application may be arranged in accordance with the breeding period of the species of fly in question which is, for example, only about 14 d at 68°F (20°C) for *Psychoda* spp. As a means of control of fly nuisance, the encouragement of film predators such as *Collembola* (spring-tails) has been found successful. High-rate filters, however, do not produce fly nuisance, because the numbers breeding in the filter are much less in proportion to the total amount of slime and the continuous dosing of beds tends to prevent the escape of flies. Total enclosure of such filters is, therefore, unnecessary and tends to restrict ventilation.

Trickling filters can maintain purification efficiency during wet weather only for flows up to about one and a half times normal dry-weather flows, except that when recirculation is in use the rate of recirculation can be reduced according to the stormwater dilution. In Great Britain, however, it is the custom to treat up to three times dry-weather flow by filtration, for which purpose it is necessary to provide larger filters. Distributors must be designed for peak loads.

In winter time, filter operation can be maintained even during heavy frost, the loss of temperature in the filter beds being relatively slight, seldom exceeding about 7°F (4°C). High-rate filters are safer than low-rate because of the higher rates of loading. Thick side walls carried at least a little above the top of the medium offer useful protection against wind, while the effects of very heavy frosts may be minimized by baffling the bottom openings, provided that adequate ventilation is maintained. However, efficiency is diminished by temperature reduction (see below). Heat produced within the filter beds is relatively insignificant.

The Degree of Purification by Trickling Filtration

Treatment of municipal sewage by primary sedimentation, followed by trickling filtration[5, 8, 16, 18, 41] with final sedimentation, normally results[41] in overall purification of about 65–95%, calculated on the reduction of B.O.D.$_5$ of influent and effluent, an average being about 87%. The higher degrees of purification are obtained commonly with low-rate filtration plants, with high-rate plants only where recirculation or multi-stage filtration is used. However, the efficiency of trickling filtration is substantially diminished at temperatures lower than normal (68°F, 20°C). According to Pöpel, reduction of the temperature from 68 to 50°F (20 to 10°C) results in a loss of efficiency of about 40%. Trickling filters are particularly adaptable for partial biological treatment because there is almost no limit to the B.O.D. loading which can be applied and they may be used successfully even for the simple aeration of water-borne wastes.

It is not possible to state generally whether low- or high-rate filtration is to be preferred. It can be said that low-rate filters have proved successful during many decades and are simple and reliable in operation when properly maintained. Their design and construction are even today improved as a result of continuing studies and experience. A higher degree of purification is achieved, while greater proportions of solid impurities are decomposed within the bed itself.

High-rate filters have some advantages. They are much smaller, and, therefore, cheaper to construct, comparatively free of odour and filter-fly nuisances and may be more adaptable for treating waste flows varying widely in volume; they also show disadvantages. The degree of purification achieved is commonly less than for low-rate filters, particularly when operated without recirculation which may be relatively costly. In addition, the costs of sludge digestion and disposal are increased because of the greater quantities of humus-tank sludge which is also more putrescible and, in larger plants, usually has a greater moisture content. Savings in construction costs are often not as high as might be expected at first sight on a comparative basis, allowing for these additional costs.

It is essential that either low-rate or high-rate filtration be decided upon, because designs aiming to be intermediate are likely to be unsuccessful. If surface loadings are not high enough but merely intermediate between low and high rates, overloading or clogging usually occurs because the flushing effect is not attained, particularly with relatively fine media and inadequate ventilation. The process of alternating double filtration, which is operated at rates higher than used for simple low-rate operation, was developed as a technique for controlling rapid slime growth when treating particular wastes and may indicate a way to control clogging in other cases at intermediate rates of surface loading.

SUBMERGED (EMSCHER) FILTERS, CONTACT AERATORS

Submerged filters, biological treatment units known also as Emscher filters[42, 43] or contact aerators,[44] are aerated filter beds in which, like trickling filters, an inert medium forms the basis for developing an active zoogloeal slime but is constantly fully submerged in the waste to be purified. The media commonly used for such units include stone, coke, timber, laths of loosely-packed cork or timber, corrugated aluminium sheet or concrete blocks, and even asbestos-cement sheets, simply spaced about 2 in (50 mm) apart standing vertically. Biological activity can only be maintained if a plentiful source of dissolved oxygen is always available and this requires artificial aeration. Without sufficient aeration, contact aerators are at least useless if not actually damaging to putrescible wastes.

Contact aerators are constructed by providing definite compartments, inside larger tanks, into which the selected medium is packed, being so arranged that air can pass freely from below upwards through the full depth, carrying with it (as in air-lift pumping systems) the waste to be purified, and can then escape freely to the surface, while the accompanying waste then travels downwards outside the medium-filled compartment, to return again to the lower section of the tank and once more circulate through the bed (Fig. 12.10). Thus a continuous circulation of waste and air through the

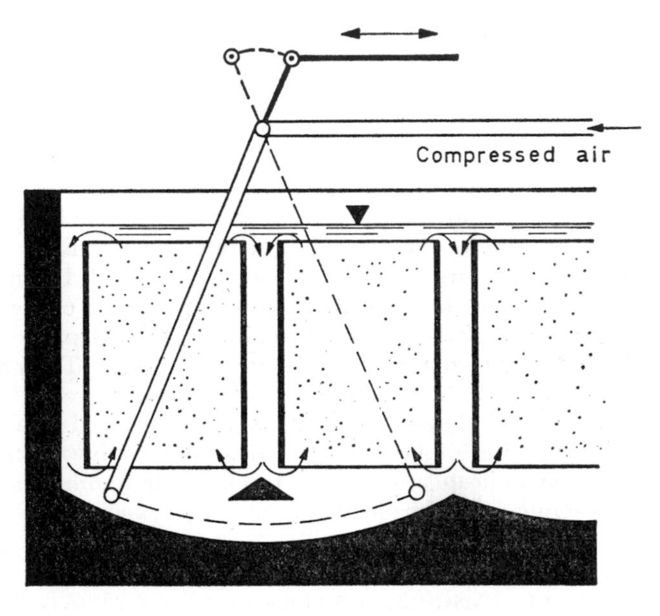

Fig. 12.10. *Submerged (Emscher) filter or contact aerator*

contact units is automatically maintained, the waste passing through the bed over and over again.

Compressed air is supplied to perforated pipes located below the contact units. These might be arranged as a grid spaced about 1 ft (0·3 m) apart. The arrangement of a swinging diffusion system, as shown, is more economical and, moreover, concentrates the air discharge so that a better flushing of the bed is ensured. This constant flushing of accumulating slime and sludge is essential to the success of the process. With fixed-air diffusers, the bottom of the tanks must be so constructed as to allow collection and discharge of sludge without interruption of treatment. This must then be removed regularly and frequently enough to prevent septicization. When swinging diffusers are used, however, the turbulence created in the bottom of the tanks would so disturb the sludge that it could not be collected properly, so such tanks are arranged to discharge the solids in the effluent from which it is separated by sedimentation, as with trickling-filter effluents. The bottom section of such a tank, below the filter medium, is constructed so that no pockets of sludge may form, providing only sufficient space for the regular movement of the swinging diffuser.

A partly submerged filter system, using rotating discs, has been developed especially for smaller plants (Fig. 12.11).[45, 46] In this process a number of thin discs of 10 ft (3 m) diam. are mounted about 0·8 in (20 mm) apart on a horizontal shaft which rotates so that the discs are almost half submerged in the waste to be treated,

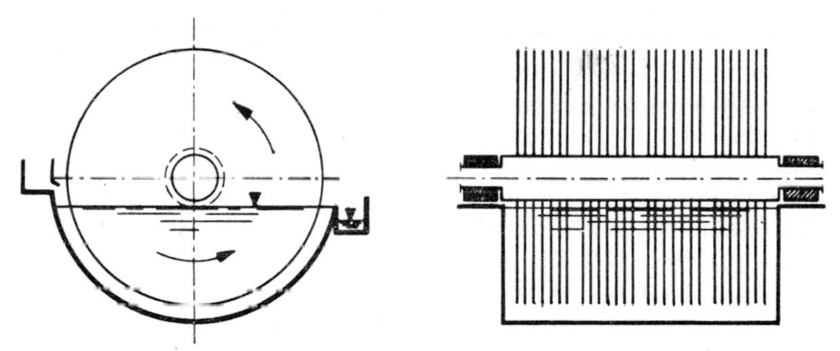

Fig. 12.11. *Partly submerged disc filter*

which flows through a suitably designed trough. Slow rotation with a speed of 2–3 rev/min has the effect of alternately submerging (feeding) and aerating the slimes which develop on the discs, so that aeration by compressed air is not required. For vertical plates, discs and other such media used for submerged filtration, the surface area required amounts to about 20 ft^2/hd (1·9 m^2/hd).

With normal municipal sewage, after sedimentation, complete biological treatment by submerged filtration produces about 0·3 gal/hd d (1·4 l/hd d) of sludge containing about 97·5% moisture; subsequent digestion, along with primary sludge, yields an additional quantity of digested sludge and causes an increase in the volume of the digested sludge derived from the primary sludge to about 0·13 gal/hd d (0·6 l/hd d) and an increase of the moisture content to about 90% or more.

Contact aerators are suitable for grease removal from wool-scouring wastes because the free fatty constituents are carried directly to the tank surface where they may then be removed by simple skimming. They are also applied favourably to the treatment of distillery and phenolic, etc. waste types in which sludge formation depends entirely on the build-up of bacterial slimes by metabolism of soluble waste constituents only, in the absence of suspended solids in the wastes themselves.

Full biological treatment by contact aerators uses about $\frac{2}{3}$ ft^3 of air/gal (4·2 m^3 air/m^3) of settled municipal sewage, partial treatment requiring smaller volumes according to the reduced surface area of the medium.

AERATION

Generally, any aeration of water-borne wastes results in some improvement of its character,[47, 48] the desirable effects depending in most cases on solution of dissolved oxygen. This reacts with hydrogen sulphide, thereby raising the redox potential (p. 42) and controlling odours, improving colour and assisting flocculation, e.g. by oxidizing iron compounds and helping to prevent the proliferation of anaerobic organisms. The undissolved gases of the air, if dispersed through the wastes causing mixing and perhaps some flocculation, also tend to float fatty and oil particles to the surface (cf. skimming tanks, p. 146). Volatile compounds may likewise be removed by aeration, being carried away with the escaping undissolved gases, after first volatilizing through the large areas of interface between the wastes and the gas bubbles or exposed surface. Thus aggressive constituents such as carbon dioxide or hydrogen sulphide may be removed from solution simply by vigorous aeration.

These various simple effects may be achieved by adequate pre-aeration of the waste liquid for short periods prior to primary sedimentation. In some cases, longer pre-aeration periods of up to 30 min may be justified. The rates of aeration are about 50 and 120 ft^3 of air/1 000 gal of sewage (0·3 and 0·75 m^3 air/1 000 l sewage) for aeration periods of 10 and 30 min, respectively. Although the wastes may be freshened or even saturated with dissolved oxygen by such means, oxygen solubility is low (see Table 4.1, p. 32) compared with the oxygen demands of most wastes, and normally the natural processes of purification require to be greatly intensified for economical application to artificial purification works (but see ponding, pp. 242–246). It is necessary to provide special engineering arrangements for the oxygen supply and suitable space for the build-up and maintenance of an active mass of zoogloeal slimes or flocs, usually called activated sludge. Such arrangements applied to intensify the purifying effects of aeration are utilized in activated-sludge plants. Aeration may cause local odour problems.

ACTIVATED-SLUDGE TREATMENT

If a waste such as municipal sewage is aerated continuously, aerobic decomposition takes place, as can be demonstrated by B.O.D. tests, but only after a period of about 20 d is purification substantially complete. In trickling filters and contact aerators, the rate of decomposition is greatly increased and the time required reduced to hours (p. 63) because an active accumulated culture of aerobic organisms is available for contact treatment of the waste. Investigations made by Ardern *et al.*[49, 50] showed that sludge particles derived from sewage can be activated by aeration so as to develop a similar active culture of aerobic organisms, in this case carried by the particles. By accumulating and circulating a sufficiently large quantity of activated sludge, contact treatment of sewage is likewise possible, and its rapid purification can be ensured accordingly. While the processes are similar to those of self-purification in natural waters, they are accelerated by the artificial means of vigorous aeration together with continuous turbulence, whereby the activated flocs are kept in suspension and the mixture of waste and sludge thoroughly mixed in the presence of sufficient dissolved oxygen at all times. The process is also applicable to wastes other than sewage if capable of supplying sufficient sludge particles of the right kind and density, or if suitable sludge can otherwise be supplied (see also p. 236).

The basic phenomena of activated-sludge treatment have been summarized by Buswell and Long:[51, 52]

> 'Activated-sludge flocs are composed of a synthetic gelatinous matrix, similar to that of nostoc or merismopedia, in which filamentous and unicellular bacteria are imbedded and on which various protozoa and some metazoa crawl and feed. The purification is accomplished by digestion and assimilation by organisms of the organic matter in the sewage and its resynthesis into the living material of the flocs. This process changes organic matter from colloidal and dissolved states of dispersion to a state in which it will settle out.'

Thus this treatment by biological processes is similar to that of others such as trickling filtration, which also depend upon adsorption into active zoogloeal slimes, and is basically different only in that no medium is used to support the active zoogloeal slime. The first stage of the process involves adsorption, which is very rapid, but the subsequent processes of decomposition and oxidation are slower. However, the ability of the activated sludge to adsorb polluting material depends on the latter, usually resulting in continued growth of active sludge. Normally bacteria are the primary agents of stabilization, protozoa acting along with other commensals, as part of a balanced biological system[52, 53] of growth and destruction.

A wide range of operating conditions is possible in activated-sludge plants[54–56] and the actual results of treatment vary accordingly, but in standard-rate plants the dry solids content of the sludge usually grown is about 50–60% of the corresponding dry weight of organic solids contained in the wastes to be treated or, say, sludge produced = 0·5 × B.O.D.$_5$ removed.[55–59] In high-rate plants the amount of sludge produced is greater, and in very low-rate, such as extended-aeration plants, smaller.

For normal operation, nitrogen and phosphorus requirements are about as follows:

B.O.D.$_5$:N, 17:1– 32:1
B.O.D.$_5$:P, 90:1–150:1

Compared with trickling filters, activated-sludge plants mostly effect a high degree of purification; they are also free of flies and odours and operate throughout the winter with almost the same efficiency as in summer. Their disadvantages are that, being somewhat sensitive to large variations in raw-waste composition, their control is less simple, and usually they produce more sludge (i.e. more solids, and with much higher water content) to be disposed of; therefore, they may be less suitable for particular cases, while for very small plants extended aeration processes, which provide for some solids destruction by aerobic digestion (p. 234), may alleviate sludge-accumulation problems and allow attendance costs to be kept low.

Plants providing for biological treatment using activated sludge usually include efficient pretreatment by sedimentation, with separate sludge digestion. In some cases, partial sedimentation at high tank loadings is found to be economical. Treatment by the activated-sludge process requires aeration tanks, in which the mixture of pretreated waste and activated sludge recirculated from the final sedimentation tanks, the 'mixed liquor', is agitated and aerated (by special arrangements described below, pp. 221–228). In these aeration tanks where the processes of biological purification operate, the mixed liquor overflows into secondary (final) settling tanks, in which the activated-sludge solids are separated from the purified water. These settling tanks are of various designs but often of the vertical-flow type, and detention periods normally range over 2–3 h. (For density currents, etc., see p. 153.) The final effluent is discharged, but a suitable proportion of the activated sludge separated in the final sedimentation tanks is continuously returned to form part of the mixed liquor, while any surplus of sludge, 'excess sludge', is treated and disposed of separately, usually along with the primary sludge, by pumping it into the raw-sewage inlet channel. Thickening of the excess sludge may be desirable, using thickening tanks (p. 259), especially if pumped separately to the digestion plant.

Activated-Sludge Characteristics

Activated sludge is actually produced in the aeration tanks and new works must be operated for about a fortnight before the necessary quantity is developed. It is essential that the sludge be kept continuously aerated and suspended in water in any tanks in which it is produced, used or stored. Its activity can only be maintained by constant and adequate supplies of oxygen, nutrients and suitable food. To grow activated sludge requires about $\frac{1}{2}$–1 lb (kg) phosphorus and about 3–8 lb (kg) combined nitrogen per 100 lb (kg) $B.O.D._5$ removed. In ordinary sewage, sufficient rates of nutrients are available, but industrial wastes may be deficient in nutrients and, therefore, may require their artificial supply (p. 314), for instance by admixture of domestic sewage. Activated sludge may become 'sick'. This can result from overloading, as shown below, or otherwise where the organic matter can consume more oxygen than is available. A similar effect can result from poisoning by toxic substances in the waste being treated, such as copper compounds, acids, phenols, mineral oils, etc., particularly from sudden discharges of industrial wastes or when the normal concentrations are suddenly increased. *Bulking* of the activated sludge then results, due primarily to the unbalanced growth of filamentous fungi. The sludge then occupies a larger volume, does not settle as readily as usual and tends to rise to the surface and form a scum or floating sludge masses, consequently spoiling the quality of the final plant effluent.

An activated sludge in good condition will settle in 30 min to a concentration of about $1\frac{1}{2}\%$ solids or more. The *Mohlman Sludge Volume Index* is the volume of sludge (in ml) which contains 1 g weight of solids; taking settled sludge to contain $1\frac{1}{2}$ g solids/100 ml, then the Mohlman Index will be $100/1{\cdot}5 = 67$ mg/l. The corresponding *Sludge Density Index* (Donaldson) is simply the percentage of solids in the settled sludge, in this case $1{\cdot}5\%$. The relationship between these two indices is

$$\text{Sludge Volume Index} = \frac{100}{\text{Sludge Density Index}}$$

Experience has shown that such figures provide a measure of the settleability of the activated sludge which is a useful control tool. A high Mohlman Index, of more than 100, points to a tendency of the sludge to bulking, but the Index varies from time to time during the day, and a range up to say 150 is common. Bulking sludges may show values as high as about 400 ml/g.

The concentration of activated sludge in the aeration tanks may be expressed as a proportion by volume. For measurement, the aeration tank contents, the mixed liquor, are settled in an Imhoff cone. Settling of a good sludge is practically complete in 10 min ($1\frac{1}{2}$ in/min (0·6 mm/s)), but the reading of sludge volume is made after 30 min. In place of the Imhoff cone test, the concentration of sludge may be determined by weighing the solids, which yields more accurate figures. Thus, if the sludge amounts to 12% of the tank volume and contains 98·5% water, the solids content of the mixture of waste and activated sludge would contain about $12 \times 1{\cdot}5/100 = 0{\cdot}18\%$ or 1 800 p.p.m. of dry activated-sludge solids. In practice, mixed-liquor solids concentrations ranging from 600 to 10 000 p.p.m. are in use (see Table 12.5, pp. 233–235).

As stated above, the activity of the sludge can only be maintained by a sufficient supply of food and oxygen. However, the separation of the sludge in the secondary sedimentation tanks does not allow aeration at the same time, so the addition of food and oxygen must be interrupted and the detention period of activated sludge in the secondary tanks should therefore be as short as possible. In consequence, it is not usually possible to complete its separation, and generally it is drawn off the secondary sedimentation tanks with a high content of effluent water, the actual volume returned amounting usually to 20–30% of the sewage dry-weather flow in the case of standard (low-rate) activated-sludge plants, but up to 50–100% if higher mixed-liquor solids concentrations are used, also in some high-rate plants providing shorter aeration periods (Table 12.5).

During this recycling of activated sludge, its quantity is continuously augmented as the organic matter and other suspended and colloidal solids contained in the waste are purified. Most of the material purification achieved results in additional sludge-solids production, although a small proportion is actually consumed by respiration within the sludge. The proportion consumed is variable, depending on the process operated and on the composition of the waste being treated, but may be only about 10% for normal sewage. This accumulation of sludge requires the continuous withdrawal from the activated-sludge system of the equivalent daily growth of sludge solids. This is done by diverting an appropriate proportion of the returned activated sludge which becomes excess sludge. Its solids content normally amounts to about 0·5–0·8 lb/lb (or kg/kg) of B.O.D.$_5$ removed in the aeration tanks. In very low-rate plants, as in extended aeration, the solids content may be only 0·3 lb/lb B.O.D.$_5$ removed or less. In high-rate plants, on the other hand, the proportion may reach 1·0 lb or more.

The age of the activated sludge is reckoned from the ratio of the total in the aeration tanks to the excess (neglecting the sludge content of the secondary sedimentation tanks because, in any case, this quantity should be kept as low as possible). Thus if the aeration period is 6 h and the tank sludge volume 30%, then the total sludge volume is $(6/24) \times 30 = 7\frac{1}{2}\%$ of daily sewage flow; if the excess sludge is, on the same basis, 3% of the sewage flow, the age of the sludge is $7\frac{1}{2}/3 = 2\frac{1}{2}$ d. Gould[60, 73] determined the age of sludge from sewage and sludge analyses by dividing the weight of suspended matter present in the aeration tanks by that of suspended matter entering the tank.

Types of Aeration Tank

The aeration tanks have to supply sufficient dissolved oxygen to the mixed liquor for the process of biological treatment. They must provide the necessary detention period of the waste flow for maintaining the activity of the activated sludge and its desired quality, especially regarding proper settleability in the final sedimentation tanks. In addition, they must be designed hydraulically in such a way that deposition of solids is prevented; therefore, the mixed liquor in them must always be in turbulent motion.

Following the first applications of the newly developed activated-sludge process early this century, various engineering techniques and designs were developed. The main development occurred in connection with the original idea of the *diffused air system* in which compressed air is introduced by means of diffuser plates, domes or pipe diffusers (pp. 214, 227), located at or towards the bottom of aeration tanks which usually are much longer than they are wide. *Ridge and furrow*, or longitudinal-furrow tanks (Fig. 12.12(*a*)) provide for aeration symmetrically from the floor, while *spiral flow* (Hurd or Manchester) tanks (Fig. 12.12(*b*)) use aeration along one side of the floor only (or, sometimes, along both sides). Vigorous overturning and turbulence of the tank contents are produced with both types, but the ridge-and-furrow types permit utilization of a larger floor area for the diffusers, which may be advantageous. The tanks are usually 10–17 ft deep (3–5 m) and commonly provide for the introduction of the air into the mixed liquor as *fine* bubbles. How-

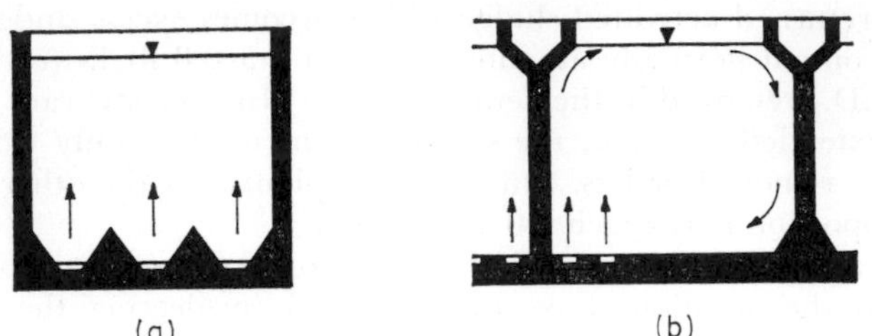

Fig. 12.12. *Diffused air activated-sludge systems* (*a*) *Ridge and furrow* (*b*) *Spiral flow*

ever, simpler diffuser systems in which the aeration is effected by discharge of medium-sized or relatively *coarse* bubbles from perforated pipe systems are also used.

The *INKA process* (Fig. 12.13) likewise employs a diffused-air system of aeration but differs appreciably from the usual arrangement.[61] The air supply, of relatively low pressure, is introduced with the aid of a stainless steel or plastic grid with perforations on the underside, submerged at a depth of about 2½ ft (0·8 m) and placed along one side of deep tank channels, in the centre of which vertical baffles are installed to assist spiral-flow movement of the tank contents. The low-pressure air is supplied by centrifugal fans at relatively high rates, say, 4–5 times higher than in diffused-air plants of more conventional type. Another variant of the simple diffused-air system is the *paddle tank with diffused air* (Fig. 12.14(*a*)), but such systems are not now commonly used, although better air utilization is possible. The air is introduced along one side of the

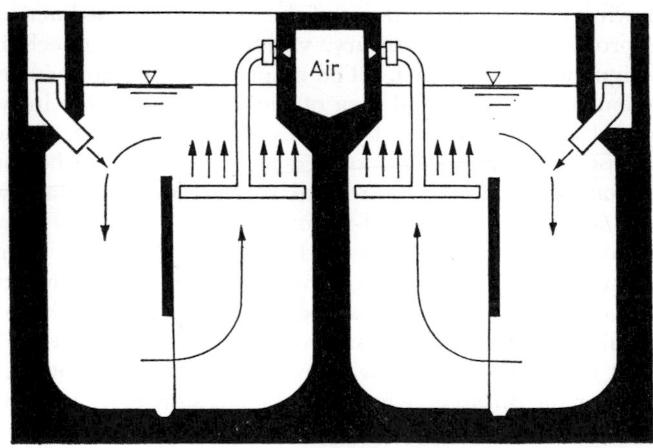

Fig. 12.13. *INKA aeration tank*

tank, in opposition to the motion of the paddles. The elongated paddles are rotated so that their peripheral speed is about 2 ft/s (0·6 m/s), the tanks usually being about 10 × 10 ft (3 × 3 m) in section.

The aeration period required depends on the degree of purification desired and normally ranges between 1 h (high-rate plants) and 10 h (low-rate plants). These are theoretical figures, calculated on the basis of total aeration tank capacity in use and the raw-waste inflow only, ignoring any recirculation of activated sludge and effluent.

Aeration tanks may also be based on *mechanical aeration systems*. These include designs where power-driven paddles (e.g. the 'Bolton Wheel') cause circulation of the mixed liquor through a system of

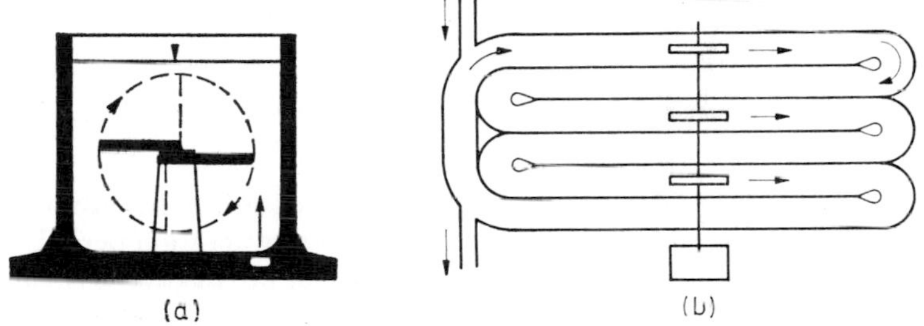

Fig. 12.14. (*a*) *Paddle-wheel activated-sludge tank with diffused air* (*b*) *Haworth or Sheffield activated-sludge system*

extended channels, using velocities of about 2 ft/s (0·6 m/s). The *Haworth* or *Sheffield* systems (Fig. 12.14(*b*)) depend to an important degree on the uptake of oxygen through the flowing liquor's surface and require extended periods of contact for moderate degrees of purification, say about 15 h for average municipal sewage. Mechanical aeration systems include in particular the *Simplex* and *Kessener Brush* systems.

The *Simplex* system (Fig. 12.15(*a*)) uses hoppered tanks equipped with a vertical pipe through which the contents are drawn upwards and thrown centrifugally outwards as a fine spray across the tank surface. This is done by means of a rapidly rotating impeller mounted across the top of the central pipe. In this way, the tank contents are circulated, aeration being confined to the intake of air during spraying (also directly into the surface). The aeration period is about 8 h for average sewage. In new plants using the 'High-Intensity Cone', circulation of the mixed liquor is increased to once every five minutes, and the aeration periods are reduced and may be about the same as for diffused aeration.

The *Kessener Brush* system (Fig. 12.15(*b*)) achieves circulation and aeration by an elongated rotating brush mounted at the water surface along one side of the tank. The brush, normally consisting of series of stainless steel combs, is usually driven at between 40 and 60 rev/min but recently up to 120 rev/min; at these speeds, the aeration tank contents are sprayed over the surface, as in Simplex

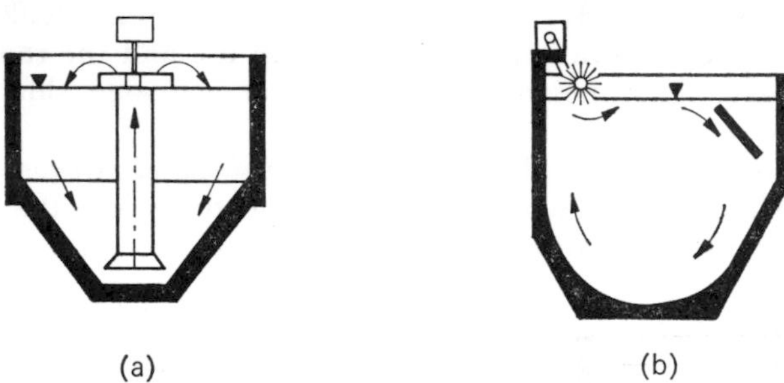

Fig. 12.15. (*a*) *Simplex activated-sludge tank* (*b*) *Kessener brush activated-sludge tank*

systems. Improved performance has been obtained by installing a baffle over the spray, allowing reduction of the aeration period, in some cases to as little as 1 h. Kessener systems are particularly suitable for smaller works. The *Oxidation Ditch* (Ring Ditch) system (Fig. 12.16) developed in Holland[62, 63] is similar to the Haworth channel flow system but uses a Kessener brush in place of a simple

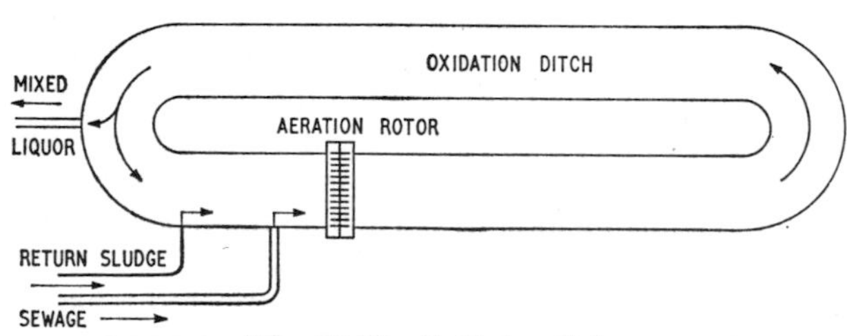

Fig. 12.16. *Oxidation ditch*

paddle. The flow velocity of the mixed liquor in the oxidation ditches is maintained at about 1 ft/s (0·3 m/s). The aeration period usually amounts to three days, resulting in a high degree of purification and highly mineralized excess activated sludge. This system is sometimes adopted for relatively small municipal sewage-treatment plants (up to about 2 500 population served), because it is comparatively simple. In some cases, final settling and excess sludge removal are provided for by special constructions and arrangements of the ditches, interrupting circulation for short periods during settling and removal of excess sludge.

Extended aeration systems make use of tanks similar to normal types and are discussed under the heading of Process Arrangements (p. 234).

Mixed-Liquor Solids Concentration

In these various aeration tanks, the mixed liquor suspended solids (MLSS) concentrations used vary between about 1 500 and 3 000 mg/l in conventional low-rate activated-sludge plants, but are usually lower in plants operated for only partial biological treatment. With some recently developed arrangements (see below), greater rates of uptake of oxygen per unit of tank volume are obtained and allow operation with much greater mixed-liquor solids concentrations, up to 8 000 mg/l or even more. Such high concentrations of activated-sludge solids are sometimes used in high-rate activated-sludge plants and commonly especially in small low-rate plants such as oxidation ditches and extended aeration systems.

Load Allowances for Activated-Sludge Plants

If a certain aeration period is accepted in the design of an activated-sludge plant, using a particular type of aeration tank, the total volume of aeration tanks required may be estimated on the basis of raw sewage inflow only, on a mixed-liquor flow basis or on raw sewage inflow plus returned activated-sludge flow. However, $B.O.D._5$ loadings calculated on total aeration tank volume and the amount of MLSS present provide a better basis for design estimates.[55–68]

Standard or normal low-rate loadings are calculated at about 35–40 lb $B.O.D._5$ input/1 000 ft^3 d, say 1 lb/yd^3 d (0·55–0·65 kg/m^3 d), of aeration tank capacity for diffused-air systems and highly efficient mechanical plants. High-rate loadings may be about 90 lb $B.O.D._5$ input/1 000 ft^3 d (1·5 kg/m^3 d), but in Europe both low-rate and high-rate loadings are usually higher, e.g. up to 110 and 220 lb/1 000 ft^3 d (1·75 and 3·5 kg/m^3 d). Concentrations of mixed-liquor solids are maintained accordingly with the aim to keep the relative daily sludge loadings at the level appropriate for the required overall plant efficiency (see below), the minimum usually being about 2 000 mg/l.

Within the normal range of MLSS concentrations, the performance of activated-sludge plants is best calculated on the basis of lb $B.O.D._5$ daily input per lb total activated-sludge solids (or kg/kg) in circulation.[65, 68] If loadings are kept below about 50 lb $B.O.D._5$ daily input/100 lb circulating sludge solids, removal of 90% or more is to be expected, but when the loading reaches higher figures, plant efficiency is reduced; e.g. if the loading is increased to 400 lb B.O.D.$_5$ daily/100 lb sludge solids, removal efficiency may be reduced to only about 50% as shown in Table 12.3. Local circumstances should be taken into account; e.g. load allowances should be reduced for small plants.

Table 12.3

ACTIVATED-SLUDGE PLANTS, B.O.D.$_5$ LOADINGS AND REMOVAL EFFICIENCY

Type of plant	*Daily sludge loading,* lb B.O.D.$_5$/lb *sludge solids* (kg/kg *sludge solids*)	*Removal of* B.O.D.$_5$ *applied* %
Low-rate	up to 0·4	>90
	0·5	90
High-rate	1·0	75
	1·5	65
	2·0	60
	4·0	50

Air Supply for the Activated-Sludge Process

The air supply to aeration tanks must be abundant enough to maintain a dissolved-oxygen concentration throughout the tanks of not less than 1–1½ p.p.m. In high-rate plants, it should be even greater, perhaps up to 3 p.p.m. Normally the consumption of oxygen is greatest near the inlet and least near the outlet, and the rate of supply of air may be varied accordingly along the tank, an arrangement which is called *tapered aeration.* Alternative methods include the introduction of the sewage load progressively along the tank maintaining a uniform rate of air supply, known as *step aeration* (see p. 230). The rate of air supply may also be varied by automatic devices measuring and controlling the dissolved-oxygen concentration of the mixed liquors.

Taking oxygen utilization at 1 lb (kg) oxygen/lb (kg) of B.O.D.$_5$ removed and allowing for the efficiency of absorption from the air supplied to range between about 5 and 15% depending on the type of aeration device used and on the depth of air inlet below the water-surface level (say 11% for fine bubble aeration at 10 ft (3 m) depth) and on the basis that 10 ft^3 (0·3 m^3) of air contain 0·175 lb (80 g) of oxygen, the air supply requirement is then $(10 \times 100)/(0{\cdot}175 \times 11) \simeq 520$ ft^3 of air/lb of B.O.D.$_5$ removed (32 m^3/kg B.O.D.$_5$).

Average domestic sewage has a B.O.D.$_5$ amounting to the equivalent of 0·12 lb/hd d (55 g/hd d), which is normally reduced by sedimentation to say 0·08 lb/hd d (37 g/hd d), equal to 1 lb (0·45 kg) of B.O.D.$_5$ daily per about 12 persons, from which it follows that an air demand for activated sludge of 520 ft^3/lb B.O.D.$_5$ removed corresponds to one of 520/12, say, 44 ft^3/hd d (1·25 m^3/hd d) of population served or equivalent. If 90% of B.O.D.$_5$ applied has to be removed, the air supply required amounts to about 0·90 $\times$ 44 $\simeq$ 40 ft^3 air/hd d (1·1 m^3/hd d) or equivalent (Table 12.4).

The capacity of air blowers should be at least 50% greater than calculated from the figures of Table 12.4, to be flexible enough to meet the variations in rates of B.O.D.$_5$ loading to be expected during

Table 12.4

AIR REQUIREMENTS OF AERATION TANKS

Size of bubbles	*Oxygen uptake efficiency,* %	*Air supply required* ft^3/lb B.O.D.$_5$ removed (m^3/kg)	*Air supply required* ft^3/hd d (m^3/hd d)
Fine	11	500 (32)	40 (1·1)
Medium, 0·06–0·12 in (1·5–3 mm) diam.	6·5	900 (57)	70 (2·0)
Large (coarse)	5·5	1 000 (64)	80 (2·3)

Note—These figures are for 90% B.O.D.$_5$ removal and for air inlets 10 ft (3 m) below water-surface level. The air volumes have to be increased if introduced at a shallower depth below surface. In the case of aeration by bubbles of medium to large sizes, the efficiency of oxygen absorption may be improved by the installation of mechanical mixers just above the air inlets.

any normal day. The data of the table apply to normal operation of normal systems. If the proportion of sludge solids is higher, additional oxygen demand must be allowed for. Sludge solids in normal systems consume about 7–12 mg oxygen/g sludge solids hourly.

The use of pure gaseous oxygen in place of air has been tried by Okun[64] who has named the technique 'Bio-precipitation'.

The units used for introducing the compressed air into the aeration tanks in the diffused-air systems were usually made up of diffuser plates about 1 ft (0·3 m) square set at floor level, but now other types of diffusers, porous and non-porous, are also used. Plate diffusers are composed of ceramic materials partly fused together to leave pores about $\frac{1}{80}$ in (0·3 mm) diameter, corresponding to No. 40 of the U.S.A. permeability gradings. This corresponds to a permeability, under dry conditions and with pressure losses of 2 in (50 mm) of water, of 40 ft^3/min of air per ft^2 (730 m^3/m^2 h). Such plates under water pass about 4 ft^3/min per ft^2 (73 m^3/m^2 h), with pressure losses between about 4 and 8 in (0·1 and 0·2 m) of water. Diffuser plates with finer pores, and permeability gradings up to 120, have been used but the pores must not be too fine, otherwise clogging occurs and too much pressure is required for operation. Other types of diffusers using porous material have the form of long tubes or mushroom-shaped domes installed at a level 1–2 ft (0·3–0·6 m) above the floor of the aeration tank. These devices are now generally preferred to diffuser plates because they can be more easily removed for cleaning. Seven-inch (180 mm) dome diffusers pass about 0·5–0·8 ft^3 air/min (0·85–1·4 m^3/h) through each dome. The air supplied to them must be free of dust, and the frames supporting the diffusers must not form rust which may cause choking.

Swinging-pipe diffusers with filters which can be swung out of the tank have the advantage that they can be cleaned regularly without shutting down and emptying the tank. Successful aeration has also been obtained by introducing the air directly through comparatively large holes in pipes located at the bottom along one side of the tanks or even through open ends of vertical pipes about $\frac{1}{2}$–$1\frac{1}{2}$ in (13–38 mm) diameter and about 2 ft (0·6 m) apart. In these cases, aeration not only depends upon air bubbles but also upon the rapid circulation of the whole contents of the tank with the aid of coarse air bubbles, whereby the exposed water surfaces are renewed in a matter of seconds and the oxygen is also rapidly absorbed from the atmosphere.

The mechanical aeration systems such as the improved Kessener and Simplex systems using modern high-intensity cones and others, have about the same power requirements as the diffused-air systems on the basis of oxygen uptake.[66, 67] The advantages of flexibility in the aeration arrangements are given particular attention for larger plants. Greater flexibility requires costlier air-supply piping systems—additional fittings, valves, air-flow meters, perhaps less efficient compressor operation, etc.—but the added costs may be justified by improved purification efficiencies. Automatic dissolved oxygen or redox potential measurements may also be provided for.

Arrangement of Works for Activated-Sludge Treatment[68–77]

Of the many arrangements which have been proposed from time to time, the following techniques for utilization of activated sludge have been proved by experience.

1. *The Normal Arrangement* as illustrated in Figs 12.17 and 12.18 provides for plain sedimentation of the waste, with the object of removing the coarser suspended matter with as much of the finer

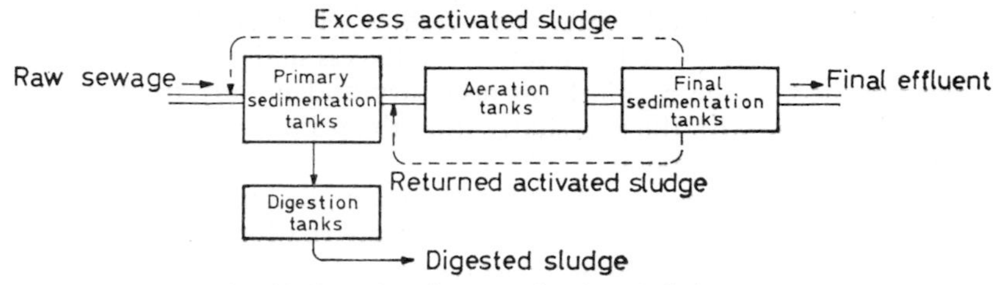

Fig. 12.17. *Flow diagram of activated-sludge process*

solids as feasible, using detention periods of at least 2 h for municipal sewage; the settled waste is then mixed with the returned activated sludge and the mixture discharged directly into the forward end of the aeration tanks. As stated above, the various systems of aeration tanks require aeration periods of 1–10 h or more; the actual period depends on the system used and the strength of the waste to be treated but should be adjusted to effect the required degree of purification and, at the same time, to keep the sludge 'alive' or activated. The effluent purified waste and activated-sludge mixture is then treated by sedimentation in final tanks (p. 237), from which

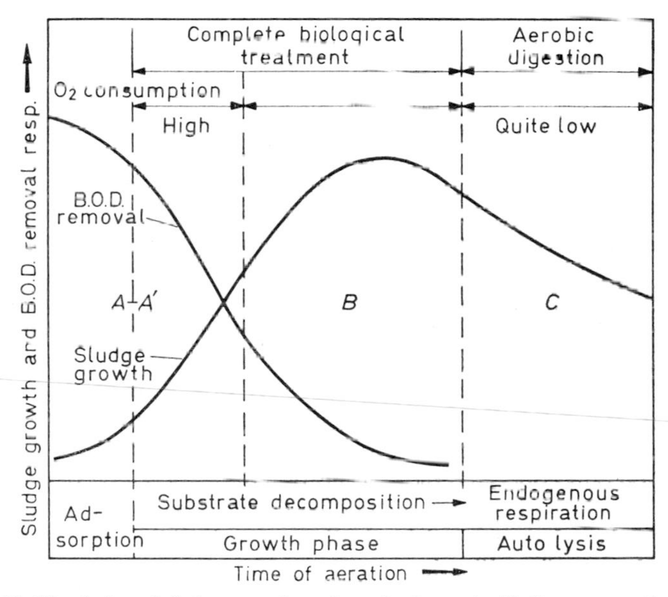

Fig. 12.18. *Activated-sludge growth and respiration. A–A' Contact aeration and high-rate oxidation; B Normal activated-sludge treatment; C Extended aeration (after Benedek*[59]*)*

the sludge is returned while the overflow of purified effluent is ready for disposal or further treatment. Excess sludge is taken from the returned sludge regularly as necessary and usually treated by digestion along with the raw sludge from the primary sedimentation tanks before final disposal.

With this arrangement, the air may be supplied uniformly along the whole length of each aeration tank; alternatively, the rate of supply may be adjusted to some degree according to the actual rates of oxygen consumption within particular zones, e.g. somewhat more air towards the inlet zone than further along the tank where the uptake may be lower. However, allowance must be made for the continuing demand of the active sludge during its separation from the effluent and return from the secondary sedimentation tanks.

2. *The Step-Aeration Process.*[60] From the Tallman's Island New

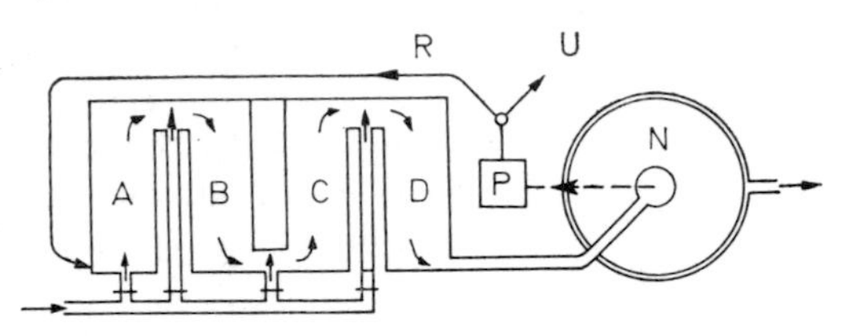

Fig. 12.19. *Step–aeration using four raw-waste feed inlets*

York works, Gould has reported more efficient utilization of the air supply by using what he has named 'step-aeration', a technique for more uniform loading of the aeration tanks, whereby the influent waste is introduced at several inlets distributed along the tank's length, instead of only at the inlet end as in the normal arrangement. Preliminary and final (N) settling tanks are provided as in the ordinary processes. As shown in Fig. 12.19 the returned sludge, R, is pumped separately to the inlet of the aeration tanks which are subdivided into four sections A–D. The settled waste also is divided into four portions (usually equal) and discharged stepwise into the tanks, one portion into each section. With this arrangement, the activated sludge is relieved of the full load at the forward end of the tank, and the tank operates more uniformly and more efficiently. Aeration tanks operated by step aeration may, therefore, be designed for relatively higher loadings per unit of tank volume, usually 25–30% higher than normal load allowances (p. 225) to achieve the same degree of purification.

3. *Treatment with Separate Reaeration of Sludge.* In this process, aeration of the waste is restricted to a period of, say, 1 h. The activated sludge, separated from the effluent, is reaerated in special

reaeration tanks, with a detention period of up to about 12 h, before returning it to the aeration tanks. Although used in place of the normal arrangement, this is more often applied in partial activated-sludge treatment processes, such as preliminary biological treatment of wastes ahead of trickling-filter installations. In another variation, primary sedimentation is omitted and the returned sludge is reaerated using four-fifths of the aeration-tank capacity, and the raw sewage introduced only into its last section. This modification, using a high rate of returned sludge, is called *biosorption*;[78] it is similar to the *contact-stabilization* process[77] which also uses reaeration of the sludge but restricts aeration of the mixture of returned sludge and sewage to the adsorption stage only (see p. 217).

4. *High-rate Activated-Sludge Processes.*[72–74, 76–81] A high-rate process developed by Gould[73, 74] consists in using less sludge in the aeration tanks, being of much less age accordingly, resulting in more economical treatment, but at the same time, of course, operating with correspondingly higher rates of sludge loadings, i.e. a considerably lower degree of purification. This process has been adopted for several new municipal plants in the U.S.A., of which the New York–Jamaica plant was the pioneer. Operating characteristics are given in Table 12.5. Primary sedimentation is not provided, preliminary treatment consisting only of screening and grit separation. The total aeration period is about 2 h. B.O.D. reduction achieved by the plant (overall, i.e. including final sedimentation) is only 75%, compared with 90–95% in normal plants. The Gould age of the activated sludge is as low as 0·37 d, only about one-tenth of the normal. The excess sludge is, accordingly, readily digestible and is reported to contain 74% organic matter (dry solids basis) and to settle easily to 6·5% dry solids.

Other high-rate processes, developed in Germany, allow higher loading rates by reducing the aeration-tank capacity to about one-third normal size or less but providing a higher rate of air supply and maintaining a higher content of activated sludge in the tanks. A B.O.D. reduction of 75% can be obtained in this way, for instance at Wuppertal[76] (see Table 12.5).

The characteristic property of high-rate processes compared with normal low-rate (Fig. 12.18, A, B) lies in the much shorter age of the activated sludge used in the aeration tanks and in the higher rates of sludge loadings (p. 225). The biological activity of the high-rate sludge depends mainly on its bacterial flora, no protozoa being present (see p. 126). The sludge is denser and settles quickly but does not develop fungi when overloaded so does not bulk readily.

5. *Extended Aeration Processes,*[82–85] first developed in the United

States, are characterized by extremely low B.O.D. loadings, say up to 15 $B.O.D._5$/1 1 000 ft^3 (0·25 kg/m^3) aeration-tank capacity per day, or 0·03–0·07 lb (kg) $B.O.D._5$ daily per lb (kg) of mixed-liquor dry solids concentration, and aeration periods of at least 1–2 d. Usually the crude sewage enters the aeration tanks without prior primary sedimentation. Because of the long aeration period, endogenous decomposition of the activated sludge solids (Fig. 12.18, C) is obtained, resulting in highly stabilized excess sludge which would not need digestion before passing it to sludge-drying beds. The rate of sludge production, relatively low, may be as little as 0·3 lb (kg) sludge dry solids/lb (kg) $B.O.D._5$ removed or even less. The required flow of air is relatively great and, with coarse-bubble

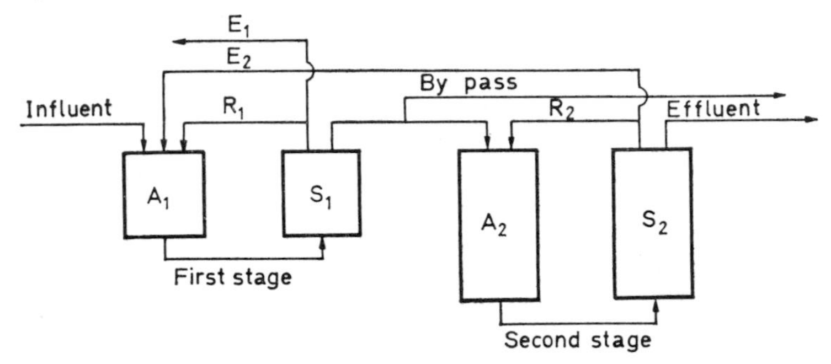

Fig. 12.20. *Two-stage activated-sludge process*
A_1 *and* A_2 = *aeration tanks* R_1 *and* R_2 = *return sludge*
S_1 *and* S_2 = *settling tanks* E_1 *and* E_2 = *excess sludge*

aeration, may amount to about 1 500–3 000 ft^3/lb $B.O.D._5$ applied (95–190 m^3/kg $B.O.D._5$). The extended aeration process is especially suitable for the treatment of sewage from small communities. *Oxidation ditches* (see p. 224) are also operated with extended aeration periods.

6. *Two-stage Activated-Sludge Processes*[86] (Fig. 12.20) may be preferable in some cases. In each stage, aeration tanks are followed by secondary sedimentation tanks, primary settling sometimes being omitted. The raw or settled sewage is mixed with return sludge from the first stage (R_1) and excess sludge from the second stage (E_2). Excess sludge from the first stage (E_1) is removed for disposal, as in single-stage treatment. Activated sludge of the first stage (R_1) is overloaded but continuously reconditioned by admixture of excess sludge from the second stage (E_2), thus making use of its residual purification capacity. Usually one-quarter of the required aeration-tank capacity is provided in the first stage. However, the

Table 12.5

ACTIVATED-SLUDGE PLANT OPERATING DATA

Location	Capacity $\times 10^6$ gal/d ($\times 10^3 m^3$/d)	Sedimentation		Aeration	Average Sludge returned	Sludge in Aeration Tanks		Daily Loading Rates		Air Supply		Overall Removal		Operating Year
		Primary h	Final h	h	%	Suspended solids mg/l	Gould age d	lb B.O.D.$_5$ /lb MLSS (kg B.O.D.$_5$ /kg MLSS)	lb B.O.D.$_5$ /1 000 ft^3 (kg/100 m^3)	ft^3/lb B.O.D.$_5$ removed (m^3/kg)	ft^3/gal (m^3/100 l)	Suspended solids %	B.O.D.$_5$ %	
Full biological treatment using conventional low-rate activated-sludge processes														
Mogden, England	60 (273)	3½	6½	8		4 600	23	0·09	24 (38)	700 (44)		96½	96½	1947[69]
Ryemeads, England	8 (36)	8	4	10	70	9 000	29	0·06	29 (46)		2 (1·2)	97	97½	1960[70]
Glenelg, Australia	8 (36)	3	4½	8	60	4 200	13	0·09	24 (38)	1 400 (87)	2 (1·2)	97	97	1963†
Ochiai, Japan	16 (73)	2·5	5½	9	50	1 200	6	0·13	10 (16)	760 (47)	⅓ (0·20)	96	93	1964†
Milwaukee E., U.S.A.	45 (205)			9½	20	3 000	3½	0·30	52 (83)	480 (30)	1¼ (0·78)	95	95	1945[71]
Milwaukee W., U.S.A.	55 (253)			10½	25	3 000	4¼	0·25	47 (75)	720 (45)	2 (1·2)	94	94	1945[71]
Chicago S.W., U.S.A.	600 (2 730)			4½	50	2 300	4¼	0·22	32 (51)	900 (56)	⅔ (0·42)	86	92	1954[72]
New York, U.S.A.	180 (819)	0·9		4·2		1 770	'Modified' aeration			380 (24)			94	1949[73]
Wards Isl., U.S.A.		Nil	370* (35 400)	2·5	28	950	3·4	Step aeration			0·43 (0·26)	95	94	1954[74]

* Overflow rate, U.S. gal/ft^2 d (l/m^2 d) † Private communication

Table 12.5 continued

Location	Capacity × 10^6 gal/d (×10^3 m^3/d)	Sedimentation		Aeration	Average Sludge returned	Sludge in Aeration Tanks		Daily Loading Rates		Air Supply		Overall Removal		Operating Year
		Primary h	Final h	h	%	Suspended solids mg/l	Gould age d	lb B.O.D.5 /lb MLSS (kg B.O.D.5 /kg MLSS)	lb B.O.D.5 /1 000 ft³ (kg/100 m³)	ft³/lb B.O.D.5 removed (m³/kg)	ft³/gal (m³/100 l)	Suspended solids %	B.O.D.5 %	
Full biological treatment using surface aeration														
Oxford, England	5·5 (25)	5	5	11	80	2 000	7½	0·17	21 (34)	High intensity cones		88	90	1958[75]
Plain aeration and partial biological treatment by high-rate activated-sludge processes														
Ann Arbor, U.S.A.	5 (23)			6½	0	150	0·4	4·00	40 (64)	600 (38)	½ (0·31)	66	63	1946[71]
Jamaica, U.S.A.	50 (228)	Nil	730* (29 700)	1·7		1 000	0·5	1·60	90 (145)	480 (30)	0·44 (0·26)	81	76	1952[73,74]
Wuppertal, Germany				1¾		2 500	1·1		150 (240)	500 (31)	0·6 (0·38)		72	1954[76]

* Overflow rate, U.S. gal/ft² d (l/m² d)

Table 12.5 continued

Partial biological treatment using contact stabilization

Nature of Waste Treated	*Analysis* mg/l		C.O.D. $\times 10^{-5}$ lb/gal (mg/l)	*Reaeration Tank*		*Contact (Stabilization) Tank*		*Overall Removal*	
	Suspended solids	B.O.D.$_5$		*Aeration period* h	*Suspended solids* $\times 10^{-5}$ lb/gal (mg/l)	*Contact period* h	*Suspended solids* $\times 10^{-5}$ lb/gal (mg/l)	B.O.D.$_5$ %	C.O.D. %
Domestic waste	—	200–250	—	0·2	2 000–3 000	1	4 000–5 000	90	
Fruit canning waste	—	400	700	$\frac{3}{4}$–$1\frac{1}{4}$	2 000–3 000	$1\frac{1}{2}$–$2\frac{1}{2}$	4 000–5 000	90	90

Full biological treatment using extended aeration[84]

				Aeration air supply ft^3/lb B.O.D.$_5$ (m^3/kg)	*Sludge produced* lb/100 lb B.O.D.$_5$ (kg/100 kg)	*Aeration period* d	MLSS p.p.m.	B.O.D.$_5$ lb/1 000 ft^3 d (kg/100 m^3 d)	*Removal* %
Domestic sewage	500	600	—	at least 1 500 (90)	20	1	8 000	up to 30 (50)	90–95

advantages which may result, such as reduction in total aeration period, may be offset by the relatively large total capacity including the two settling tanks required.

Two-stage processes are more flexible in operation. Some first-stage effluent may be bypassed around the second stage if necessary to protect the second stage from undue variations in the pollutional load reaching the works. Continuous bypassing may be applied for *partial biological treatment* and allows further reduction in the sizes of the tanks. Chasick[87] obtained similar effects during a trial run in which two groups of activated-sludge units were operated in parallel, excess sludge of one group being returned into the aeration tanks of the other; he called the process 'activated aeration'.

7. *Activated-Sludge Processes using Mineral Additives*. Processes have been developed using various kinds of mineral solids as carriers for sludge activation. Normal experience has shown that, during rainstorms, the presence of semi-colloidal inorganic constituents introduced by the stormwater improves the settleability of the activated sludge, and even the addition of raw sludge along with the returned activated sludge may have a similar effect. In one special case of low nitrogen content, Kraus obtained an improvement in activated-sludge plant performance by adding digested sludge and supernatant liquor from digestion tanks to the activated sludge in the aeration tanks, increasing the aeration period to obtain a proper activiated sludge of high nitrate content.[88, 89] Adjustment of the nitrogen or phosphorus contents in such a way is desirable if the levels are otherwise low enough to limit purification efficiencies.

Iron is also used for sludge improvement. In the Guggenheim process, iron salts (usually ferrous sulphate) are used. Phelps and Bevan[81] found bacterial growth to be stimulated by using iron contents of 5–10 p.p.m. When activated sludge is enriched with iron in this way, the capacity for absorbing organic matter is increased together with the oxidation capacity, and the sludge settles more readily. The aeration period may be lowered accordingly. The process is also less sensitive to industrial waste discharges.

In the Niers process cast-iron scrap is placed in a preaeration tank through which the incoming (settled) waste is passed and aerated, before discharge into the activated-sludge aeration tanks, carrying dissolved ferrous hydroxide forward and improving the character of the activated sludge accordingly.[90]

Another such process is that of Zigerli in which asbestos is added to the activated sludge, at 1 p.p.m. (reckoned on the total inflow of untreated waste). Asbestos is known to be a good medium for supporting bacterial cultures; it has the advantage of being inert without oxygen demand and is unaffected by toxic substances. An aeration period of about one hour is used. The purification effected by such processes is somewhat less than by normal systems.[91]

Activated-sludge processes may be modified to yield only *partial biological purification*. However, if the aeration period is reduced or the air supply restricted, not only is the degree of purification reduced but the activated sludge itself tends to become sick, and accordingly the process cannot be maintained. In general, it is essential that a high degree of purification be achieved for the activity and settle-ability of the activated sludge to be sustained. Full biological treatment must result, with a high degree of B.O.D. reduction yielding a stable effluent. However, nitrification may not occur to any appreciable extent, as this depends not on the purification of the waste but on the extent of decomposition of the sludge itself (second stage of biochemical oxidation). Many of the earlier activated-sludge plants were designed for longer aeration periods, a greater air supply and the return of larger volumes of activated sludge (greater sludge age) and as a result achieved not only a higher degree of purification (greater B.O.D. reductions) but also yielded effluents with a high content of nitrate. Just as in low-rate filtration (see p. 198), so in these plants the nitrate results from a more extensive decomposition of the activated sludge. The surplus sludge from these plants is denser and less in quantity; against this, however, it is noted that costs are relatively high. The more recent activated-sludge plants are, consequently, not designed to produce nitrates, thereby reducing costs significantly without loss of stability of effluents. There is nothing to be gained by nitrification, the nitrogen remaining usefully in the sludge and entering the digestion tanks, etc., with the surplus sludge. The overall degree of purification remains relatively high and is regarded as full biological treatment.

Partial biological treatment by means of activated sludge can nevertheless be provided by processes such as those described which depend on the use of reaeration, reduction of sludge age or adding inorganic substances such as iron or asbestos. In these ways continuous processes may be maintained to yield lower degrees of purification, say only 65–85% instead of 90–95% representing full purification

Secondary or Final Sedimentation Tanks

The effluent of the aeration tanks is passed into secondary or final sedimentation tanks to separate the activated sludge from the final effluent. The sludge must not remain long in these tanks because there is no supply of air by aeration, and the oxygen carried in solution represents only a poor supply compared with the sludge-respiration rate. The aerobic sludge has a tendency to rise in the tank, especially when the activated-sludge process is operated so as to effect substantial nitrification, because gaseous nitrogen results from the reduction of nitrates (denitrification, pp. 58, 210) and tends to cause carryover of sludge particles by flotation.[96]

In general, the average storage period of activated-sludge solids in the final sedimentation tanks should not exceed 6 h. To achieve this, the sludge has to be drawn off from the tanks continuously at a high rate, along with recirculation of effluent, up to 100% by volume calculated on the average dry-weather raw-sewage flow into the plant (p. 220).

The types of settling tanks best suited for activated sludge are vertical-flow tanks (p. 164). The detention period should be 2–3 h, calculated on maximum flow of raw sewage entering the plant. The upward rate of flow, equal to the surface loading rate (p. 152), should not exceed about $1\frac{1}{2}$ in/min (0·6 mm/s), or about 1 100 gal/ft^2 of tank area daily (2·3 m^3/m^2 h). Generally, activated-sludge solids are flocculent in character, and upward-flow final settling tanks operate commonly as 'sludge-blanket'-type clarifiers. Mechanical tanks are common, continuously operating mechanisms being preferred. In addition to upward-flow tanks, however, rectangular longitudinal-flow and circular radial-flow types are used. Special designs were referred to in Chapter 10; recent designs also include types in which the radial flow is directed inwards. Multiple-weir systems are commonly used to keep weir overflow rates low, say within 5 000–10 000 gal effluent/ft length of weir (70–150 m^3/m) daily (see p. 162); sludge-storage hoppers, sometimes of relatively substantial capacity, are normally provided as part of the underflow withdrawal system. Special designs have been developed to take advantage of density currents and of the unusually large proportion of underflow characteristic of the final tanks of an activated-sludge plant.

Surplus Activated Sludge

Excepting extended aeration and aerobic sludge-digestion[84] processes, only a small proportion of the polluting organic matter absorbed by the activated sludge is actually decomposed; consequently, the remainder forms additional sludge and must be disposed of daily in order to prevent its accumulation. This additional quantity varies; under average conditions (p. 257), it amounts to about $\frac{1}{2}$ gal/hd d (2·5 l/hd d) of contributing municipal population (calculated on a moisture content of about 98·5%). It is withdrawn from the returned activated sludge, which is pumped from the final sedimentation tanks, and is consequently more diluted than the settled sludge. Taking, as before, average conventional low-rate plants as example, the excess sludge usually amounts to about 1 gal/hd d (5 l/hd d) or about 3% of the daily D.W.F.; with very weak municipal sewage, it may be only 1% or less. The returned activated sludge normally amounts to about 30% of the total inflow of sewage in municipal activated-sludge plants, so that about 10% of the returned sludge is diverted to the sludge disposal works as excess sludge.

Special provision is required for the disposal of surplus sludge. If surplus activated sludge is discharged directly to drying areas, a considerable odour nuisance will arise, but it is readily digested along with raw sludge in digestion tanks. However, pretreatment of the surplus activated sludge in order to reduce the total volume may be justified. Thickening tanks for this purpose are usually provided for a loading of about 10 lb sludge dry solids/ft^2 d (50 kg/m^2 d) of tank surface and for settling periods of 12–24 h. The separation of the water may be facilitated by slowly rotating, vertically mounted bars which tend to dislodge gases from the sludge particles. Chlorine should be applied to delay decomposition of the sludge which otherwise disturbs settlement, about 0·4–1·1 $\times$ 10^{-3} lb/hd d (0·2–0·5 g/hd d) or 1 lb (0·45 kg)/1 000 persons daily being required. This technique reduces the moisture content of the sludge to as low as 96%.

A simpler method consists in the diversion of the surplus sludge to the primary sedimentation tanks influent channel (Fig. 12.17) so that it is recovered as a mixture of raw and surplus sludge. Mixed raw and surplus sludge drawn from the primary sedimentation tanks with this arrangement of municipal sewage-treatment works contains about 95·5% water. It has been shown that, provided the surplus sludge is discharged into the raw-sewage influent channel continuously in proportion, the efficiency of the pretreatment by sedimentation is improved.

These mixed sludges are easily digested in the ordinary way, with a corresponding increase in gas production. Usually, however, the digested mixed sludge has higher moisture content than raw sludges (Table 13.1, p. 257). The required capacities for sludge digestion and sludge-drying beds are about twice those required for primary sludge only. Surplus activated sludge may be partly dewatered mechanically by means of vacuum filters, and the filter-caked sludge may be dried finally by heat and sold as fertilizer. It is more economical, however, especially for smaller works, to combine the surplus sludge with raw sludge either in the primary sedimentation or in the sludge-digestion tanks and to filter the mixed sludges either before or after digestion because, as pointed out above, the moisture content and, hence, the sludge volume is then relatively much smaller. Another technique is to digest the raw sludge separately and to mix the digested and the (undigested) fresh surplus sludge before filtration. This may be preferable in some cases, because although the production of digester gas is diminished, the humus content of the dried sludge is better and the sludge may be more easily dried.

Operation and Maintenance of Activated-Sludge Plants

When activated-sludge plants are first brought into operation, a ripening period, without additives, about 14 d, is required for the development of the activated sludge. During this period, foam forms on the surface of the aeration tanks; once the sludge is ripened, continuous day and night aeration is essential. During later normal operation, foam formation may indicate that the proportion of activated sludge returned is too small. Foaming may also occur with higher concentrations of detergents. Spraying with water[93] or the addition of mineral oil[94] (about 0·1 p.p.m.) or anti-foaming chemicals may then diminish or minimize foaming, which otherwise can be very troublesome within the works and on windy days even far beyond the works area.[95]

Activated sludge may become 'sick' (see p. 219), resulting in bulking.[96] Bulking sludge under water occupies a greater volume and does not settle readily; eventually it floats in large aggregations on the secondary sedimentation (humus) tanks and passes over into the effluent in proportions sufficient to spoil its quality. Plant operators may recognize such a change in the character of the sludge readily by simple inspection, while the behaviour of the sludge in the Imhoff cone also shows such changes clearly. The following means are effective in rectifying such changes:

1. Reduction of the rate of sludge return and rejection of a large quantity of excess sludge. If it is due to a single unusual disturbance it is better to reject a large part of the sludge and to commence a new period of ripening.
2. Increased aeration.
3. If aeration cannot be increased, reduction of the amount of waste load discharged into the aeration tanks. This may require discharge of waste after sedimentation-tank treatment only until the sludge recovers. Load reduction usually ensures its rapid recovery.

4. Dilution of the contents of the aeration tanks using fresh water or even stormwater if available.
5. Addition of coagulants such as ferric chloride to the final sedimentation (humus) tanks.
6. Application of chlorine either to the influent, the humus tanks or the returned sludge; a rate of about 5 mg/l may be sufficient.
7. Where normally practised, the discharge either of excess activated sludge or of digestion-tank supernatant liquor to raw-waste inflow to the primary sedimentation plant should be interrupted.
8. Step aeration or sludge reaeration should be brought into operation where possible.
9. The addition of nutrients (nitrogen, phosphates) if the waste flow is deficient, in some such respect.

When activated-sludge plants are inadequate in capacity, it may be necessary to bypass more or less of the settled waste without biological treatment in order to maintain a healthy sludge, but if arrangements for sludge return and for aeration of sludge and mixed liquor are flexible enough, considerably increased loadings may be handled temporarily and satisfactorily without bypassing.

Pretreatment of the settled waste using trickling filtration or contact aeration is helpful for combating erratic discharges of industrial wastes which may otherwise prove damaging to the plant and is also helpful to overloaded plants. Provision of reaeration tanks in which overloaded activated sludge may be rested and reactivated by several hours' reaeration serves a similar purpose, and the same is achieved by two-stage operation.

Apart from raw and settled sewage, final effluent and digester testing, operating supervision of larger activated-sludge plants involves the following tests and measurements:

1. Quantities of waste flow, returned sludge and excess sludge.
2. Settling tests of the contents of the aeration tanks at outlet end.
3. Determination of suspended solids contents of the aeration tank, returned sludge and final effluent.
4. Dissolved oxygen contents in the aeration tank.
5. B.O.D. of the effluent.
6. Depth of sludge in the final sedimentation tanks.

The machinery required for activated-sludge plants includes sludge pumps, air compressors and motors for mechanical aeration or circulation where used. The power consumption may be about 1½ h.p. (say 1 kW)/1 000 persons or equivalent for average municipal sewage, or about 2 h.p. (1½ kW) for weak sewage, but actual power consumption figures vary greatly according to the system and type of tanks used.

TREATMENT BY PONDING

In Chapter 6 it was shown that ponds or lakes may greatly increase the self-purification capacity of streams because of their large surface areas and increased detention periods. In much the same way, artificial ponds or lakes[97–106] provide means by which organic wastes can be treated biologically and, at the same time, mechanically by sedimentation.[97] They may be constructed on a small or large scale, as required, and are capable of yielding highly purified effluents, in many cases of a quality comparable with that of natural waters. Such effluents may, however, be aggressive on account of the carbon dioxide and dissolved-oxygen contents characteristic of purified wastes, while fertilizing constituents such as nitrates and phosphates may also encourage excessive algal growths.

Ponds may be either essentially aerobic or anaerobic. Aerobic ponds (oxidation ponds or sewage lagoons)[98] are designed and loaded with a view to the continued maintenance of a positive oxygen balance throughout the whole volume. Some sludge deposition and accumulation are inevitable, and these bottom-sludge layers become anaerobic, as with stream deposits (p. 65). Wastes should be discharged into ponds so as to avoid uneven sludge deposition and to achieve 'natural' mixing and recirculation of its waters. Mechanical mixing, recirculation by pumping or mechanical aeration within the pond provide supplementary aids for achieving higher rates of treatment.

Such ponds are usually designed for long average periods of detention[99]—commonly 10–40 d—with loadings as low as 1–2 lb or less of B.O.D.$_5$ equivalent/1 000 ft^2 pond area daily (0·5–1 kg/100 m^2) (depth, say 3–4 ft (about 1 m) or less). The basic principle of their operation is that normal water and soil bacteria cause oxidation of these organic solids, producing some carbon dioxide. Its utilization by photosynthetic planktonic green plants (algae) releases oxygen which then is available for further aerobic fermentation, and so on. This is symbiosis, a mutual living together.

Under circumstances favouring maximum bacterial–algal symbiosis, detentions of two or three days with loadings up to 5 lb/1 000 ft^2 d or more (2·5 kg/100 m^2) become feasible, but mechanical mixing and circulation are then essential. Solar radiations are the main determinants of total purifying capacity because of their control of maximum feasible algal efficiencies. Effluents from such high-rate pond systems are usually dark green in colour, owing to high concentrations of algae. The total content of unstable organic matter may, accordingly, differ little from, or exceed, that of the influent; nevertheless, there are many environments into which such effluents can be discharged without nuisance. The algae may serve as food for fish, directly or indirectly, or, if not, eventually die. Their ultimate fate will determine whether or not pond treatment is suitable for any particular situation.

Effluents from low-rate ponds usually contain only relatively small numbers of pathogens (or the associated coliform indicator organisms), but bacterial reductions are usually not uniformly reliable. Suspended solids contents may also be variable, mainly because of variations in algal content. Where ponds are used for 'tertiary' treatment of biologically treated wastes,[104] the lower content of organic matter, etc., commonly restricts the development of algae to relatively low levels. In summer, however, algal cell concentrations in pond effluents may reach the equivalent of 100–200 mg/ml of (dry) suspended solids.

Anaerobic ponds[105, 106] become established when water-borne wastes carry organic matter into open bodies of water at such a rate that the water remains continuously deoxygenated despite uptake of oxygen through the surface. Algae normally cannot grow in or on the pond because the anaerobic conditions are too inhibitory.

Loading rates commonly are as high as 50 lb of organic solids/1 000 ft^2 pond area d (24 kg/100 m^2) or even more, corresponding to average detentions of 1–3 d for raw (or settled) sewage. Digesting sludge accumulates on the bottom of the pond and acts to seed the whole pond continuously with anaerobes of the type suitable for rapid septic decomposition and precipitation of organic solids added (see septic treatment, p. 246).

The area of pond required for waste purification may be computed by means of the oxygen balance, allowing a suitable number of days' detention. Taking complete biological purification at 90% B.O.D. reduction, a detention period of 10 d is required (temperature about 68°F (20°C)), with a loading of about 1 lb of B.O.D.$_5$/150 yd^2 d (0·4 kg/100 m^2), equal to 400 persons/acre (1 000/ha) for full treatment of settled municipal sewage. Dilution by fresh (river) water is usually necessary, except for weak sewage. Loss of water by leakage or evaporation should be minimized, because concentration of the polluting matter is unfavourable to efficient purification. Net evaporation amounts to about 2–3 ft/year (0·6–0·9 m/year) in moderate climates, but up to 10 ft/year (3 m /year) in warm, dry climates, about 20% of this loss occurring in winter and 80% in summer. This may be compared with the evaporation from soils with plant cover of about 1½–2 ft/year (0·4–0·6 m). Cooling towers lose about 1% of the water cooled for each 12°F (7°C) of cooling.

Generally, sludge should not be discharged into ponds, and wastes such as raw municipal sewage must be pretreated by primary sedimentation. However, where pond treatment is used in two stages, the first being an anaerobic pond, such pretreatment is often omitted. Sludge deposits accumulate gradually on the bottom of the pond and, if not carried away by floods, must be removed by dredging or emptying and cleaning. Since sludge deposits affect the oxygen balance, allowance must be made accordingly (p. 65 et seq.). The waste should be discharged into the pond in such a way that full use is made of the surface area available; usually, several points of discharge are provided which may be used all at once or alternately, according to circumstances.

It is also desirable to reduce siltation to a minimum, e.g. by treatment in grit chambers prior to ponding. Bypasses should be constructed to prevent siltation by floodwaters and may be useful also for temporary bypassing of part or all of the wastes when necessary, as to relieve temporary overloading in drought periods or in any other circumstances where oxygen depletion of the pond is imminent. If the pond is not flushed by flood flows from time to time, allowance must be made for sludge accumulations.

For satisfactory operation, a sewage pond should carry fish life.[104, 107–109] This ensures among other things that the development of mosquitoes is prevented or controlled, smaller fish such as minnows and sticklebacks being especially suitable. Ducklings also feed on mosquito egg rafts. For fish life, the dissolved oxygen of the pond waters must exceed 3 p.p.m. everywhere. Local depletions of the oxygen content may be tolerated if they occur only for short periods, provided that the fish may find a suitable volume of adequately oxygenated water where they can rest. Suitable conditions may be achieved in some cases simply by discharging the oxygen-consuming wastes some distance downstream from the upper end of the pond, so that a portion of it remains always quite clean.

Sewage fishponds provide a means of biological purification yielding marketable food products. The sewage or other similar non-toxic waste must first be treated by sedimentation, designed to remove all settleable solids. The settled waste is then discharged through a suitable number of inlets into a pond stocked with carp and tench, a total diluting flow of clean natural surface water, amounting to at least five times the waste flow, being discharged separately into the pond. The depth of the pond is restricted to about 2 ft (0·6 m), and the loading is about 800 persons or equivalent per acre (2 000/ha), the volume corresponding to about two days' detention. In climates with a cold winter, the fish are taken out in the autumn, yielding about 400 lb/acre (450 kg/ha) per half year, and the ponds are left unused throughout the winter and not available for treatment of wastes again until late spring.[109] Additional means of treatment must, therefore, be provided for proper disposal of wastes during winter periods.

Computations of the oxygen balance for impounding reservoirs are illustrated by Examples 2 and 3 (pp. 71–74), and a similar technique is applicable to aerobic ponds or lagoons.

Overloading of ponds may be offset more or less by additional treatment such as:

1. Preliminary, partial or complete biological treatment.
2. Chlorination.
3. Artificial aeration of the ponds.
4. Adding nitrate.

Ponding has been applied successfully in the U.S.A., particularly in the states of California and Texas, and has proved useful for putrescible wastes from canneries, refineries, distilleries and strawboard mills, also for small towns and military camps. Its simplicity and reliability often compares more than favourably with technically more elaborate mechanized sewage treatment works of today. Preliminary treatment by plain sedimentation is usually provided, after which the detention period in the ponds is 20–30 d. The loading rate is 3–5 p.p.m. of daily B.O.D.$_5$ computed on the total pond volume for ponds about 4 ft (1·2 m) deep, the equivalent rates being about 0·8–1·3 lb B.O.D.$_5$/1 000 ft^2 (4–6 g/m^2 d), or 400–700 persons/acre (1 000–1 750/ha). If anaerobic conditions occur, the addition of sodium nitrate may be helpful, using an amount containing oxygen sufficient for about 20% of the B.O.D.$_5$. Excessive use of nitrate should be avoided. When ponds are shallower than 4 ft (1·2 m), plant growth usually spreads through the pond and must be cleaned out from time to time. Small fish are introduced for mosquito control, *Gambusia affinis* being commonly used. Bacteriological purification by ponding is usually more efficient than biological treatment processes such as activated sludge or trickling filters.

In addition to aerobic ponding processes, two-stage processes, using a primary-stage anaerobic (septic) pond and a secondary-stage oxidation pond, have been developed in the U.S.A. and Australia.[105, 110] Detention periods of only a few days have been used, yielding effluents comparable to those of activated-sludge or trickling-filter treatment plants. Loadings may be 2–5 times as high as ordinary oxidation ponds.

ANAEROBIC OR SEPTIC BIOLOGICAL TREATMENT

Under any conditions by which oxygenation is restricted, wastes such as raw or settled municipal sewage rapidly become septic. Such wastes contain large numbers of anaerobic bacteria, so that lack of oxygen allows their active growth, and this results in the septic decomposition of the organic substances present in the waste, including both dissolved and suspended material. Sewage becomes septic in sedimentation tanks, particularly if the settled sludge is not removed soon enough, the rate of septicization depending on temperature (it is very rapid in hot climates) and on the nature of the sewage itself. Anaerobic decomposition within the settled sludge sets in at once, and eventually active decomposition of the whole of the contents of the tank ensues. After several weeks (sometimes months) of operation, such a tank becomes a fully ripened septic tank and is capable of effecting a considerable decomposition of organic matter carried into the tank either in suspension or in solution. Compared with aerobic decomposition, septic-tank treatment has the great disadvantage that the effluent almost invariably contains hydrogen sulphide, is, therefore, foul-smelling and appears black or dark in colour. Even if the organic matter is largely decomposed, such effluents exert a large direct (chemical) oxygen demand and, consequently, are unsuitable in many cases for direct disposal into natural waters or onto land as this may give rise to odour nuisance. They are also not in a favourable condition for secondary treatment by aerobic processes.

Septic tanks were commonly used in earlier days for the treatment of municipal sewage, even for relatively large communities, but today they are seldom employed, except for small installations such as single dwellings or institutions (see Part D). In these cases, the disadvantages are offset by simplicity of construction and because the tanks provide a satisfactory degree of purification with a minimum of attention and without regular maintenance. The older municipal septic tank works were designed for a minimal detention period of between 12 and 24 h, while for small domestic installations a detention period basis of one day or more is customary (see p. 370). According to Viehl,[111] biological purification by septic processes is practically complete in about three weeks.

Details of the construction, operation and maintenance of septic tanks may be taken from the appropriate sections of the following chapter on sludge digestion; the rules for uniform temperature control and proper mixing and seeding also apply to septic tanks. Likewise, it is important to retain as large an amount of sludge as possible in septic tanks so as to prolong the digestion period, thereby increasing the effective decomposition and purification per unit of tank volume. This may best be achieved by introducing the waste at the bottom of the tank so as to ensure good mixing and seeding, also by recovering any sludge carried over in the effluent and using secondary settling tanks from which it is returned to the septic tank. With such improvements, septic tank treatment may be applied successfully for the treatment of many kinds of industrial wastes.[112–114] Loadings may reach 1 lb total dry solids/ft^3 tank d (15 kg/m^3 d). The carbon : nitrogen ratio in the wastes should not exceed about 20 : 1 for successful treatment. As in older municipal plants, septic tank effluents may be treated on trickling filters, although they are not in a favourable condition for efficient treatment.

A two-stage fermentation process, using successively acid fermentation (p. 260), liming and alkaline fermentation (p. 260), has been applied successfully, e.g. in the treatment of beet-sugar wastes[115] (p. 323).

BIBLIOGRAPHY

ECKENFELDER, W. W., JR. *Theory and Design of Biological Oxidation Systems for Organic Wastes* (1965) University of Texas

ECKENFELDER, W. W., JR and O'CONNOR, D. J. *Biological Waste Treatment* (1961) Pergamon, London

FAIR, G. M., GEYER, J. C. and OKUN, D. A. *Water and Wastewater Engineering*, Vol. 2, *Water Purification and Wastewater Treatment and Disposal* (1968) Wiley, New York

ISAAC, P. C. G. (Ed.). *Waste Treatment* (1960) Pergamon, Oxford

MCCABE, BRO. J. and ECKENFELDER, W. W., JR (Eds). *Biological Treatment of Sewage and Industrial Wastes*, Vol. 1, *Aerobic Oxidation* (1956) Reinhold, New York

REFERENCES

1. BOARD, R. G. 'Biological Aspects of Organic Waste Treatment', *Wat. Pollut. Control* **67** (1963) 614

2. HARKNESS, N. 'Bacteria in Sewage Treatment Processes', *J. Proc. Inst. Sew. Purif.* (1966) Pt 6, 542: see also BAARS, J. K., *J. Proc. Inst. Sew. Purif.* (1965) Pt 1, 36.
3. IMHOFF, K. 'Was bedeutet hoher Nitratgehalt im biologisch gereinigtem Abwasser?', *Gesundheitsingenieur* **64** (1941) 14, 632
4. IMHOFF, K. 'Bodenfilter', *Gesundheitsingenieur* **64** (1941) 405
5. BRUCE, A. M. and MERKENS, J. C. 'Recent Studies of High Rate Biological Filtration', *Wat. Pollut. Control* **69** (1970) 113
6. GOLDTHORPE, H. H. and NIXON, J. 'A report upon the treatment of sewage on percolating filters at Huddersfield', *J. Proc. Inst. Sew. Purif.* (1942) 101: see also REYNOLDSON, T. B., *J. Proc. Inst. Sew. Purif.* (1942) 116
7. HUNTER, A. and COCKBURN, T. 'Operation of an enclosed aerated filter at Dalmarnock Sewage Works', *J. Proc. Inst. Sew. Purif.* (1944) 12
8. BAKER, J. M. and GRAVES, Q. B. 'Recent Approaches for Trickling Filter Design', *Proc. ASCE* **94** (SA1) (1968) 65
9. Water Pollution Control Federation, *Manual of Practice No. 8* '*Sewage Treatment Plant Design*', Chapter 11, 'Trickling Filters' (1961) Washington (cf. Am. Soc. civ. Engrs *Manual of Practice No. 36* (1959))
10. FOX, G. T. J. 'The Percolating Filter', *J. Br. Granite Whinstone Fed.* **2** (1962) 29
11. EDEN, G. E., BRENDISH, K. and HARVEY, B. R. 'Measurement and Significance of retention in Percolating Filters', *Wat. Pollut. Res. Rep. No. 462* (*J. Proc. Inst. Sew. Purif.* (1964) Pt 6, 513)
12. ALLEN, L. A., TOMLINSON, T. G. and NORTON, I. L. 'Effect of treatment in percolating filters on bacterial counts', *Survr munic. Cty Engr* **103** (1944) 585
13. MONTGOMERY, J. A. 'High-capacity trickling filters', *Wat. Sewage Wks* **93** (1946) 35, 119, 198
14. DAVISS, M. R. V. 'Alternating double filtration at works of the Birmingham Tame and Rea District Drainage Board', *Sewage ind. Wastes* **23** (1951) 437
15. OSBORN, D. W. 'Operating experiences with Double Filtration in Johannesburg', *J. Proc. Inst. Sew. Purif.* (1965) Pt 3, 272
16. BRUCE, A. M., TRUESDALE, G. A. and MANN, H. T. 'The Comparative Behaviour of Replicate Pilot Scale Percolating Filters', *Instn publ. Hlth Engrs J.* **66** (1967) 151
17. SOLBE, J. F. DE L. G., WILLIAMS, N. V. and ROBERTS, H. 'The Colonization of Percolating Filters by Invertebrates and their effect on Settlement of Humus Solids', *Wat. Pollut. Control* **66** (1967) 423
18. GALLER, W. S. and GOTAAS, H. B. 'Analysis of Biological Filter Variables', *Proc. Am. Soc. civ. Engrs* **90** (SA6) (1964) 59; discussion SCHULZE, K. L. and HERNANDEZ, J. W., JR **91** (SA3) (1965) 129
19. SHEIKH, M. I. 'Retention Time in Biological Filters', *Ph.D. thesis* (1966) Univ. Newcastle on Tyne
20. EDMONDSON, J. H. and GOODRICH, S. R. 'The cyclo-nitrifying filter', *J. Proc. Inst. Sew. Purif.* (1943) 57

21. 'The Comparative Performance of Plastic and Conventional Media', *A. Rep. Wat. Pollut. Res. Bd* (1966) 84
22. CHIPPERFIELD, P. N. J. 'The Work of the Brixham Research Laboratory of Imperial Chemical Industries Ltd.: Recent Investigations of Biological Treatment Processes', *J. Proc. Inst. Sew. Purif.* (1964) Pt 2, 105
23. 'The Ecology of Invertebrates on Trickling Filters', *A. Rep. Wat. Pollut. Res. Bd* (1966) 90
24. JAMES, A. 'The Bacteriology of Trickling Filters', *J. appl. Bact.* **27** (1964) 25
25. SCHULZE, K. L. 'Experimental vertical screen trickling filter', *Sewage ind. Wastes* **29** (1957) 458
26. TRUESDALE, G. A. 'The measurement of sewage flow using radioactive tracers', *J. Proc. Inst. Sew. Purif.* (1953) Pt 2, 97
27. MOHLMAN, F. W. *et al.* 'Sewage treatment at military installations, Report of sub-committee', *Sewage Wks J.* **18** (1946) 789–1028
28. MOHLMAN, F. W. *et al.* 'Sewage treatment at military installations —summary and conclusions', *Sewage Wks J.* **20** (1948) 52
29. HAWKES, H. A. and JENKINS, S. H. 'Comparison of four grades of media in relation to purification, film accumulation and fauna of sewage percolating filters operating on alternating double filtration', *J. Proc. Inst. Sew. Purif.* (1958) Pt 2, 221
30. LUMB, C. and EASTWOOD, P. K. 'The recirculation principle in filtration of settled sewage—some notes and comments on its application', *J. Proc. Inst. Sew. Purif.* (1958) Pt 4, 380
31. IMHOFF, K. 'Wie hoch macht man biologische Tropfkörper?', *Gesundheitsingenieur* **74** (1953) 41
32. TOMLINSON, T. G. and HALL, H. 'The effect of periodicity of dosing on the efficiency of percolating filters', *J. Proc. Inst. Sew. Purif.* (1955) Pt 1, 40: see also Ref. 37 below
33. IMHOFF, K. 'Dic Bodenlüftung bei biologischen Tropfkörpern', *Gesundheitsingenieur* **63** (1940) 262
34. HALVORSON, H. O., SAVAGE, G. M. and PIRET, E. L. 'Some fundamental factors concerned in the operation of trickling filters', *Sewage Wks J.* **8** (1936) 888; also MITCHELL, N. T. and EDEN, G. E., *Wat. Waste Treat. J.* **9** (1963) 366
35. HURLEY, J. and WINDRIDGE, M. E. D. 'Enclosed aerated filters', *J. Proc. Inst. Sew. Purif.* (1938) Pt 1, 221
36. 'Some recent observations on percolating filters', *D.S.I.R. Notes Wat. Pollut. No. 14* (1961)
37. HAWKES, H. A. 'The effects of methods of sewage application on the ecology of bacteria beds', *J. Proc. Inst. Sew. Purif.* (1960) Pt. 4, 478 (cf. *Ann. appl. Biol.* **47** (1959) 339)
38. CAROLLO, J. A. 'Control of trickling-filter flies with D.D.T.', *Sewage Wks J.* **18** (1946) 208
39. TOMLINSON, T. G. and JENKINS, S. H. 'Control of flies breeding in percolating sewage filters', *J. Proc. Inst. Sew. Purif.* (1947) Pt 2, 94

40. DEN OTTER, C. J. 'A physical method for permanent control of Psychoda pests at wastewater treatment plants', *J. Wat. Pollut. Control Fed.* **38** (1966) 156
41. VELZ, C. J. 'A basic law for the performance of biological filters', *Sewage Wks J.* **20** (1948) 607
42. BACH, H. 'Phenolhaltige Abwässer und ihre Reinigungsmöglichkeit', *Z. angew. Chem.* **39** (1926) 1093
43. BACH, H. 'The Tank Filter for the purification of sewage and trade wastes', *Wat. Wks Sewer.* **84** (1937) 389, 446
44. WILFORD, J. and CONLON, T. P. 'Contact Aeration Sewage Treatment Plants in New Jersey', *Sewage ind. Wastes* **29** (1957) 845
45. HARTMANN, H. 'Entwicklung und Betrieb von Tauchtropfkörpern', *Gas- u. WassFach* **101** (1960) 281
46. PÖPEL, F. 'Aufbau, Abbauleistung und Bemessung von Tauchtropfkörpern', *Schweiz. Z. Hydrol.* **26** (1964) 394
47. SEIDEL, H. F. and BAUMANN, E. R. 'Pre-aeration and the Primary Treatment of Sewage', *J. Proc. Inst. Sew. Purif.* (1960) Pt 4, 444
48. NOGAJ, R. and HURWITZ, E. 'Making Supernatant Behave!', *Wastes Engng* **34** (1963) 230
49. ARDERN, E. and LOCKETT, W. T. 'Experiments on oxidation of sewage without the aid of filters', *J. Soc. chem. Ind., Lond.* **33** (1914) 523
50. 'Pioneers of Activated Sludge: Ardern, Lockett and Fowler', *Survr munic. Cty Engr* (17, 6, 1967) 28
51. BUSWELL, A. M. and LONG, H. L. 'Microbiology and Theory of Activated Sludge', *J. Am. Wat. Wks Ass.* **10** (1923) 309
52. DIAS, F. F. and BHAT, J. V. 'Microbial Ecology of Activated Sludge', *Appl. Microbiol.* **12** (1964) 412: see also CURDS, C. R., COCKBURN, A. and VANDYKE, J. M. 'An Experimental Study of the Role of the Ciliated Protozoa in the Activated Sludge Process', *Wat. Pollut. Control* **67** (1968) 312
53. TENCH, H. B. 'Sludge Activity and the Activated Sludge Process', *Wat. Pollut. Control* **67** (1968) 408
54. SAWYER, C. N. 'Milestones in the development of the activated sludge process', *J. Wat. Pollut. Control Fed.* **37** (1965) 151
55. LESPERANCE, T. W. 'A Generalized Approach to Activated Sludge', *WatWks & Wastes Engng (U.S.)* (1965) 44, 52; 34, 52; 37, 52; 40
56. STEWART, W. J. 'Activated Sludge Process Variations—the Complete Spectrum', *Wat. Sewage Wks* **3** (1964) 153, 246, 295
57. WASHINGTON, D. R. and HETLING, L. J. 'Volatile Sludge Accumulation in Activated-Sludge Plants', *J. Wat. Pollut. Control Fed.* **37** (1965) 499
58. SIMPSON, J. R. 'The Biological Oxidation and Synthesis of Organic Matter', *J. Proc. Inst. Sew. Purif.* (1965) Pt 2, 171
59. BENEDEK, P. 'New Developments in Activated-Sludge Process', *Proc. 2nd int. Conf. Wat. Pollut. Control, Tokyo* **2** (1964) 351; also *Gas- u. WassFach* **105** (1964) 773

60. Gould, R. H. 'Economical practices in the activated-sludge and sludge-digestion processes', *Sewage ind. Wastes* **31** (1959) 399: see also Gould, R. H., *Munic. Sanit.* **10** (1939) 815 and *Sew. ind. Wastes* **31** (1959)
61. Fischerström, N. C. H. 'Low pressure aeration of water and sewage', *Proc. Am. Soc. civ. Engrs* **86** (SA5) (1960) 21
62. Walker, P. G. W. 'Rotor Aeration of Oxidation Ditches', *Wat. Sewage Wks* **109** (1962) 238: see also Ref. 85 below
63. Baars, J. K. 'The use of oxidation ditches for treatment of sewage from small communities', *Bull. Wld Hlth Org.* (1962) 465
64. Pfeffer, J. T. and McKinney, R. E. 'Oxygen-enriched air for biological waste treatment', *Wat. Sewage Wks* **112** (1965) 381: see also Okun, D. A., *Sewage Wks J.* **21** (1949) 764; F. W. Mohlman, 792; also *Civ. Engng, Easton, Pa* **18** (1948) 32
65. Haseltine, T. R. 'A rational approach to the design of activated-sludge plants', *Biological Treatment of Sewage and Industrial Wastes*, Vol. 1 (Eds McCabe, B. J. and Eckenfelder, W. W.) (1956) 257 Reinhold, New York
66. Downing, A. L. 'Aeration in the Activated-Sludge Process', *Inst Publ. Hlth Eng. J.* **59** (1960) 80
67. Von der Emde, W. 'Die Technik der Belüftung in Belebtschlammanlagen', *Schweiz. Z. Hydrol.* **26** (1964) 338
68. Pöpel, F. 'Determination of the Size of Tanks for the Aerobic and Anaerobic Degradation of Organic Wastes', *Int. J. Air Pollut.* **7** (1963) 199
69. Townend, C. B. 'West Middlesex Main Drainage—Ten Years' Operation', *Proc. Instn civ. Engrs* **27** (1946–7) 351
70. Balfour, D. R., Manning, H. D., Cripps, T. and Drew, E. A. 'Design Construction and Operation of the Middle Lee Regional Drainage Scheme', *Proc. Instn civ. Engrs* (1960) 283
71. Stanley, W. E. 'Factors affecting the Efficiency of Activated-Sludge Plants', *Sewage Wks J.* **21** 625
72. Anderson, N. E. 'Sewage Aeration Practice in the Sanitary District of Chicago', *Proc. Am. Soc. civ. Engrs* **79** (1953) 310
73. Gould, R. H. 'Sewage Aeration Practice in New York City', *Proc. Am. Soc. civ. Engrs* **79** (1953) 307
74. Torpey, W. N. and Chasick, A. H. 'Principles of Activated-Sludge Operations', *Biological Treatment of Sewage and Industrial Wastes* (Eds McCabe, B. J. and Eckenfelder, W. W., Jr) (1955) 284, Reinhold, New York: see also *Sewage ind. Wastes* **26** (1954) 1059
75. Lewin, V. H. 'The First Year's Operation of The City of Oxford Sewage Treatment Works', *J. Proc. Inst. Sewage Purif.* (1959) Pt 1, 32
76. Möhle, H. 'The High-rate Activated-Sludge Process at the Sewage Treatment Works at Wuppertal-Buchenhofen', *J. Proc. Inst. Sew. Purif.* (1956) Pt 3, 297
77. Zablatzky, H. R., Cornish, M. S. and Adams, J. K. 'An Application of the Principles of Biological Engineering to Activated-Sludge Treatment', *Sewage ind. Wastes* **31** (1959) 1281

78. ULLRICH, A. H. 'Experiences with the Austin, Texas, Biosorption Plant', *Wat. Sewage Wks* **104** (1957) 23
79. SAWYER, C. N. 'Activated-sludge modifications', *J. Wat. Pollut. Control Fed.* **32** (1960) 232
80. KEHR, D. and VON DER EMDE, W. 'Experiments on the high-rate activated-sludge process', *J. Wat. Pollut. Control Fed.* **32** (1960) 1066
81. PHELPS, E. B. and BEVAN, J. G. 'A Laboratory study of the Guggenheim biochemical process', *Sewage Wks J.* **14** (1942) 104
82. TAPLESHAY, J. A. 'Total oxidation treatment of organic wastes', *Sewage ind. Wastes* **30** (1958) 652
83. MCCARTY, P. L. and BRODERSEN, C. F. 'Theory of extended aeration activated sludge', *J. Wat. Pollut. Control Fed.* **34** (1962) 1095
84. SIMPSON, J. R. 'Extended Sludge Aeration Activated-Sludge Systems', *J. Proc. Inst. Sew. Purif.* (1964) Pt 4, 328
85. GUIVER, K. and HARDY, J. D. 'Operational Experiences with Extended Aeration Plants', *Wat. Pollut. Control* **67** (1968) 194.
86. IMHOFF, K. 'Two-stage operation of activated-sludge plants', *Sewage ind. Wastes* **27** (1955) 431
87. CHASICK, A. H. 'Activated aeration at the Ward's Island sewage treatment works', *Sewage ind. Wastes* **26** (1954) 1059
88. KRAUS, L. S. 'Digested Sludge—an aid to the Activated-Sludge Process', *Sewage Wks J.* **18** (1946) 1099, and **20** (1948) 989
89. KRAUS, L. S. 'Use of digested sludge and digester overflow to control bulking of activated sludge', *Sewage Wks J.* **17** (1945) 1177
90. JUNG, H. 'Erfahrungen bei der chemischen Abwasserreinigung', *Gesundheitsingenieur* **69** (1948) 305
91. ZIGERLI, P. 'Das Z-Verfahren als neuer Beitrag zur Abwasserreinigung', *Gesundheitsingenieur* **60** (1937) 499
92. SAWYER, W. T. and BRADNEY, L. 'Rising of activated sludge in final settling tanks', *Sewage Wks J.* **17** (1945) 1191
93. IMHOFF, K. 'Spray channels eliminate foaming at Neersen, Germany', *Wastes Engng* **23** (1952) 414
94. MCNICHOLAS, J. 'A method for the control of foam produced in activated-sludge plants', *Wat. sanit. Engr* **2** (1952) 416
95. LEDBETTER, J. O. 'Air Pollution from Aerobic Waste Treatment', *Wat. Sewage Wks* **111** (1964) 62
96. HATTINGH, W. J. 'Activated Sludge Studies 3. Influence of Nutrition on Bulking', *Wat. Waste Treat J.* **9** (1963) 476: see also pp. 380, 424
97. PARKER, C. D. 'Food Cannery Waste Treatment by Lagoons and Ditches at Shepparton, Victoria', *Proc. 21st ind. Waste Conf., Purdue Univ.* (1966) 284; cf. *Proc. 2nd Fed. Conv., Aust. Wat. Waste Wat. Ass.*, Melbourne (1966)
98. *Proc. Symp. Waste Treat by Oxidation Ponds* (Oct. 1963), Centr. publ. Hlth Engng Res. Inst., Nagpur and Inst. Engrs, India
99. MERON, A., REBHUN, M. and SLESS, B. 'Quality Changes as a function of Detention Time in Wastewater Stabilization Ponds', *J. Wat. Pollut. Control Fed.* **37** (1965) 1657

100. O'CONNOR, D. J. and ECKENFELDER, W. W., JR. 'Treatment of organic wastes in aerated lagoons', *J. Wat. Pollut. Control Fed.* **32** (1960) 365 with discussion by ORFORD, H. E.
101. ISAAC, P. C. G. and LODGE, M. 'The use of algae for sewage treatment in oxidation ponds', *J. Proc. Inst. Sew. Purif.* (1960) Pt 4, 376
102. HERSHKOVITZ, S. Z. and FEINMESSER, A. 'Sewage reclaimed for irrigation in Israel farm oxidation ponds', *Wastes Engng* **33** (1962) 405
103. COLLOM, C. C. 'The Manukau sewerage scheme, Auckland, New Zealand', *Survr munic. Cty Engr* **123** (1964) 47
104. GAILLARD, J. R. and CRAWFORD, J. 'The Performance of Algae Ponds in Durban', *J. Proc. Inst. Sew. Purif.* (1964) Pt 3, 221; also HUNT, M. A. and WESTENBERG, H. J. W., p. 230 and WILLIAMS, R. K., p. 238
105. PARKER, C. D., JONES, H. L. and GREENE, N. C. 'Performance of Large Sewage Lagoons at Melbourne, Australia', *Sewage ind. Wastes* **31** (1959) 133
106. OSWALD, W. J., GOLUEKE, C. G., COOPER, R. C., GEE, H. K. and BRONSON, J. C. 'Water Reclamation, Algal Production and Methane Fermentation in Waste Ponds', *Proc. 1st int. Conf. Wat. Pollut. Control, Lond.* **2** (1962) 119
107. FALCK, T. 'Die Entwicklung der Abwasserreinigung in Fischteichen', *Gesundheitsingenieur* **58** (1935) 6
108. KISSKALT, K. and ILZHÖFER, H. 'Die Reinigung von Abwasser in Fischteichen', *Arch. Hyg. Bakt.* **118** i. (1937) 1
109. WADDINGTON, J. I. 'Munich Fish Ponds', *J. Proc. Inst. Sew. Purif.* (1963) Pt 3, 214; also *Chem. Zentbl.* **129** (1958) 4293; **131** (1960) 15867
110. PORGES, R. 'Industrial waste stabilization ponds in the United States', *J. Wat. Pollut. Control Fed.* **35** (1963) 456
111. VIEHL, K. 'Über die Reinigung von häuslichem Abwasser durch Ausfaulen', *Gesundheitsingenieur* **65** (1942) 391
112. GEHM, H. W. and BEHN, V. C. 'High-rate anaerobic digestion of industrial wastes', *Sewage ind. Wastes* **22** (1950) 1034
113. DAGUE, R. R., McKINNEY, R. E. and PFEFFER, J. T. 'Anaerobic Activated Sludge', *J. Wat. Pollut. Control Fed.* **38** (1966) 220
114. BUSWELL, A. M. 'Methane Fermentation', *Proc. 19th ind. Waste Conf., Purdue Univ.* **1** (1964) 508
115. NOLTE, E. 'Drei Jahrzehnte Erfahrungen mit Zuckerfabrikabwässern', *Beitr. Wass- Abwass.- u. Fisch. Chem.* **1** (1946) 24

13

SLUDGE TREATMENT

TREATMENTS of wastes involving sedimentation—such as grit removal, skimming, primary (raw waste) sedimentation and secondary or final sedimentation following trickling filtration or activated-sludge treatment—result in the separation of solid or pseudo-solid matter in the form of sludges and/or scums containing less, but still considerable, amounts of water. The origin of sludges or scums may normally be recognized by appearances, including colour and odour, which depend also on the particular state or condition of the sludge or scum.

Primary Sludges, separated from fresh municipal sewage in primary sedimentation tanks, are mostly grey to yellowish-grey. They contain small or larger pieces of faecal matter, paper, matchsticks, waste vegetable residues from kitchens and other readily recognizable particulate matter. Such sludges have characteristically sour objectionable odours, do not filter readily, and have high moisture (water) content, usually 94–97%. The solids not readily recognizable include bacteria, viruses, parasitic ova and other organisms and constituents, some of which are pathogenic or otherwise harmful. Chemically precipitated sludge has a different colour and appearance, varying according to the coagulants used, etc.; otherwise it may be similar to primary sludge but is usually less odorous.

Humus Sludges, separated from the effluent of trickling filters in humus tanks, are brown and flocculent, and in a fresh state are not objectionably odorous.

Activated Sludges, separated from the mixed liquors in aeration tanks in the final sedimentation tanks of activated-sludge plants are also flocculent in character and have a relatively very high moisture content. Their colours are brownish, and odours in a fresh state not unpleasant, but they decompose readily producing objectionable odours. (Sludges from extended aeration plants are usually not unstable.) *Digested Sewage Sludges* are more homogeneous and have a moisture content usually appreciably less than that of the fresh sludges. They are commonly jet black (due to iron sulphide) and have a not unpleasant tar-like odour. Contrary to most raw (fresh) sludges, they may be drained or filtered readily, because the particles are distinctly granular.

Sludges from Industrial Wastes may be very different in nature and appearance, varying, for example, from completely inorganic sludges, such as from inorganic chemical industries, or from metal industry wastes (metal pickling, galvanizing, or electroplating) to sludge almost wholly organic, such as from fermentation industries or from meat industries. Sludges settled from coal-mining wastes usually are granular and contain relatively little water, but the sludges from metal industry wastes usually include both crystalline and gelatinous solid matter with water content nearly as great as normal sludges from sewage treatment. Sludges from abattoir wastes are often yellow to brown in appearance due to a high content of paunch manure. Sludges from wool-scouring wastes may be intermediate regarding water content and include considerable vegetable and animal residues, partly fibrous, so that they often choke drains and sewers.

From the engineering viewpoint the moisture content of sludge is its most important characteristic. Thus a 90% sludge containing 10 lb (or 1 kg) of solids and 90% by weight water has a volume of about 10 gal (10 l), whereas an 80% sludge containing 10 lb (or 1 kg) solids and 80% water has a volume of about 5 gal (5 l) only, and similarly a 97% sludge containing 10 lb (or 1 kg) solids occupies about 33 gal (33 l). The moisture content is usually determined indirectly by evaporating a given weight of sludge on a waterbath and drying the residue to constant weight in an oven at a temperature of about 230°F (110°C). The weight of residue gives the proportion of dry solids by weight, and the difference is reckoned as moisture.

Ignition of these dry solids leaves an ash-like residue, which is called the 'fixed residue', and the loss on ignition is taken as the organic content of the dry solids, more strictly described as the volatile matter. The dry solids content (total solid residue on evaporation and drying) of fresh raw municipal sewage solids consists normally of 70% organic matter and 30% mineral matter (fixed residue). During digestion about two-thirds of the organic matter is transformed into gaseous or other volatile substances so that the proportion of organic matter in fully digested sewage sludge is only about 45% and the fixed residue 55%. For separate sewerage systems the inorganic matter is less in proportion, while the addition of stormwater to sewage, of course, increases the relative mineral content of sludges.

The digestibility of sludge and the expected yield of digester gas may also be determined by simple bottle experiments,[1] using an apparatus such as that shown in Fig. 13.1. For such experiments two parts of sludge to be tested are mixed (for 'seeding') with one part of fully digested sludge taken from a digestion tank which has been in operation at the same temperature as the experimental temperature. Such tests should preferably be duplicated, and blank tests should also be run in parallel so that the gas production due to the digested (seed) sludge may be allowed for.

Another significant property of sludge is its drainability, which may be measured by simple laboratory tests using either filter-papers or sand-beds. Measuring characteristics such as the specific

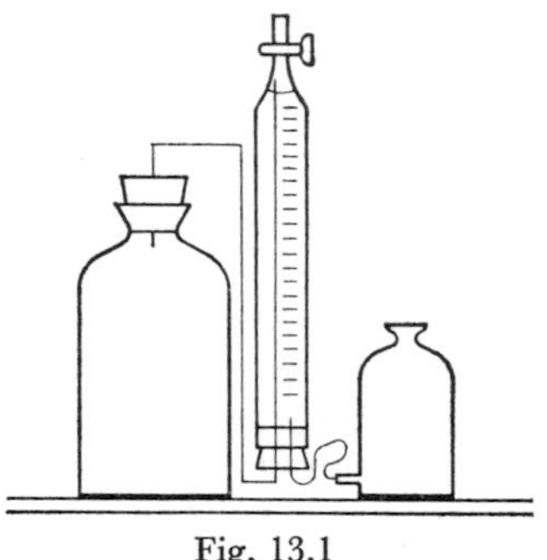

Fig. 13.1

resistance to filtration[2, 3] makes it possible to decide upon treatment before disposal; whether simply to drain before or after digestion, after elutriation, or use vacuum filters, and whether the use of chemical coagulants is desirable or more economical.

For considering the agricultural utilization of sludge it is necessary to determine the content of fertilizing compounds, including available and total nitrogen, phosphates, humus, potash, lime and magnesia. Other analyses are commonly made from time to time for cyanides, arsenic, toxic metals, grease and calorific value.

QUANTITIES OF SLUDGE IN SEWAGE TREATMENT

Fresh sewage sludge is composed of the organic and inorganic matter separated during sewage treatment by sedimentation and is roughly proportional to the number of persons contributing sewage, while the amount of sludge per person (usually measured as gal/hd d or l/hd d, the solids content as lb/hd d or g/hd d) depends on the degree of treatment, being the larger the more thoroughly the water is treated. Thus, by primary sedimentation alone, only the settleable solids are separated in the sludge, while biological treatment of settled sewage results in the formation of more settleable solids including those which in the untreated sewage were dissolved or

Table 13.1

AVERAGE QUANTITIES OF SLUDGE PRODUCED FROM MUNICIPAL SEWAGE

Type of sludge		(a) *Dry Solids* lb/ hd d	(a) *Dry Solids* g/ hd d	(b) *Dry solids* %	(c) *Moisture content* %	(d) *Quantity of sludge* gal/hd d	(d) *Quantity of sludge* l/hd d
A. *Primary sedimentation*							
Fresh sludge from hoppered tanks		0·12	55	3·5	96·5	0·33	1·5
Fresh sludge after thickening		0·12	55	5·0	95·0	0·24	1·1
Wet digested sludge		0·07	32	8·0	92·0	0·09	0·40
Dried digested sludge in bulk		0·07	32	50·0	50·0	(0·03)	(0·14)
B. *Trickling filtration*							
I Low-rate							
II High-rate							
Humus sludge	I	0·03	14	6·0	94·0	0·05	0·23
	II	0·04	18	6·0	94·0	0·07	0·30
Fresh humus sludge mixed with fresh primary sludge	I	0·15	68	5·0	95·0	0·30	1·4
(thickened)	II	0·16	73	5·0	95·0	0·32	1·5
Wet mixed sludge	I	0·09	40	8·0	92·0	0·11	0·50
after digestion	II	0·10	45	8·0	92·0	0·12	0·55
Dried mixed sludge after digestion,	I	0·09	40	50·0	50·0	(0·04)	(0·18)
in bulk	II	0·10	45	50·0	50·0	(0·04)	(0·18)
C. *Activated-sludge process*							
I Standard (low) rate							
II High-rate							
Fresh surplus (excess)	I	0·08	36	0·7	99·3	1·10	5·1
sludge as pumped	II	0·06	27	1·5	98·5	0·40	1·8
Surplus sludge after half an hour of	I	0·08	36	1·5	98·5	0·50	2·4
settling	II	0·06	27	2·0	98·0	0·30	1·4
Fresh surplus sludge mixed with fresh	I	0·20	90	4·5	95·5	0·43	2·0
primary sludge	II	0·18	82	4·5	95·5	0·40	1·8
Wet mixed sludge	I	0·12	55	7·0	93·0	0·17	0·8
after digestion	II	0·11	50	8·0	93·0	0·13	0·6
Mixed sludge after digestion and drying,	I	0·12	55	45·0	55·0	(0·06)	(0·27)
in bulk	II	0·11	50	45·0	55·0	(0·05)	(0·23)

Note—The above table is intended as a guide for the design of sludge-disposal works. When making such designs, consideration must be given especially to particular factors which may alter the above figures, such as type of sedimentation tank, operating schedules and procedure, etc., the character of the municipal sewage treated, etc., sludge handling and treatment details, and so on. (See also text, pp. 254, 255, 304).

otherwise too fine to settle out. Table 13.1 sets out the average quantities of sludge to be expected for normal domestic sewage treatment works in dry weather. Higher living standards may result in higher figures, say up to 20%. Where appreciable amounts of sludge-forming solids are discharged into the municipal sewers, e.g. with industrial wastes, special allowances must be made for the additional sludge quantities. In appropriate cases these can be allowed for by using population equivalents (see p. 309). During wet weather, with combined sewerage systems, considerable quantities of solids derived from roadways, yards, etc., also enter the sewerage system and the sludge quantities may differ accordingly. However, when sewage overflows come into operation some of the sewage solids also are carried away in the overflow. It is not possible, therefore, to give a general rule for the adjustment of quantities in such cases, but for combined systems with stormwater tanks at least 20% additional solids are likely to enter the sludges separated in the treatment works.[4] In the U.S.A., for example, the quantities for separate systems are mostly about the same as the average figures given in Table 13.1, but for combined systems the average quantities for ordinary wet seasons are 50–100% higher. Of course the characteristics of the sludges are also altered by stormwater or in some cases by groundwater infiltration. The proportion of inorganic matter is usually increased, and the raw sludges are usually denser with a lower moisture content.

PURPOSE AND METHODS OF SLUDGE TREATMENT[5,6]

Treatment is usually necessary to make sludge disposal easier and safer, especially regarding hygienic and aesthetic aspects. Economic considerations play an important part too, because the costs of sludge disposal (including treatment) are usually greater than the return and must normally be kept as low as possible. Utilization of fertilizing or other valuable properties (e.g. humus-forming capacity)are other possibilities (cf. p. 14).

Many processes have been developed for the pretreatment of sludges before final disposal, but anaerobic digestion is most commonly applied. As indicated above, digestion alters the fundamental character of the organic matter present in the raw sludge. Treatment by digestion accordingly simplifies disposal problems because of:

1. Substantial overall reduction in the volume of sludge finally to be disposed of

2. Improved drainability
3. A marked degree of 'stabilization' of the organic matter so that it does not continue to decompose rapidly
4. Greatly reduced odour.

Although digestion does not eliminate all dangers of harmful organisms, an appreciable measure of control is achieved, partly by actual destruction during the process, also indirectly by allowing a cleaner separation of the solids and the associated water. Digestion followed by draining and drying, although still not eliminating all possible dangers, nevertheless provides a high degree of protection. The processes of sludge digestion, therefore, are dealt with later in detail, referring also to the aspects of conservation and waste recovery by using the solids and the gases produced during digestion.

Sludge Thickening[7–9]

It is sometimes desirable, before proceeding with digestion or other processes, to provide for a preliminary concentration of the sludge solids. This may be effected in *sludge thickening tanks*, with or without chemicals, using either continuous flow, or fill and draw systems. Efficiency may be improved by up to 20%, particularly with gassy sludges, by slowly revolving stirrers, commonly of the type known as a picket fence.[8] Chlorination of the sludge before thickening has been used to delay decomposition; suitable dosage rates with activated sludges are 40–120 mg/l of chlorine.

Generally thickening tanks may not be compared directly with settling tanks for raw wastes (which are discussed in Chapter 10), because in primary tanks, for the most part, the concentration of raw waste solids is low enough for each particle to settle freely and independently of the other particles, the different particles falling through the tank at different rates according to individual shapes, dimensions and densities. In thickening sludges, however, particle concentrations are already high enough to hinder the independent motion of individual particles (both downwards and upwards).

Continuous-flow tanks for sludge consolidation (thickening) are usually deep circular tanks with central feed and overflow at the circumference. Usually either a detention period of 6 h or more is provided, or the tank is designed for a surface loading rate of not more than about 400 gal/ft^2 (20 000 l/m^2) tank area per 24 h. The use of chemicals may allow twice these loadings. Operation on the fill and draw system requires two or more tanks used alternately, each with a capacity of at least half the daily sludge volume. Thickening tanks are equipped with suitable inlet and supernatant draw off or overflow arrangements with provision for skimming as well. Withdrawal of the thickened sludge is usually from a central collecting hopper, in continuous systems by pumping at suitable adjustable rates.

Thickening of primary or mixed sludges in such consolidation tanks may reduce the sludge volume by 40–50%, or even more, compared with the total volume of the sludges as drawn from the sedimentation tanks.

SLUDGE DIGESTION

Acid Fermentation[10]

If fresh faecal matter is allowed to decompose under water, but without addition of sufficient seeding material, a process of digestion proceeds almost immediately with the production of hydrogen and carbon dioxide along with other gaseous and volatile compounds, some of which are strong smelling and responsible for very objectionable odours. Within one day normal sewage sludge will produce acids enough to reduce the pH from 7 to 6, or even 5, and digestion will continue steadily in this way until all the acid-producing compounds have been decomposed. The pH of the sludge may fall even to about pH 4, depending on the concentrations of the various organic and inorganic compounds originally present. This fermentation process is commonly referred to as *acid fermentation.* It is of little value and is undesirable from the engineering viewpoint because the reduction in sludge volume is only slight, the sludge does not dewater readily, and moreover, the sludge produced stinks strongly and offensively. Such sludge has a distinctly yellowish or yellowish-grey colour and is relatively viscous, and tends to retain gas bubbles and form scum accordingly.

Methane or Alkaline Fermentation[11–15]

This is the type of process normally used for sludge digestion and is completely different from acid fermentation. It may be demonstrated simply as shown on p. 255 in the test for gas production, and yields methane and carbon dioxide as the principal gaseous products of digestion. The pH value of sludges during digestion is normally above pH 7, up to about pH 7·6, and the sludge has only a slight but characteristic odour not commonly described as objectionable. This process is normally called *alkaline fermentation*, but it must be considered as proceeding in two stages which, however, should operate concurrently.[14–15] The first stage yields organic acids which are then decomposed further (second stage) with the production of carbon dioxide and methane; the second stage is essential to maintain alkaline conditions. If, because of the introduction of too high a proportion of fresh sludge, or following a sudden fall in temperature, the second stage is overwhelmed, then acids accumulate and the bacteria essential to the second stage are inhibited so that acid fermentation then ensues. If the content of volatile acids, calculated as acetic acid equivalent, exceeds 2 000 mg/l there is a danger that acid fementation will develop.[16] Reducing the loading or increasing the temperature (slowly) can usually overcome such a tendency.

The elements essential for cell growth and for enzyme production and functioning, such as nitrogen and phosphorus, are usually present in domestic sewage sludges. For other sludges, for example, textile wastes consisting mainly of cellulosic or other nitrogen-free organic matter, etc., which may not contain all the essential elements in sufficient proportion, suitable additions of chemicals, or admixture with other wastes, may be necessary.

The result of alkaline digestion is a considerable reduction in sludge volume, yielding finally a uniformly dark-coloured dense sludge which may be dewatered readily and which does not stink offensively, and the process, therefore, is widely used for waste engineering practice. It is applicable generally to any relatively concentrated fermentable organic wastes (see p. 285) in much the same way as municipal sewage sludge; however, it is not so suitable for relatively dilute wastes. This chapter is devoted, therefore, to a full description of the processes of alkaline fermentation plants and the more important details of their design, maintenance and operation. Acid fermentation processes are not discussed because their usefulness is limited to special industrial applications and is seldom of value in waste treatment.

The first important question is that of *ripening*. As already shown, if a digester is charged only with fresh sludge the process of acid fermentation follows automatically. Eventually the rate of acid fermentation becomes less than the rate of alkalinization, and from then on conditions of fermentation approach more and more to those of normal alkaline fermentation. At 60°F (15°C) a period of about six months is required for this ripening of fresh municipal sewage sludge in ordinary large-scale digesters, the period being longer or shorter according to temperatures (see Fig. 13.2). Sufficient amounts of septic sludge, from septic tanks, cesspits or even dammed sewers or unusually long rising mains, can be expected to reduce the period required for ripening. It is customary to make special provision for ripening digesters rapidly when first brought into use. Usually a new digester is ripened artificially by first introducing seeding sludge from a digester which is already operating normally, using an amount large enough to ensure alkaline fermentation of the raw sludge input. If insufficient seeding sludge is available the rate of addition of fresh sludge must be limited until a sufficient quantity of properly digesting sludge is present in the digester. Seed sludge is best at the stage where active fermentation is almost completed, subsequent storage impairing its value for seeding. The addition of other suitable organic matter, such as activated carbon, can be of assistance; rotting leaves have been used successfully as seeding material. During winter periods heating the contents of the digester also hastens ripening, and this is done most

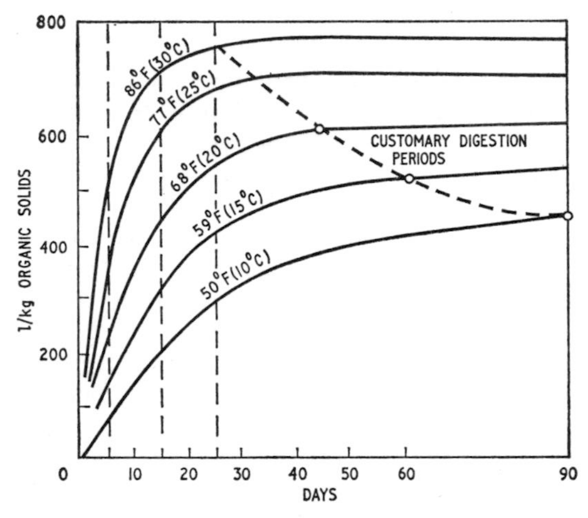

Fig. 13.2. *Gas production, per kg of dry volatile solids, from seeded raw sludge during digestion at different temperatures (after Fair and Moore)*

readily by filling the digester with water of temperature about 82°F (28°C), say 77–86°F (25–30°C). The efficiency of alkaline digestion is illustrated by the rapid destruction of most forms of organic matter which are so decomposed within one or two weeks at ordinary temperatures as to be quite unrecognizable (although keratinous material such as hair remains practically unaltered) while the volume occupied also is markedly reduced (p. 264). Most plant seeds are destroyed completely, although some resist digestion for long periods, tomato seeds, pumpkin seeds and passion-fruit seeds being among the most resistant. Many pathogenic organisms (including parasitic ova) are destroyed also (see p. 115), tubercle bacteria being among the most resistant.

The *temperature* of digesting sludge is of the greatest importance because it has a profound effect on all the details of the digestion processes. This is demonstrated by Fig. 13.2, showing the different[17] volumes of gas produced from fresh sludge organic solids digesting at different temperatures.[14] Not only is decomposition accelerated at higher temperatures (up to 90°F (32°C)) but the total weight of gas produced is very much greater. This is shown clearly by the figures for total gas yields, given in Table 13.2, which might be expected from the digestion of sewage solids only if the digestion period is long enough (see also Table 13.3).

Even small variations of only a few degrees may result in a significant decrease in the amount of gas produced, through disturbance

Table 13.2

GAS YIELD BY ALKALINE FERMENTATION

Temperature of digestion	50°F (10°C)	59°F (15°C)	68°F (20°C)	77°F (25°C)	86°F (30°C)
Total gas produced, measured at s.t.p., ft^3/lb organic solids (m^3/kg)	7·2 (0·45)	8·5 (0·53)	9·8 (0·61)	11·4 (0·73)	12·2 (0·78)

Digestion maintained constantly at the given temperature

of the digestion processes, especially at lower temperatures, 50–60°F (10–15°C). It is important, therefore, that digestion tank temperatures be kept as constant as possible. The figures of Table 13.2 apply only to digestion tanks maintained constantly at the temperature given.

Temperatures of 86–95°F (30–35°C) are normally regarded as optimum—measured by gas production and rate of digestion—the process depending on the activity of mesophilic bacteria (that is, bacterial types living at normal temperatures, usually about 50–95°F (10–35°C)), and it is still common practice to use temperatures within the mesophilic range. However, it has been found that thermophilic bacteria active at temperatures above 95°F (35°C) are also useful. The optimum temperature for these bacterial types is about 130°F (54°C) (range 125–140°F (52–60°C)) and some modern digestion plants have been designed accordingly. Rates of digestion are shown graphically in Fig. 13.3, Curves I and II being applicable respectively to mesophilic and thermophilic digestion.

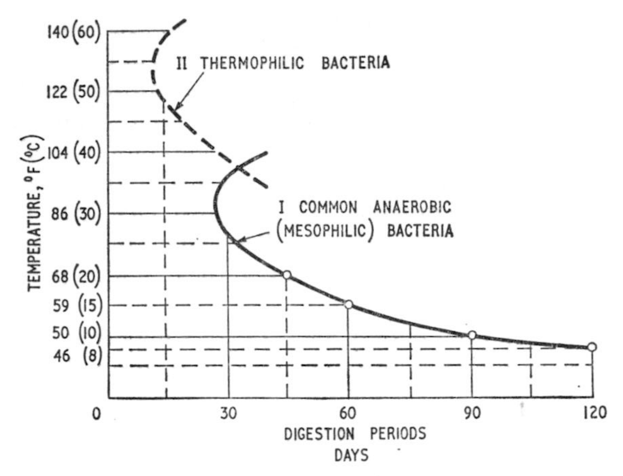

Fig. 13.3. *Customary digestion periods for ripe digestion tanks*

The digestion periods usually provided in municipal sewage treatment plants, depending of course on the temperature of operation, are shown in Table 13.3.

The rate of digestion (that is, anaerobic decomposition) is slow compared with aerobic decomposition processes as measured by the rate of B.O.D. reduction. The relative rate of digestion of 8%/d at 68°F (20°C) (Fig. 13.2) is very much less than the corresponding rate of aerobic decomposition (more than 20%, see pp. 35, 63 et seq.).

Table 13.3

NORMAL PERIODS FOR SLUDGE DIGESTION PLANTS (MESOPHILIC RANGE)

Temperature of digestion	50°F (10°C)	59°F (15°C)	68°F (20°C)	79°F (26°C)	90°F (32°C)
Digestion period (d)	90	60	45	32	25

The above figures are applicable to ordinary types of digestion tanks of municipal sewage treatment works containing solids and liquor in the usual proportions (water content averaging between about 92 and 98%).

The marked decrease in sludge volume as a result of digestion is of great practical importance for the final disposal of the sludge, by whatever means. This decrease is due partly to gasification and to a minor extent liquefaction, of the solids, but mainly to the settlement of the sludge to a lower moisture content. Taking, for example, the figures given in Table 13.1 (p. 257), raw and digested sludges may contain on an average 0·12 and 0·07 lb/hd d respectively (55 and 32 g/hd d) of dry solids, moisture content being 96·5% and 92%. Thus the reduction in total solids by digestion is about 40%; but the specific volume per lb of raw solids also is reduced from 2·9 to 1·3 gal of water, say to one half the original specific volume. The overall reduction in volume from 2·9 gal of water per lb of raw solids (29 l/kg) to 0·7 gal of water per 3/5 lb of the digested residual equivalent solids, is about three-quarters. In any particular works the moisture and organic matter content of the sludges may differ considerably from the average figures given in Table 13.1. Usually, the higher the proportion of organic matter to inorganic the greater the moisture content; also a higher moisture content of the raw sludge usually produces a higher moisture content in the digested sludge. Thus where excess activated sludge is digested along with raw sludge, the final digested sludge has a much higher moisture content. Despite its relatively low moisture content digested sludge flows more easily and is more easily pumped than raw sludge, and the latter must be relatively dilute to facilitate movement through pump systems or gravity lines. Furthermore, whereas fresh sludge does not drain readily and stinks offensively while drying, well-digested sludge is readily drained and dries without offensive odour.

DIGESTION TANKS

Three classes of digestion tank are in common use; septic tanks, two-storey tanks such as Imhoff tanks and separate sludge-digestion tanks.

Septic Tanks

These combine the processes of sedimentation and sludge digestion in the same compartment and are, therefore, to be regarded as waste-treatment tanks and not merely as sludge-digestion tanks. They are dealt with in more detail in pp. 370–372 and pp. 382, 383. Since the process of sludge digestion by alkaline fermentation is essential for proper sewage treatment in septic tanks, the technical requirements for such sludge digestion must be observed in their design and operation.

Two-storey Sedimentation-digestion Tanks[18–20]

This type of tank is illustrated in Fig. 13.4 and shows a sedimentation compartment located directly above a much larger sludge digestion compartment. Sludge settled out in the upper section of the tank falls by gravity through an elongated slot directly into the digesting sludge below, the slots being arranged so that they are protected against direct access of gas bubbles rising from the digesting sludge (Fig. 13.4). This protection is provided by special construction, either by extending the floor of the sedimentation compartments from one of the sides down below the slots or by placing a special baffle under the full length of the slots, thereby deflecting any rising gas or sludge solids. The slots are usually about 9 in (250 mm) wide, and to ensure that deposited sludge slides freely into the sludge digestion compartment the bottom partitions of the sedimentation compartments (suspended floors) must be set with a slope of not less than 1·2 vertical to 1 horizontal, except that where two sloping floors intersect, the minimum satisfactory grade then is

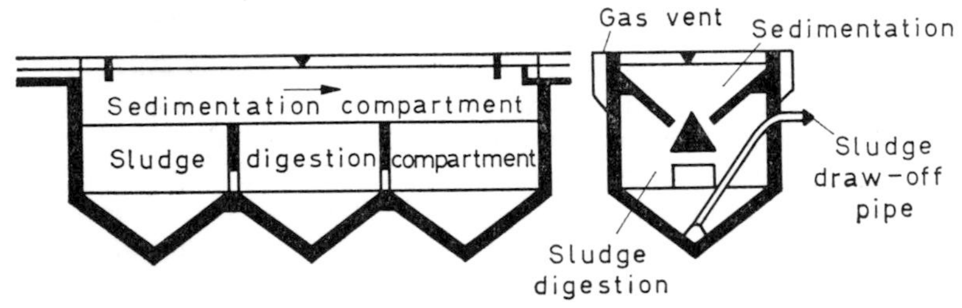

Fig. 13.4. *Two-storey sedimentation and sludge-digestion tank (Imhoff tank)*

2 vertical to 1 horizontal. If the walls and bottom partitions of the sedimentation compartment are squeegeed regularly and the slots remain free of sludge accumulations, deflection of the digester gases and sludge flocs across the slots ensures the absence of sludge from the sedimentation compartment, out of which the solids gravitate continuously, and the waste flowing through remains fresh. Simultaneously the excess water, separated as digestion proceeds, passes automatically upwards through the slots with the exchange water which is displaced from the sludge compartment in proportion to the volume of raw sludge falling through the slots. When digested sludge is withdrawn from the bottom of the tank, a corresponding volume of raw waste is drawn into the sludge compartment through the slots as exchange water. Normally the digested sludge withdrawn has a low moisture content (see Table 13.1, p. 257). This type of tank construction has the disadvantage that it is not feasible to provide heating of the sludge compartment because of the direct interconnection with the flowing waste (see also p. 270). Such tanks normally are elongated, and more sludge settles near the inlet end

of each sedimentation compartment than settles near the outlet end. The loading of the sludge compartment may be made uniform by reversing the inlet and outlet of the sedimentation compartment, for example, from week to week, provided that the tanks are not unduly elongated; alternatively, the sludge compartment may be enlarged at the inlet end of the tank.

The construction necessary to support the (upper) sedimentation compartment walls is simplified by subdividing the (lower) sludge compartment by structural partitions. Interchange of sludge within the sludge compartment is obtained by providing large openings towards the bottom of these walls, while the short circulation of incoming sewage through the sludge compartment is obviated. Where two or more parallel but separate sedimentation compartments are provided it is essential that their respective sludge compartments also be separated completely by longitudinal partitions, otherwise unequal distribution of the influents results in some continuous direct flow through the digestion to the sedimentation compartments and stinking septic water and solids than interfere with normal sedimentation.

Each section of the sludge compartment should be provided with a sludge draw-off pipe equipped with a valve, and each pipe should discharge freely into an inspection chamber so that the discharged sludge may be examined visually. The size of pipe usually is at least 8 in (200 mm) in diameter, and provision for flushing with fresh water is useful, both at the inlet (inside the tank) and along the pipe from outside. For free flow of sludge in pipelines the hydraulic gradient should be 1 in 8, and for open channels 1 in 40; if the sludge is abnormally thick or viscous then higher gradients may be necessary. Air-lift pumping is sometimes used and is successful even with sandy sludge.

Scum accumulating in the digestion compartment requires special attention. It is produced by rising sludge particles carried up to the surface by gas bubbles, some of which particles do not settle again, even after release of the gas, and form a floating scum. A number of materials have a characteristic tendency to form scum, e.g. fibrous particles such as hair, kernels and such-like matter. However, where ample digestion tank space is provided and kept at a high enough temperature, for example, by warm sewage, little scum formation is to be expected. If the waste flow contains saline compounds, iron salts, carbonates or coal-dust, or other matter which so modifies the character of the sludge as to reduce its adhesion or make it denser, the scum is also reduced. Sludges such as those from domestic sewage, and industrial wastes containing much wool grease, hairs or fibres, tend to form greater scum layers, especially when the available capacity is relatively inadequate or temperatures are low. Scum is kept soft by the escape of gas bubbles from the digesting sludge below, provided that sufficient gas is generated to disturb all of the scum surface in this way. Thus, whereas the surface available for gas release may be sufficient to avoid too much swelling of the scum, scum tends to build up also if the area is too great, especially in tanks which are freely ventilated. Excessive scum layers should be broken down or the scum removed from the tank because scum occupies available capacity in a digestion tank and digests more slowly than sludge or not at all. Scum must be kept wet otherwise it dries out and cakes. In two-storey tanks it is also a danger because if accummulation continues scum will extend eventually down to the slots and enter the sedimentation compartment and proper operation of the tank will become impossible. Scum can be broken down by water jets (see Fig. 13.5), by jets of well-digested sludge or supernatant liquor (excess water separated from the digesting sludge) or by mechanical stirring apparatus similar to that used for separate digestion tanks. As it is useful occasionally to remove all or part of the scum, means should be provided to facilitate removal, say by the provision of draw-off pipes or gates (Fig. 13.5); it may be advisable also to provide for secondary digestion of such scum before disposal to avoid complaint of odour or other nuisance. In small plants it is usually practicable to remove scum by hand. If the surface area available for gas release is inadequate the scum tends to be blown up and out of the containing walls.

In large plants it is usual practice to install gas collector domes above the digestion compartments to collect and meter the digester gas. Construction costs may be small because the dividing walls between sedimentation and digestion tanks may be used for gas collection. Since gas production provides a direct measure of the rate of digestion, the metering of the gas and a daily record of gas production provide data which make operation of the digestion process much simpler. The water seal required for gas domes is 12 in (300 mm). Scum may be kept out of gas domes by means of porous plates of concrete or timber (Fig. 13.5).

Apart from the area required for release of digester gas (gas vent area) the whole surface of the Imhoff tank is available for the sedimentation compartment, and in principle the sludge slots should be located as high up as feasible, thereby utilizing the depth of the tank as far as may be for digestion and the surface for sedimentation. In small plants for which attention and control are insufficient it is advisable to provide more space for scum. Special arrangements are shown in Fig. 13.6.

The most important duty in operating Imhoff tanks, assuming good sedimentation and digestion, is to keep the slots between sedimentation and digestion compartments free from sludge accumulation. Scum must not be allowed to enter or block the slots nor must the digesting sludge accumulate until the sludge layer rises up to the slots; it must be withdrawn to prevent this, even if digestion is not complete enough. Usually the appearance of bubbles in the sedimentation compartment along a line above the slot is an indication of this and, accordingly, sludge must be withdrawn from the tank. The sludge and scum layers may be investigated readily by a pump and long section hose, lowering or raising the hose while

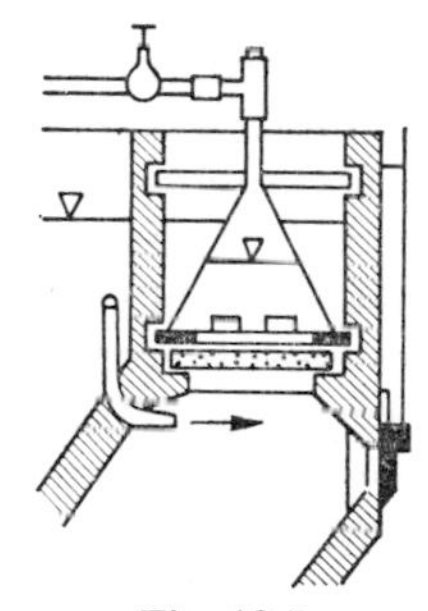

Fig. 13.5

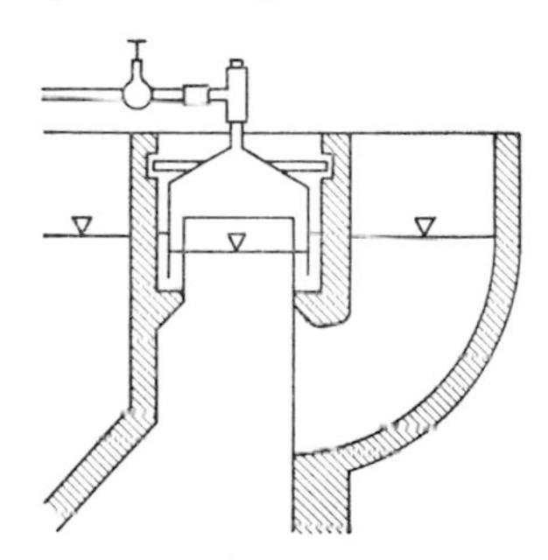

Fig. 13.6

pumping continuously until the discharge from the pump shows that the inlet of the section hose has entered sludge or scum layers. It is also very important to note that too much sludge should not be drawn from the bottom of the sludge compartment because enough must always remain there to ensure automatic seeding of the incoming fresh sludge and to maintain a well-balanced alkaline fermentation; otherwise, the digestion tank must be ripened again (p. 261).

Foaming of Imhoff tanks can occur when too many 'soapy' or surface-active substances accumulate in the supernatant liquor of the digester, with the result that the escaping gas bubbles and scum solids form a persistent foam; this is quite different from ordinary scum. Foaming commonly occurs during ripening and so can also result from overloading. The concentration of foaming agents may be reduced by the introduction of fresh water or other diluent, and foaming may thus be overcome. Judicious removal of some bottom sludge may effect the same result by drawing fresh waste into the digestion compartment and diluting the foaming agents.

The depth required in two-storey Imhoff tanks may be reduced by making the partitions between the sedimentation and the sludge digestion compartments practically horizontal instead of steeply sloping. This necessitates the use of sludge-collecting mechanisms (similar to the mechanisms, for example, of Fig. 10.13, p. 165) to move the settled sludge to slots through which it falls into the digester below.

Separate Digestion Tanks[21, 22]

Because of the relatively great depth required for Imhoff tanks it is often more economical to provide separate tanks for sludge digestion with no direct connection with the wastes flowing through the sedimentation tanks. The principal advantage of separate digestion tanks, however, is that control of the digestion processes can be better maintained and that the contents can be heated to allow digestion at optimum temperatures. Generally these advantages have proved so important as to outweigh the disadvantages and so separate digestion tanks are preferred; there are, however, disadvantages which have to be taken into account. First, the continuous automatic separation of sludge from sedimentation must be replaced by arrangements requiring manual attention or experienced supervision of mechanical apparatus; often this cannot be arranged for continuous operation. Second, the excess water (supernatant liquor), which in Imhoff tanks is automatically separated from the sludge and carried out, must, in separate tanks, be separated by special means requiring an experienced operator. Daily attention is required for this. Third, because the temperature in the separate tanks is not controlled by the temperature of the raw waste in the sedimentation tanks (which is relatively constant) separate digestion tanks, unless heated, are subject to much wider variation in temperature from season to season, and the much lower temperatures prevailing in winter considerably reduce the efficiency. In addition, the seeding of the fresh sludge in Imhoff tanks is also automatic, whereas in separate tanks special provision for seeding is necessary, and requires personal attention for the maintenance and operation of this equipment.

These disadvantages of separate tanks are overcome by providing the appropriate digestion tank capacity, by good design to facilitate maintenance and operation, by careful operation, and commonly also by arrangements for heating the sludge to control temperatures.

Where ample space is available it may be sufficient to construct open digestion tanks, using earthen or even concrete basins for sludge digestion. Such basins should be as deep as possible, commonly 10–20 ft (3–6 m), with the working level raised sufficiently to allow the discharge of digested sludge on to drying beds or into pumps by gravity; sludge draw-off pipes equipped with valves are, therefore, necessary. It is essential to provide for seeding of the fresh sludge from sedimentation basins by mixing with ripe (digested) sludge—at least 50% by volume—before discharge into the digestion basin to prevent acid fermentation. This mixture of fresh and seed sludge should be introduced at several places in turn in order to load the basin as uniformly as possible. Effluent pipes from different levels provide for removal of supernatant liquor and sludge separately as required. In open tanks or basins, the scum layer should not be disturbed because it is important for odour control. Often raw sludge, if not thoroughly mixed and seeded with digesting sludge, rises through the scum layer to the surface and fly nuisance then may occur, particularly in warmer climates. Usually the accumulation of silt and consolidated sludge on the bottom necessitates interruption of the process after several years, the basin being cleaned out for a fresh start after removal of the deposits which would otherwise reduce effective capacity and endanger operation by clogging sludge withdrawal pipes. If such basins are constructed as emergency or temporary digesters they can be used later as a secondary stage, using fully equipped covered concrete tanks for a primary stage.[23]

Fig. 13.7 shows a typical layout of a reinforced concrete tank of the usual covered digestion type, into which fresh sludge mixed with seeding sludge—using a proportion of one or two to one by volume—is discharged near the top, and from which the digested sludge is withdrawn at the bottom and supernatant liquor from above.

Such digestion tanks are covered in order to reduce heat losses from the surface and so conserve heat which keeps the scum layer more fluid. Three types of cover are in use:

1. The fixed submerged cover, such as that shown for Imhoff tanks in Fig. 13.5, has the advantage that the working range of the water-levels in the sludge tank is always above the cover. Gas collection is then safer because air cannot gain entry under the cover.
2. The fixed cover placed above the working level usually is flat or better cone-shaped with a slope of about 30–45° (Fig. 13.7). It is protected against the ingress of air when connected with a separate gas-holder which maintains a positive pressure of gas inside the tank above sludge- and water-levels so that air does not enter when sludge or supernatant liquor is withdrawn.

As the operator must know the levels of sludge and liquor inside the tank, indicators must be provided.

3. The floating cover (Fig. 13.8) can also be made safe against the entry of air under the cover. Enlarged covers of this type can serve also as gas-holders, in which case additional gas storage may not be necessary (p. 280).

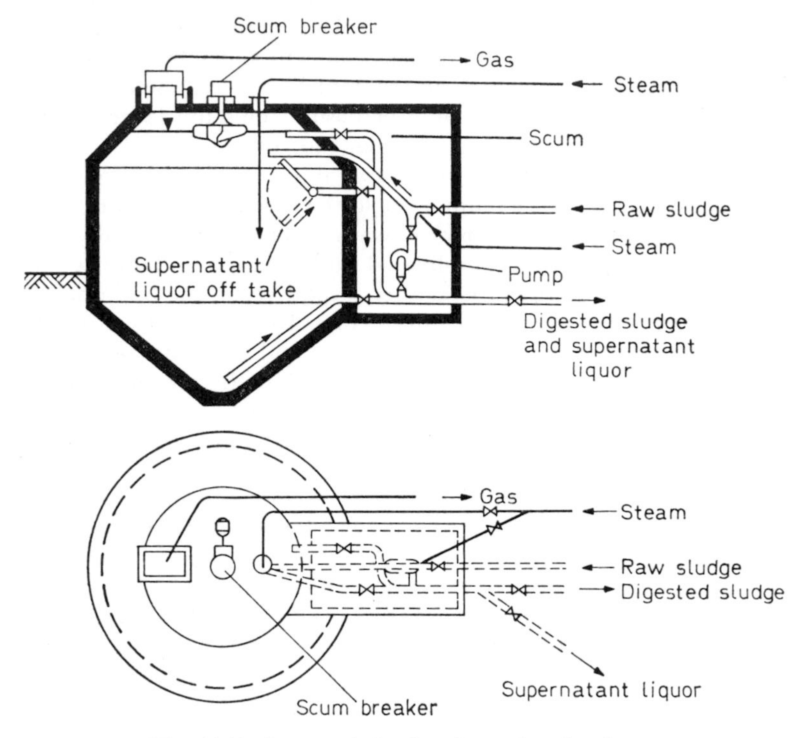

Fig. 13.7. *Separate sludge digestion tank and equipment*

Since the gas produced by digestion consists principally of methane which has a high calorific value it is highly explosive when mixed with air. To prevent accidents care must be taken to exclude air from gas-holders, pipelines, etc., otherwise explosions and expensive structural damage may occur.

The bottom of such tanks is usually made hoppered (Fig. 13.7). If the tank bottom is flat or only gently sloping it is necessary to provide mechanical scrapers to move the digested sludge to the draw-off pipe.

In covered and heated digestion tanks excessive scum formation must be prevented not only because it occupies space which can be better used, but also because it does not come within reach of the heating apparatus since it does not move freely. It is, therefore, necessary to break up the scum continuously or at least once daily. The following techniques are practicable and in use:

1. Spray jets directed into or onto the scum; water is sometimes used, but it is better to use supernatant liquor which carries bacteria and enzymes appropriate to alkaline fermentation and which, moreover, is already at digestion tank temperature.
2. Spraying the seeded fresh sludge over the top of the tank's contents, dispersing the fresh sludge widely and ensuring

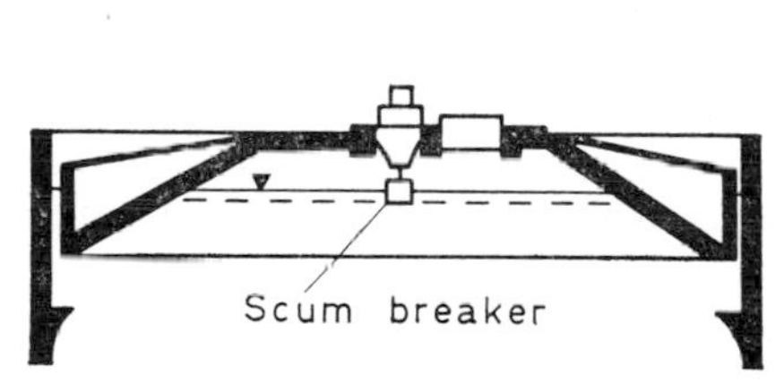

Fig. 13.8

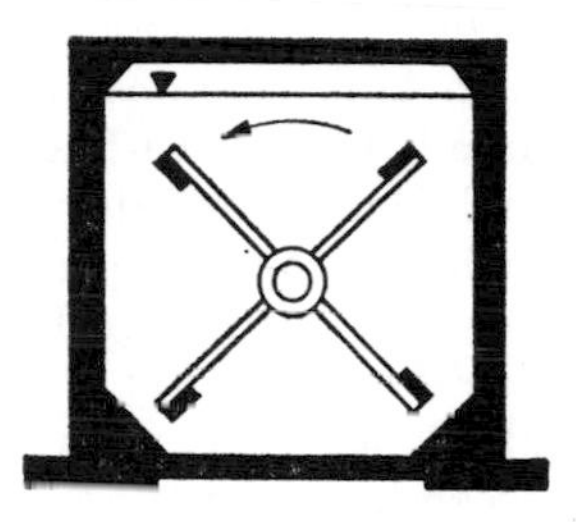

Fig. 13.9

thorough wetting of the scum layer. Scum formation is, thus, either greatly reduced or prevented.
3. Circulating the contents of the tank using an external pump, which discharges sludge from below onto or into the scum to achieve mechanical breakdown of the scum layer.
4. Mechanical means installed inside the tank; such means include either a centrifugal type of mechanism which causes circulation and wetting of the scum layer or stirring mechanisms as shown in Figs 13.8 and 13.9.
5. Blowing sludge gas into the digesting sludge.[24, 25]

In spite of these provisions, floating matter gradually accumulates at the top of the tanks, especially where material such as garbage or other mixed residues is being digested (p. 285) in order to increase gas production or to manufacture special fertilizers. As for two-storey tanks (p. 265) draw-off pipes or gates (see Fig. 13.6) are, therefore, necessary and water jets should be installed at these openings. It should also be made possible to remove floating matter manually by providing suitable openings in the cover.

TWO-STAGE DIGESTION

The whole process of sludge digestion can be carried through as a single-stage process, using only a single tank, or two or more tanks operating in parallel. On the other hand, two-stage (and even multiple-stage) digestion has also been developed.

The sludge compartment of an Imhoff tank might be used as a first-stage tank from which partially digested sludge may be transferred to a second-stage separate digestion tank for further digestion. This arrangement combines the simplicity of operation inherent in the Imhoff tank with the advantages of separate digestion-tank systems, the dimensions of the Imhoff tank thus being reduced. The partly digested sludge which is transferred to the separate second-stage tank contains much less moisture than raw sludge and is already seeded; with a second-stage tank it may be heated and the final digestion of the sludge hastened and improved accordingly. Such a two-stage procedure is specially useful for older plants, including overloaded Imhoff tanks, which require additional sludge digestion capacity; this is best achieved by the construction of additional tank capacity as separate digesters for second-stage (secondary) digestion. In any case, designs for new Imhoff tanks should provide separate capacity for separate digestion of scum or floating matter drawn off from either sedimentation or digestion compartments.

For works using separate digestion tanks only, two-stage digestion has come more and more into favour. In plants using heated tanks, particularly, the temperature of digestion commonly results in marked turbulence in the digesting sludge, with consequent difficulty in separating clear supernatant liquor with a low content of suspended sludge solids. If supernatant liquor cannot be separated properly the effective detention period of the sludge in the digestion tank is too short for proper digestion unless excessive additional capacity is provided. Where biological treatment follows sedimentation it is customary to return the supernatant liquor to the raw waste influent channel, but if the liquor contains significant amounts of digestion tank solids the treatment processes are adversely affected and the pollution passing through to the receiving waters is correspondingly increased; where the digester gas is used some of the useful digester-gas production is also diminished accordingly. In such cases the arrangement of two-stage digestion may be helpful. The first stage of digestion, the primary tank, is equipped for gas collection, heating and with scum-breaker mechanisms. With a digestion tank temperature of 86°F (30°C), for example, gas production amounts to two-thirds in 5 d and nine-tenths in 14 d (Fig. 13.2, p. 262). Normally no supernatant liquor free of sludge solids can be removed from this primary tank; separation and removal of supernatant liquor take place in the secondary tank into which sludge is discharged after primary tank digestion and is kept free of turbulence and without any agitation. Heating apparatus is usually not required for second-stage digestion since little advantage is gained thereby, nor is scum-breaking necessary; also, since gas production is relatively small, the provision of gas-collecting covers is not usually justified. The secondary tank may, therefore, be of simple construction, even an open tank or earthen basin being adequate if provided with proper draw-off pipes (p. 271). Apart from further digestion, and the separation of supernatant liquor, secondary tanks also serve as balancing tanks for the sludge-drying works; they may be enlarged to provide for storage of digested sludge, e.g. during winter when open sludge-drying beds become less effective. Such secondary tanks do not depend upon seeding and the working level need not be held constant.

Two-stage digestion plants are also more flexible in relation to sedimentation plants. In operating sedimentation tanks equipped with mechanical sludge-collecting apparatus, Schulz found a digestion plant was used to better advantage by discharging the denser sludge, settled in the hoppered inlet section of the sedimentation tank, into the primary digestion tank, while the much less dense sludge scraped from the bottom of the latter section of the sedimentation tank was collected separately and discharged to the secondary digestion tank. In this way the gas yield per cubic foot of primary tank may be increased because a denser sludge is digested therein. Genter and Kennedy have suggested that the overall efficiency of two-stage digestion can be improved by elutriation (washing) of the sludge withdrawn from the primary tank so as to reduce the moisture content before transfer to the second stage. In the same way, digested sludge dries more readily after elutriation.[26]

The supernatant liquor separated from the sludge during digestion processes is mostly returned to the raw waste influent usually just ahead of the sedimentation tanks. As stated above, if the liquor carries excessive amounts of sludge solids due, for example, to the great activity of the digestion tank, the raw waste may be so contaminated and its B.O.D. so much increased that the operation of the whole treatment works may be seriously disturbed. In such cases it is advisable to pretreat the liquor in a small settling tank (returning the sludge to the digestion tank) which may be part of the digestion tank structure. Sometimes it is pretreated chemically or filtered on special sand filters and discharged separately. As suggested by Kraus, the sludge-contaminated supernatant liquor may be first aerated in a separate aeration tank and so converted to an activated sludge which may then be returned to the activated-sludge plant. This often improves the carbon : nitrogen balance.

The expected amounts of supernatant liquor may be calculated from the sludge quantities tabulation (see Table 13.1, p. 257). In municipal sewage plants the daily volume is normally between 0·15 and 0·3 gal (0·7 and 1·4 l), averaging about 0·2 gal/hd (0·9 l/hd). Relatively clear supernatant liquor usually contains at least two or three times as much suspended solids in p.p.m. of liquor as raw sewage, and has a B.O.D. usually at least several times that of raw sewage.

CALCULATION OF SLUDGE DIGESTION TANK CAPACITIES[27, 28]

The tank capacity required may be estimated from the quantities of sludge to be digested and to be expected after digestion (see Table 13.1, p. 257) and from the necessary digestion period appropriate to the given temperature corresponding to figures of Table 13.3 (p. 264) or of Fig. 13.3. For example, for a standard activated-sludge plant with a fresh surplus sludge mixed with fresh primary sludge of 0·43 gal (2·0 l) and after digestion of 0·17 gal/hd d (0·8 l/hd d), averaging to, say, 0·3 gal/hd d (1·4 l/hd d) (assuming a linear rate of reduction per unit of time) and a digestion period of 30 d at 86°F (30°C), we may compute the required digestion tank capacity to $0{\cdot}3 \times 30 = 9$ gal, or say, 1·5 ft^3/hd (40 l/hd).

Otherwise the required tank capacity may be calculated on the basis of the organic matter or dry solids content of the sludge thereby avoiding a direct determination of the digestion period. A basis commonly used for heated tanks of primary sedimentation plants is 0·8 ft^3/hd (0·022 m^3/hd), or the equivalent of say 1·2 persons/ft^3 (42·4/m^3). The corresponding loading rate per tank capacity on the basis of average living standard (see Table 3.2, p. 19) amounts to $1{\cdot}2 \times 0{\cdot}09 \simeq 0{\cdot}11$ lb/ft^3 (1·8 kg/m^3) daily of settleable organic matter, or accordingly about 0·15 lb total dry solids/ft^3 (2·4 kg/m^3) daily. For the design of Imhoff tanks the capacity of the digestion compartment should be at least 2 ft^3/hd (60 l/hd) if no secondary digestion is provided. Where sludge from industrial wastes is also included the population equivalent of the solids contribution must be allowed for and the digestion tank capacity increased accordingly.

Design figures on a population served basis are given in Table 13.4. In practical designs digestion tanks are sometimes made larger than these figures because ample space is helpful in making operation safer, easier and more flexible. On the other hand, the required capacity may be lessened by careful operation, and by ensuring always the most favourable conditions for the most efficient loading

Table 13.4

AVERAGE REQUIREMENTS OF DIGESTION TANK CAPACITIES FOR MUNICIPAL SEWAGE TREATMENT WORKS

Sewage treatment processes from which the raw sludges are derived	*Required digestion tank capacity* (ft^3/hd (l/hd) *or equivalent*)		
	Imhoff tank sludge compartment	*Separate digestion tanks*	
		Digestion in unheated tanks (1)	*Digestion at* 90°F (32°C) *in heated tanks* (2)
Plain sedimentation	2 (60)	up to 5 (140)	0·5–1·5 (15–40)
Chemical precipitation	—	up to 8 (230)	1–2 (25–55)
Sedimentation and low-rate filtration	3 (85)	up to 6 (170)	0·75–2 (20–60)
Sedimentation and high-rate filtration	4 (110)	up to 9 (250)	1–3 (30–80)
Sedimentation and standard activated-sludge treatment	6 (170)	up to 10 (280)	1·5–2·5 (40–70)
Sedimentation and high-rate activated-sludge treatment	4 (110)	up to 8 (230)	1·5–3 (40–80)

(1) Loadings dependent on climate (temperatures) and living standards.
(2) Lower figures applicable only for large plants under most favourable circumstances.

of the tanks and for continuously high rates of fermentation,[28, 29] including:

1. Addition of fresh sludge after thickening, either continuously or in relatively small batches, preheated and properly seeded.[30]
2. Continuous efficient mixing of the digester contents and minimizing scum formation.
3. Maintaining a constant temperature within the tanks.
4. Two-stage digestion, separating supernatant liquor as much as practicable, especially in the first stage.

In such cases it may be possible to increase the load allowances to 0·20 or 0·25 lb of organic dry solids, or, accordingly, to 0·25 or 0·35 lb total solids of raw sludge/ft³ digestion tank capacity daily (3·0 or 4·0 kg organic dry solids, 4·0 or 5·5 kg total solids/m³).

Allowance must be made for the working temperatures of the tank from time to time throughout the year. Where heating of the tanks is practised the temperature is practically uniform, but in unheated tanks it varies considerably, usually from season to season. Fig. 13.10 shows, for example, the conditions which actually occurred with Imhoff tanks at Munich, Germany, where temperatures and gas production varied according to seasonal effects.[30] In

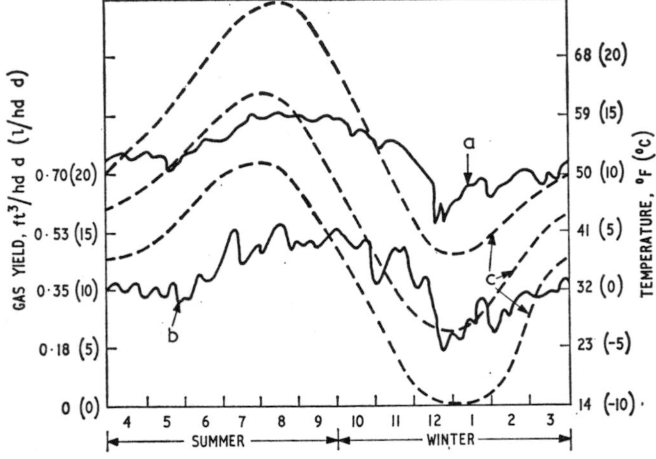

Fig. 13.10. *Sewage temperatures (a), and gas yields (b), in the Imhoff tanks at Munich, Germany, and air temperatures (c) (max, avge, min)*

such cases digestion tank capacities must be adjusted according to the operating temperatures which will be experienced.

In many instances additional digestion tank capacity must be provided to meet special local requirements. Under certain climatic conditions, as in winter periods of colder climates, digested sludge cannot be drawn off the digestion tanks regularly for longer periods thus requiring some sludge storage capacity which might be allowed for in the digestion tanks. With smaller plants serving populations, say under 5 000 persons, the allowances must be about 50% greater —in small treatment works serving less than 300 persons (Chapter 18) even greater in proportion—because attention usually is less frequent. About 20% additional allowance is necessary where stormwater tanks also are included in the sedimentation plants.

GAS PRODUCTION AND UTILIZATION

Sludge gas evolved from digestion tanks should be measured regularly, preferably continuously, and analysed for methane content,[31] its calorific value also being determined. Less frequently, the contents of carbon dioxide, hydrogen, nitrogen and sulphuretted hydrogen may be determined.

The sludge gas produced in two-storey tanks (Imhoff tanks) consists normally of about 70–80% methane and 30–20% carbon dioxide with a calorific value of about 700–800 Btu/ft³ (26 050–28 500 kJ/m³). The gas from separate digestion tanks, however, contains only about 65–70% methane and 35–30% carbon dioxide, and has a calorific value of about 650 Btu/ft³ (24 300 kJ/m³). The difference is due principally to the exchange of liquid between settling and sludge compartments of the Imhoff tanks, soluble carbon dioxide being carried away in the effluent.

Fresh sewage sludge solids digested at ordinary temperatures in the usual way yield about 8 ft³ of sludge gas per lb of dry organic matter (500 l/kg). At 86°F (30°C) gas production is greater, yielding about 12 ft³/lb of dry organic matter from settled sewage sludge (780 l/kg) (Fig. 13.2, Table 13.2, p. 263). From Table 3.2 (p. 19), allowing about 95% removal of settleable solids, we have up to about 0·09 lb/hd d (or 40 g/hd d) of organic solids in fresh raw sludge, and gas yields may reach about 1 ft³/hd d (28 l/hd d) in heated tanks. For Imhoff tanks the quantity of sludge gas is less because of the retention of carbon dioxide in the effluent, but the volume of methane and the equivalent total calorific value are about the same.

The gas yield of 1 ft³/hd d applies only to the digestion of sedimentation-tank solids. Digestion of humus-tank solids from trickling-filter plants or activated-sludge plants yields only a quarter or half as much gas per lb of organic matter because these solids are already partly decomposed. It can be taken that the total yield of gas from plants with secondary biological treatment amounts to 1¼–1½ ft³/hd d (35–42 l/hd d), or about 30%, more than for primary sludge alone. These are, of course, average figures and will be increased, for example, if industrial wastes or sludge due to stormwater are present. On the other hand, gas production may be inhibited by certain industrial wastes or will be reduced by lower temperatures or because of particular local factors of design or operation.

Gas collection is profitable not only in large plants but also in many municipal or industrial waste treatment plants of medium or even small size. For economy, then, the digestion tanks should be constructed with relatively small surface area so that the gas collecting cover may also be small and the tank therefore made deeper. Gas collection is relatively inexpensive in Imhoff tanks (p. 268) because it can be arranged economically, using the essential structures of the tank. In separate digestion tanks, covers may be either fixed or floating, according to design of the tank (see p. 271).

It is important to ensure that no air is allowed to mix with the sludge gas (e.g., during desludging or the withdrawal of supernatant liquor) because of the increased explosion risks, and because of lowered calorific value of the mixture. Imhoff tanks have a practically constant working level, even during sludge withdrawal, and are, therefore, safer, and it is sufficient to allow a water seal of only 1 ft (300 mm) at minimum working level. With separate digestion tanks, however, it is impracticable to maintain a constant level during charging and withdrawal of sludges and liquor and usually the variations in working level cover a wide range. Hence floating covers or covers fixed below the minimum working level are safest. When fixed covers are located above working level, gas-holders connected with the tank are essential so that gas is returned below the cover automatically when the liquid level falls, keeping the whole system under slight but positive pressure. The required gas-holder capacity is at least equal to the daily sludge intake into the digester. Gas-holders may be of the same types as those used in the gas industry, i.e. water-sealed or dry-piston types, or alternatively a system of large pressurized cylinders may be provided. Propane-butane gas storage then may be used for stand-by instead of oil storage. Where the gas is also utilized outside the treatment plant, e.g. when used for driving cars, it is customary to provide gas storage capacity equal to about one-quarter of the maximum daily gas production. Because of carbon dioxide and hydrogen sulphide gas-holders must be protected against corrosion.

Handling the gas from digesters requires great care because it is often practically odourless and not readily detected. It can be rendered more safe by adding odorous gases such as mercaptans. The digester gas is explosive when mixed with air in the proportions of about 1 volume gas : 5–15 volumes air. Pipelines carrying gas to burners or gas-engines must be protected against explosion by, for example, a gravel filter. Manually operated drip traps should be installed at all low points in the gas-piping system to provide for safe removal of condensate. Waste gas burners, located away from buildings, should be used to burn excess gas and so avoid possible odour nuisance.

Sludge gas normally contains some hydrogen sulphide, which is very poisonous. Hydrogen sulphide is readily detected in air in concentrations of the order of 1 p.p.m. (0·0001%), but above about 100 p.p.m. it is less noticeable; 1 000 p.p.m. (0·1%) is immediately fatal. Usually it must be removed from digester gas before the gas is used because of its corrosive effect on metals. It may be removed by 'scrubbing' with iron oxide, using oxide beds as used by the gas industry. It is also largely removed in 'water scrubbers' (washing towers) which are used in some plants to reduce the carbon dioxide content of the gas.

Digester gas may be utilized within the treatment works and for purposes external to the works. It may be burnt in boilers for the heating of the digestion tanks and treatment works buildings, for sludge drying, or for incinerator plant for screenings or sludge; it may be used in gas-engines from which the power requirements[32] of the works may be taken; it may be used as fuel for vehicles powered by internal-combustion engines using the gas under low pressure or compressed in cylinders; or it may be added to the gas supply of a public utility. In this last case allowance must be made for the relatively high calorific value and the different combustion characteristics of the sludge gas compared with normal town gas, but it is generally profitable to use the sludge gas in this way if the distance to the gasworks is not too great.

UTILIZATION OF SLUDGE GAS IN AUTOMOBILES

Sludge gas has been used successfully as fuel for automobiles, especially in Germany and England.[33] According to practical experiments in Stuttgart, 1 000 ft³ methane is practically equivalent to 8·4 gal petrol (100 m³ methane is equivalent to 135 l petrol). Calorific values of digester gas compared with town gas are:

	Calorific Values Btu/ft³	kJ/m³
Digester gas with 70% methane	670	25 000
Pure methane	950	35 400
Town gas	450–550	16 800–20 500

It has proved uneconomical to produce liquid fuels from the methane of sludge gas and the gas is used, therefore, either at ordinary pressures or as compressed gas. However, the carbon dioxide present in the digester gas is usually washed out so that the gas used consists largely of methane and contains little or no carbon dioxide.

Uncompressed gas is used for fuel in such vehicles as motor-buses, where it is carried in collapsible bags above the roof. Small bags have been used for cars, agricultural tractors and other small vehicles of this type.

Compressed gas is more commonly used and is stored in special containers (bottles) charged at about 350 atmospheres pressure (5 000 p.s.i., $3{\cdot}4 \times 10^7$ N/m^2). It is customary to wash the digester gas first, under considerable pressure, using about 1 volume of water to 10 volumes of gas, thereby removing practically all the carbon dioxide and hydrogen sulphide. The 'bottles' used for storage may be, for example, 16 ft (5 m) high by 2 ft (0·6 m) diameter, capacity 35 ft^3 (1 m^3) and tare 5 ton (5 080 kg). The gas bottles or cylinders carried by vehicles such as motor-cars are about 5 ft (1·5 m) long by 9 in (0·22 m) diameter, capacity 1·9 ft^3 (0·05 m^3) and tare 140 lb (63 kg) and are charged to a pressure of about 200 atmospheres (2 800 p.s.i., 2×10^7 N/m^2) so that about 420 ft^3 (12 m^3) methane is carried, which is equivalent to about $3\frac{1}{2}$ gal (16 l) petrol. The effective storage available in the large 35 ft^3 (1 m^3) bottles is then about 3 500 ft^3 (100 m^3). In working installations high-pressure storage is provided equal to about one day's consumption, and in addition, low-pressure storage of similar capacity.

HEATING OF DIGESTION TANKS[34]

Heating is necessary if the temperature within the digestion tanks is to be held above that of the raw sewage or sludge, say 18–45°F (10–25°C) higher, for a digester temperature within the range say 86–95°F (30–35°C) (see p. 264), requiring compensation also for losses from the heated digestion tanks to ground and atmosphere. These heat losses depend on climatic factors as well as on the heat transfer coefficients of the digestion tank structures. In Germany, heat losses of properly insulated digestion tanks operated at 86°F (30°C) have been found to reach 34 Btu/ft^3 d (1 268 kJ/m^3 d) of digestion tank capacity.

Heating of digestion tanks may be arranged: (1) by introducing hot water or steam directly into the digestion tanks, or (2) indirectly by heat-exchanger systems internal or external to the tanks.

1. The simplest method is the addition of hot water to the diges-

tion tanks, by heating fresh water or supernatant liquor outside the digester, and introducing the heated water into the bottom of it. The heated water then rises through the digesting sludge, losing its heat progressively and thereby heating the contents of the digester, its volume being added to the supernatant liquor. This simple system has the obvious disadvantage of diluting the contents of the digester, but this is often not important and can be lessened by the use of steam instead of hot water. Steam heating may be used successfully either by injecting steam into the digester at the bottom, or by heating the raw sludge with steam before discharging it into the digestion tank.[35, 36] Local overheating of sludge by steam does no harm so far as can be observed in practice. One ton (1 000 kg) of steam adds about $2{\cdot}4 \times 10^6$ Btu ($2{\cdot}5 \times 10^6$ kJ) to the sludge. Raw sludge is usually heated to the digestion tank temperature, but, in addition, more heat must be added to the contents of the tank to balance the heat losses.

2. Indirect heating is achieved by heat exchangers located outside the digestion tanks or by hot-water circulating systems located inside the tanks. In either case hot water is obtained from boilers fired by the digester gas or from the water-jacket of gas engines.

Hot-water coils and other types of heat exchanger located inside the tanks may be arranged in a number of ways. Horizontal pipe coils are installed in the form of a spiral located around the internal periphery of the tank through which the hot water is circulated. In order to avoid incrustation of the exposed surface of the coils, water temperatures should not exceed 140°F (60°C). More recently vertical heating pipes and other types of heat exchanger mounted from the top of the tanks have been preferred because they are less prone to collect sludge or to form scale than horizontal coils. They can be removed for inspection and maintenance.

In larger plants external heat exchangers are preferred. In these installations the sludge is pumped with a velocity of about 4–5 ft/s (1·2–1·5 m/s) through a system of pipes of at least 5 in (125 mm) diameter. The pipes are surrounded by hot water (or even steam) circulating counter-current-wise. Higher heat-transfer rates and, therefore, higher heating efficiencies are obtained than with internal heating coils. These installations must also provide for easy and regular cleaning of the sludge piping.

The overall heat transfer rates are about as shown in Table 13.5.

According to the required digestion tank temperatures, the heating may be applied as follows:

1. Heating during the winter only, to about 60°F (15°C), so that

Table 13.5

Heating system	*Heat transfer rates* Btu/ft^2 h °F (J/m^2 s °C)
Horizontal internal heating coils (peripheral)	about 30 (170)
Vertical internal heating pipes	about 60 (340)
External heat exchangers	about 180 (1 020)

summer conditions are simulated during winter; this provision is generally suitable for open digestion tanks.

2. Restricted heating to maintain temperatures of about 75°F (24°C) may be useful sometimes as a temporary measure to increase digestion capacity.
3. Heating to maintain the optimum temperature for digestion (about 90°F (32°C)); the capacity of digestion tanks required is much less than at 60°F (15°C), while digestion is more reliable, scum formation is lessened and the supernatant liquor is of better quality and more readily separated.
4. Superheating to maintain a temperature of about 130°F (54°C) to suit thermophilic bacteria (p. 263); this allows further reduction of the required capacity of digestion tanks but the process of thermophilic digestion has not proved altogether satisfactory for municipal sewage plants because the digested sludge is not odourless and the costs of heating are too high.[37, 38]

The amount of heat available from the digester gases may be calculated as follows: the average yield of gas from primary sludge (p. 279) is about 1·0 ft^3/hd d (28 l/hd d) with a calorific value of 600 Btu/ft^3 (22 400 kJ/m^3), corresponding to a heat content of 600 Btu/hd d (663 kJ/hd d). The amount of raw sludge averages about $\frac{1}{3}$ gal (1·5 l), or say 3·3 lb (1·5 kg) water equivalent daily, so that the heat available from the gas is sufficient to increase the temperature by 600/3·3 = 180°F (82°C). The efficiency of the boiler may be about 70% only and the heat loss in the digester may be about 40% daily equivalent on an average in temperate climates, so that the practical availability of heat may correspond to a temperature rise of $0{\cdot}7 \times 0{\cdot}6 \times 180$ = about 74°F (41°C). The heat available from sludge gas would be sufficient to raise the sludge temperature from, say, 50°F (10°C) to 124°F (51°C). This is considerably more than is required normally for mesophilic digestion.

Digester gas may also be used to generate power, 18 ft^3 (0·5 m^3) gas being sufficient for at least 1 hp (0·75 kW) for an hour. Taking activated-sludge plants as a good example, gas production in municipal sewage works amounts normally to $1\frac{1}{4}$–$1\frac{1}{2}$ ft^3/hd d (say 35–42 l/hd d) (see p. 279). Taking about 40% to be needed for digestion tank heating, the balance may be used to generate say (1 000 × $1\frac{1}{3}$ × 0·6)/(18 × 24) or about 1·9 hp (1·4 kW) per thousand of contributing population. The activated-sludge plant power requirements are about $1\frac{1}{2}$–2 hp (1·1–1·5 kW) per thousand (see p. 242) and therefore no external power is needed if sludge gas is utilized for power production.

These calculations demonstrate the importance of the utilization of sludge gas, especially within larger sewage treatment works. Gas yields generally are sufficient to heat the digester to the optimum temperature and to generate all the power required in the works. This is especially true if the raw sludge sent to the digestion tanks has a low moisture content and if the heat available from the gas is used efficiently, e.g. by using waste heat from the power plant for heating the digestion tanks.

DIGESTION OF SLUDGES AND OTHER WASTE SUBSTANCES FROM PROCESSES OTHER THAN MUNICIPAL SEWAGE TREATMENT

The processes of anaerobic fermentation may be applied usefully for the treatment of many other waste materials, and, likewise, other such materials may be useful for increased gas production.[39] Reference to Table 13.6 gives an idea of the wide application possible for industries and for municipal economies, but only a brief review can be given in these pages. Concerning the figures quoted in the table, it must be noted that more highly subdivided and dispersed material generally yields more gas than dense, hard and coarse material, and that fresh, green material is better than old, woody and rotten material. The figures given in Table 13.6 are only approximate, the data having been derived from laboratory experiments.[40] The half-digestion period given is defined as that digestion period in days required at 86°F (30°C) to obtain half of the ultimate gas yield from laboratory digesters using sewage sludge seed. Corresponding data for municipal sewage sludges, included for comparison, show that grass, potato plants and some wastes from abattoirs, are even more readily digestible than sewage sludge solids.

By comparison with water-borne wastes such as sewage sludge the digestion of solid wastes such as municipal garbage requires less space, because less is occupied by unnecessary water. For these wastes the load on digestion tanks may be as high as 0·2 lb dry solids/ft³ d (3·2 kg/m³ d) of digestion tank capacity, and the gas yield may be as high as ½–1 volume daily per unit of digester volume.

For wastes having a low nitrogen content, e.g. vegetable wastes in which the carbon : nitrogen ratio[41] is high, say over 15 or 20 : 1, it is necessary to add nitrogen, for example, in the form of ammonium

Table 13.6

GAS YIELDS FROM ANAEROBIC DIGESTION OF WASTE SOLIDS

Nature of solid waste	*Total gas yield* ft³/lb dry solids (m³/kg)		*Methane in gas*	*Half-digestion period*
	Total solids basis	*Volatile solids basis*	%	d
Municipal sewage sludge	6·9 (0·43)	9·7	78	8
Municipal sewage skimmings	9·1 (0·57)	10·0	70	—
Municipal garbage only	9·7 (0·61)	10·0	62	6
Waste paper only	3·6 (0·23)	4·1	63	8
Municipal refuse (combined, free of ash)	4·4 (0·28)	4·9	66	10
Abattoirs waste:				
cattle paunch contents	7·4 (0·47)	8·4	74	13
intestines	1·4 (0·09)	1·4	42	2
cattle blood	2·5 (0·16)	2·5	51	2
Dairy wastes, sludge	15·6 (0·98)	16·4	75	4
Yeast wastes, sludge	7·8 (0·49)	12·7	85	—
Paper wastes, sludge	4·0 (0·25)	—	60	—
Brewery wastes (hops)	6·8 (0·43)	7·1	76	2
Stable manure (with straw)	4·6 (0·29)	5·5	75	19
Horse manure	6·3 (0·40)	6·9	76	16
Cattle manure	3·8 (0·24)	5·0	80	20
Pig manure	4·1 (0·26)	6·6	81	13
Wheaten straw	5·6 (0·35)	5·9	78	12
Potato tops	8·4 (0·53)	9·7	75	3
Maize tops	7·8 (0·49)	8·3	83	5
Beet leaves	7·3 (0·46)	8·0	85	2
Grass	7·9 (0·50)	8·9	84	4
Broom (1 in (25 mm) cuttings)	7·0 (0·44)	7·1	76	7
Reed (1 in (25 mm) cuttings)	4·6 (0·29)	5·0	79	18

Based on data from laboratory digesters operated at 86°F (30°C)

sulphate or ammonium phosphate. This is commonly experienced in composting, and may be demonstrated readily in bottle experiments (see p. 255).

The special case of sewage sludge digestion is dealt with on p. 264. Digestion of garbage and screenings is referred to briefly on pp. 115 and 116.

As with other biological treatment processes, anaerobic digestion processes are susceptible to interference by the ions of metals such as copper, nickel or zinc, by arsenic compounds and by organic compounds[42] such as bactericides and fungicides. The concentrations present in domestic or municipal sewages of normal composition are generally too low to inhibit or interfere with sludge digestion although the concentrations found in the digestion tanks are often many times greater than occur in the raw sewage or waste. In many cases also other factors act to offset the potential toxic effects of such contaminants.[43] Inhibition of sludge treatment as a result of toxic wastes reaching a sewage works in large enough proportional amounts, however, is at least likely to prove costly if not intolerable, because the sludge disposal works must continue to operate practically continuously if other processes such as sedimentation, trickling filtration, etc., are to be continuously effective.

SLUDGE DRYING

Solids removed as sludge are accompanied by large quantities of water (see Table 13.1, p. 257). Sludge dewatering is generally not used when sludge is disposed of into the ocean, on to farmlands in liquid form as fertilizer, for use in composting of solid wastes such as garbage, nor perhaps even where used for land-fill. Usually, however, the sludge must be drained and dried before it is disposed of finally. While industrial types of drier have been adapted to waste treatment, ordinary sand-beds similar to the beds illustrated in Fig. 13.11 are in most common use for draining and drying.

Sludge-drying Beds

The sludge at the bottom of digestion tanks is always under pressures higher than atmospheric and contains correspondingly large volumes of gases in solution and entrained under pressure. When such sludge is withdrawn from the tank and exposed to the atmosphere, considerable expansion and foaming occur. This is even more so when the sludge is run on to drying beds and a relatively large surface from which gases escape more rapidly is exposed. The foaming carries the solids to the surface and the water consequently drains away more readily from underneath and passes rapidly away through the sand which forms the surface of the drying beds; consequently dewatering is finished quickly, commonly within one day. In summertime gas bubbles are also produced during drying, and dewatering continues to a further degree. At this stage the upper foaming sludge layer is left resting on the sand-bed and contains still about 80–85% moisture (by weight). The final drying requires a much longer period (under favourable weather conditions about two weeks) after which only about 55% moisture should remain.

On the other hand, if the sludge run on to drying beds no longer contains such gases then the process of draining and drying is different, for the water which separates rises to the surface and does not drain away readily through the denser sludge layer but must be evaporated for the most part.

Drainage rates are influenced mainly by properties inherent in the sludges (cf. drainability, p. 256), and by the nature of the sand-bed, etc. Evaporation, on the other hand, is controlled mainly by external (climatic) factors, particularly those affecting the heat balance. Cloud humidity and wind velocity play a large part, along with the direct influence of rainfall. Under favourable weather conditions sludge run on to sand drying beds may dry sufficiently in about two weeks. However, average drying periods are mostly much longer because of wet, damp or cold weather periods and in winter sludges may not dry sufficiently, even after three months or more. Generally, digested sludge dries much faster than fresh and sludges from primary sedimentation faster than sludges from biological treatment processes.

The appearance of the drying sludge on the beds often gives an indication of its nature and its behaviour. *A few fine cracks* only on the surface are characteristic of a well-digested and readily de-watered municipal sludge. With such sludges the volume occupied by the water is partly replaced by gas bubbles without much change otherwise in its composition, the water separating accordingly to the bottom. *Numerous medium-width cracks* proliferating over the whole surface of the bed are rather characteristic of sludges with a higher moisture content, due, for example, to admixture with activated sludge or chemical precipitation sludges, and which must, therefore, lose most of their moisture content by evaporation rather than by drainage. *A small number of wide cracks* are characteristic of more cohesive sludges, such as badly digested sludge, which usually dry relatively slowly. Often undigested sludge is characterized also—apart from odour—by attracting large yellow hover-flies (*Eristalis tenax*) which settle on such beds but are never seen to settle on fully digested sludge.

Drying beds (Fig. 13.11) are usually constructed of coarse materials such as slag or crushed stone or gravel overlaid by sand. The material is graded in three sizes and laid about 10 in (250 mm) deep with the coarsest on the bottom, the finest on top and over this a 4 in (100 mm) layer of sand or coke breeze. In the U.S.A. sand is chosen with an effective size between 0·3 and 0·5 mm (0·012 and 0·020 in) and a uniformity coefficient less than five. Drains are laid at the bottom of the beds, with a fall of about 1 in 150 discharging freely into open drains or pipes, from which the drainage from the beds is discharged into the supernatant liquor disposal system (p. 276). The beds usually are in units up to about 15 ft (say 5 m) wide and any convenient length. When sludge is removed after drying it is inevitable that some sand is also removed, and fresh sand must be provided as necessary; sand from grit chambers is sometimes found to be suitable for this purpose. To facilitate removal of the sludge it is customary to provide for trucks or sledges to be moved centrally along the length of each bed unit. Alternatively, in large plants travelling dredging machines may be used for dried sludge

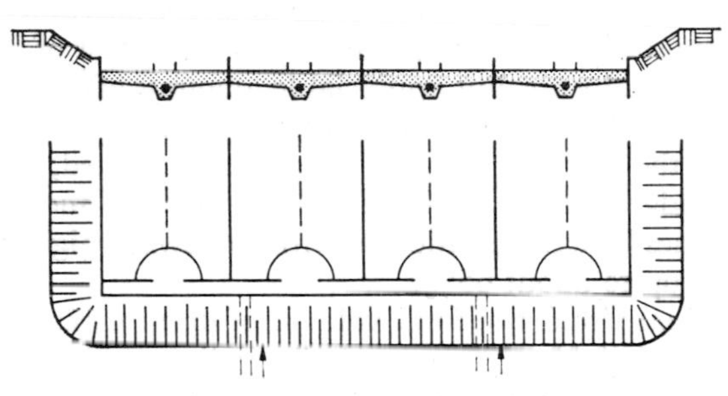

Fig. 13.11. *Sludge-drying beds*

removal.[44, 45] Strips of timber, brick or concrete located in the sand layer are often used to facilitate either manual or mechanical shovelling or dredging of the dried sludge without undue loss of sand.

Each unit of the drying bed area is isolated by concrete partitions at least sufficient in height for a full charge of sludge. Experience shows that sludge to be drained and dried should mostly be run on to beds to an initial depth of about 8 in (200 mm), and that the bed can be cleared and refilled about 5–9 times (say, commonly, six times) annually, depending on climatic conditions, equivalent to a loading of about 4 ft/year (1·2 m/year). Assuming (see p. 257) that the amount of digested sludge is 0·09 gal or about $\frac{1}{70}$ ft³/hd d (0·4 l/hd d), the surface area required for municipal sewage works is therefore about $\frac{1}{70} \times 365 \times \frac{1}{4} =$ about $1\frac{1}{3}$ ft²/hd (0·12 m²/hd), one acre for 33 000 persons (or 83 000 persons/ha). This corresponds to a daily loading of about 0·05 lb of dry digested solids/ft² d (0·25 kg/m² d). For municipal works including trickling filters, sludge-drying beds should provide about $1\frac{1}{2}$ ft²/hd (0·15 m²/hd), while for activated-sludge plants the allowance should be about $2\frac{1}{4}$ ft^2/hd ($0{\cdot}2\ m^2/hd$). Six fillings annually depend upon drying and filling on an average every eight weeks, but if this cannot be attained, then, of course, additional area is required. This may be due to prolonged winters, wet seasons or humid climates, lack of attendance in small plants, or other causes. The beds are sometimes covered to keep off rain, snow or heavy dew and in some cases are heated; such beds can be about 25–50% smaller. The sludge-drying bed area must be larger where poor sludge digestion is obtained because of badly designed or overloaded tanks. Draining and drying here may be hastened significantly by pre-conditioning of the sludge—for example, by using a coagulant such as alum or ferric chloride, or by elutriation.

Dried sludge is very porous and includes considerable gas space in its volume so that it is comparatively light. Most of this porous space is occupied by air, but appreciable amounts of digester gas are also present, especially in poorly digested sludge, and tend to be driven out if the partly dried sludge is again moistened, e.g. by rain. Normally sludge is dried to about 50–55% residual moisture but it may sometimes be easily spadeable with somewhat higher moisture content, say up to about 65%.

In municipal works the final volume of dried digested sludge is about $1\frac{3}{4}$ ft^3/hd ($0{\cdot}05\ m^3/hd$) yearly for sedimentation only, about 2 ft^3/hd ($0{\cdot}06\ m^3/hd$) yearly (low-rate process) or $2\frac{1}{2}$ ft^3/hd ($0{\cdot}07\ m^3/hd$) yearly (high-rate process) for sedimentation plus trickling filtration, and about 3–$3\frac{1}{2}$ ft^3/hd ($0{\cdot}09$–$0{\cdot}10\ m^3/hd$) yearly for sedimentation plus activated-sludge treatment.

SLUDGE LAGOONS

These are earthen basins in which raw sludge is discharged for digestion and subsequently allowed to drain and dry. They are equipped with under-drains (in similar fashion to settling beds, p. 168) which are fitted with valves. The valves are kept shut until the sludge is sufficiently digested, after which they are opened and the digested sludge is allowed to drain and dry. The capacity of such lagoons is normally reckoned on the basis of one charge of sludge say every two or three years. If charged with digested sludge operation is practically odourless and clear excess water may be drained off at any time. Such lagoons may be used also where the sludge can remain permanently in the lagoon, and may serve too as sludge storage ponds to help over winter periods.[46]

These lagoons are also used for the treatment of raw municipal sludge, but some odour is to be expected here; it can be lessened if the sludge is kept covered with water. Raw sludge seeded with at least 50% of digested sludge by volume should be discharged preferably below the liquor surface of the lagoon fills. Scum usually forms freely and should not be disturbed. The arrangement is thus similar to open sludge tanks such as earthen digesting basins (p. 270) except that sludge and liquor are not drawn off regularly. After filling the lagoon with raw sludge, the charge must remain undisturbed for some months until digestion is completed, when the drains are opened and the sludge is drained and allowed to dry.

Lagoons are also used for dewatering digested sludge.[47] They may be shallow or deep, the total capacity usually being sufficient for the total sludge output of several years.

MECHANICAL SLUDGE DEWATERING AND DRYING

Mechanical means for dewatering or drying sludges include centrifuges, filter presses or vacuum filter (Fig. 13.12) and rotary driers.[48]

Sludge Centrifuges[49, 52, 53]

The type of centrifuge normally used depends on differences in specific gravity of sludges and water, similar in principle to cream separators. Sludge and water are separated during centrifuging and are discharged separately. The process serves rather as a sludge concentration process, because if the water is to be relatively clean, the moisture content of the sludge remains comparatively high.[54] If spadeable sludge of 60% moisture content is required, this can be obtained with these centrifuges only if the separated liquor also contains a high concentration of sludge solids, thus forming a dilute sludge which may not easily be disposed of. This separated liquor may be filtered on sludge-drying beds after pre-aeration of the liquor. The process is now undergoing much development.

Filter Presses and Vacuum Filters[50, 51, 54, 55]

These utilize some form of filter cloth to separate water from the sludge, retaining the solids on the cloth. This practice requires that the filter medium should not clog with the sludge particles, otherwise water cannot be forced readily through the filter cake which is formed by the sludge solids. In the special case of municipal sewage sludge, for example, this requirement necessitates chemical treatment of the sludge prior to filtration or it may be achieved sometimes by elutriating the sludge with fresh water (p. 275). Lime or ferric chloride, less frequently alum, is commonly used for this 'sludge conditioning'.

Sludge presses consist simply of plate and frame filter presses of the ordinary types commonly used in industry. Before dewatering in such presses lime is added at the rate of about 0·5–0·8% by weight or, say, about 60–100 lb/1 000 gal (6–10 kg/m³) using even more lime if the fat content of the sludge is high. On the other hand, if the content of fats is very high it may be more helpful to treat the sludge by hot pressing whereby fats and greases may be separated. In practice, raw sludge is pumped into the presses, the pressure held to about 100 (even 200) p.s.i. (7 or 14 atmospheres) for 1–2 h, after which the plates and frames are separated and the sludge removed.[50]

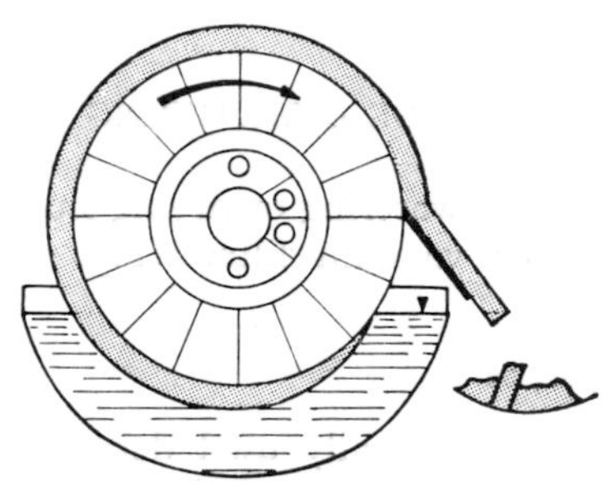

Fig. 13.12. *Vacuum filter*

The moisture content of raw municipal sludge is usually thus reduced to about 65% (from about 95%), which corresponds to a reduction of 7 : 1 or more in volume. The filtrate is highly polluted although relatively clear, and usually must be returned to the influent waste. If sufficient fat or grease is contained in the sludge the cake may be extracted readily. Generally, fresh sludge is more suitable for pressing than is digested sludge, but chemical-precipitation plant sludge and sludges from certain industrial wastes are especially suitable for filter pressing.[50]

In vacuum filters[51, 55] (Fig. 13.12) the cloth filter medium is stretched around a cylindrical drum which is mounted with its axis horizontal so that the bottom segments are continually immersed in the sludge or slurry to be filtered. The drum is segmented internally so that the segments may be subjected each independently to internal vacuum or air pressure as desired. In operation the drum rotates slowly (1·5–9 min/rev), the sludge being sucked on to the cloth and forming a continuous filter cake having a thickness about 0·2–

0·4 in (5–10 mm). The segments which have emerged are kept under vacuum so that air is sucked through the cake to displace some of the remaining water; the cake is then loosened by applying light air pressure in place of the vacuum and finally scraped off automatically with a fixed knife-edge mounted close against the filter cloth. The cleaned cloth, free of cake, then passes again into the sludge and is used for another cycle. From surplus activated sludge, for example, such filters produce a sludge cake containing about 82% moisure, from raw municipal sewage sludge about 65% and from digested municipal sludge only about 62%. Commonly, sludge is conditioned by the addition of ferric chloride before vacuum filtration, using up to about 2½% reckoned on the weight of dry solids contained in raw or digested municipal sludges, but up to 7% for activated sludge. In addition, as much as 7–10% lime may be required. If digested sludge is first elutriated the amount of ferric chloride required may be reduced by as much as 80%, lime additions are not necessary and the final moisture content is usually much less. The effective capacity of such vacuum filters varies from about 2–6 lb dry sludge solids/ft^2 (10–30 kg/m^2) filter cloth per hour, the figure being higher the better digested the sludge and the lower its moisture content.

Heat Drying of Sludge[51]

Filter pressing or vacuum filtration is sufficient for the dewatering and final drying of sludge only for well-digested or odourless sludges. Sludge cakes from other sludges, such as those from municipal sedimentation, chemical precipitation or activated-sludge plants, produce objectionable odours when stored after filtration, and such sludge cakes need to be dried finally by some other means. *Rotary driers* are suitable for this purpose. Sludge cake is fed through the driers counter-current to a flow of hot air, the inlet air having a temperature of about 1 100°F (600°C) and the escaping air about 600°F (300°C). The moisture content of the entering sludge should not exceed about 50% which necessitates recycling and admixture of sufficient finally dried sludge to reduce the average moisture of the sludge to be dried. If the finally dried sludge leaving the drier has a moisture content less than about 10% it may be ground and sold as fertilizer for spreading. It can be burnt readily and alternatively may be disposed of finally as an inorganic ash.

Treatment of sludges using heat and pressure include 'wet combustion' processes[56] (the Zimmermann process[56–58]) and direct autoclave treatment (Porteous process[59, 61]). In the former, sludges are heated first in heat exchangers and finally with steam and oxygen (air) at high pressures and temperature so that organic compounds are oxidized (burnt) to carbon dioxide and water, etc., and the residue may be almost entirely inorganic; in the latter, proteins, etc., are flocculated and the sludge deodorized and sterilized. It is then dried in filter presses of the plate and frame type.

SLUDGE INCINERATION [51, 52, 62–64]

The heat available from combustion of raw municipal sewage sludge (measured by its calorific value) is sufficient to evaporate the moisture which it still contains after drying until spadeable (about 55% moisture). It is feasible, therefore, to dry and burn such sludge[64, 65] without an external source of fuel, provided the efficiency of drier and incinerator is high enough. About half the calorific value of raw municipal sludge is lost as a result of digestion but may be substantially recovered by collecting the whole of the digester gas evolved during digestion. If the digester gas is not otherwise used and so can be made available to assist the process, incineration without using external fuel sources is also feasible for wet digested sludge, and is usually made easier because the moisture content of digested sludge is lower and is more readily drained and partly dried than is raw sludge. On the other hand, if raw sludge is dried and incinerated the costs of digestion and gas collection are saved. Nevertheless in practice, the incineration of digested sludge is preferred because odour nuisance is obviated or may be more certainly prevented. Furthermore operation of the sludge incinerators requires the provision of some storage to provide for the inevitable variations in sludge quantities discharged, also to provide for interruptions in operation. Digestion tanks automatically provide storage and may be regarded as storage units for final disposal of the sludge.

In the U.S.A. two types of furnace have been applied to sewage sludge incineration, namely, those developed by Raymond and by Nichols-Herreshoff. In the *Raymond system* the sludge cake is mixed with dry sludge of about 10% moisture content to produce a fluffy mixture which does not cake. The mixture is admitted to a cage mill into which hot gases from a furnace are discharged. The sludge particles are dispersed and carried to a cyclone separator where the dry sludge is separated. The gases are returned to the furnace. The dry sludge is discharged as a powder and may be used as a fertilizer or taken to the furnace as fuel while some of it has to be returned to the intake of the sludge to be treated and mixed with sludge cake as before. *Nichols-Herreshoff incinerators* are multi-storey furnaces through which the charge of sludge is moved progressively from top to bottom, emerging finally as cooled ash. Such a furnace of about 13 ft (4 m) diameter and 16 ft (5 m) high can incinerate 50 ton (51×10^3 kg) of spadeable sludge daily. On an average, about $\frac{1}{2}$ gal of fuel oil is required per ton of sludge ($2\frac{1}{4}$ l/tonne) for maintaining operation.

Both these techniques may be used to handle sludge from vacuum filters and are free of nuisance from odours and from smoke. The final ash from municipal sewage sludge amounts to about 15% by weight of the original spadeable sludge but varies according to the contents of moisture and fixed residue. It is possible to extract and recover the iron which then may be used for sludge conditioning.

FINAL DISPOSAL OF SLUDGE SOLIDS

The purpose of the various sludge treatment processes finally is to facilitate dispersal of the solids (contained in the sludges) to obviate public nuisance by odour, dust, etc., and to ensure that natural waters and soils are not polluted or harmfully contaminated in any way by organisms or toxic substances derived from the sludges. The final disposal of sludges, from sewage or other wastes, usually interferes to some degree with other uses of the disposal medium, waters or soil, unless adequate precautionary measures are applied. Generally this is best achieved if the disposal includes some sort of utilization of the constituents of the sludge, directly or indirectly, at the time of disposal or over a period of time. If, for instance, the sludge be discharged on to land or on to soil or into waste areas, although serving some useful purpose such as land-filling, it nevertheless might become harmful to man directly, or may interfere with some other activity. Likewise, when the sludge is disposed of into natural waters (including the ocean), the solids may be useful in stimulating biological activity: thus they may promote fish growth and ultimately benefit the economy of the original or other countries but the sludge may create local nuisance or even harm. For sludge disposal, therefore, as well as waste-water (effluent) disposal, care must be taken to balance the costs for conservation and for dispersal, including treatment costs, at the lowest level consistent with overall communal requirements.

The possibilities of sludge disposal may be grouped as follows:

1. By agricultural utilization
2. On to or into soil or as land-fill

3. Into natural waters
4. By industrial utilization.

Wherever feasible the aim should be to dispose of sewage sludge by agricultural utilization, particularly if suitable areas of land are available close enough to the sludge digesters (pp. 112, 114). Liquid sludges may be applied to arable land about once every three years, or to grassed land once each year. In both cases about 200–300 persons digested sludge equivalent/acre (500–750/ha) defines the required area. Farmlands irrigated by ploughing in digested liquid sludges (after some drying out) may need dressings with agricultural lime to avoid acidity.

Most sludges are utilized as fertilizer by drying or partly drying the digested sludge before spreading (pp. 114–116). The area required then, depending on the soil and on the nature of the sludge, may average 400–700 persons or population equivalent/acre (1 000–1 750/ha). Transport costs for dried sludge then may be economical at distances say up to 30 miles (50 km), compared with 10 miles (15 km) for pumping of liquid sludge.

Incineration of sludges yields finally an earthy ash which may also contain some nutrients valuable for agricultural land, even though the organic matter has been 'burnt off' and so dispersed into the atmosphere. Sludge ash may be transported economically over greater distances than may dry unburnt sludge residues and so areas more distant from the treatment works may benefit. In some cases this may make processes of incineration more economical.

In other cases it might be preferable to compost, for example, sewage sludge alone or mixed with other available and suitable waste material such as garbage etc. (see p. 115), as has been practised in some countries for many centuries, finally applying the composted material to agricultural areas. However, in the modern world, the costs of composting usually are relatively high and often outside economically acceptable limits.

Sometimes, where direct application of the sludge for agricultural use is not economical or practicable, the sludge may be buried, using trenches, say about 2–3 ft (0·6–0·9 m) wide and up to 3 ft (0·9 m) deep, covered finally with soil; or used for land-fill, either after drying, or by ponding the liquid sludge on the area to be filled and allowing it to drain and dry *in situ*—taking care to avoid contamination of surface or groundwaters. Tipping should be restricted to 6 ft (1·8 m) in depth, because sludges with high proportions of organic matter may heat up to 150°F (65°C) in heaps. Spontaneous ignition may result and the sludge may smoulder for many weeks or months. Disposal areas ultimately may be reclaimed as parks or for horticultural or agricultural use. Then the sludge is utilized at least partly.

In instances where agricultural or other such utilization, direct or indirect, is not feasible or cannot be undertaken for special reasons, other practicable means of final disposal must be looked for and may include, for example, dispersal in the ocean. In suitable cases it may be permissible to mix the sludge with the final effluent from the waste treatment works and so discharge it also into the receiving waters, either after digestion or as the residual ash from sludge incinerators. Then allowance should be made for any content of B.O.D. or for the suspended matter, etc., which is thereby discharged.

The disposal of sewage sludge into the ocean is still the practice of a number of coastal cities in England and America.[66, 67] Special vessels are used to carry the sludge—either raw or after partial or full digestion—which is dumped at selected places about 15 miles (25 km) or more from the shore lines. Special attention is given to the prevention of odour nuisance during loading or while traversing public waterways. Transport costs are reduced if the sludge is dewatered as far as practicable before loading.

The discharge of digested sludge or of industrial waste sludges into streams may sometimes be permissible during flood flows where, for example, the sludge discharged is carried downstream to the ocean without creating any nuisances along the course of the stream, its estuary or the adjacent ocean shores.[67]

Transport of Sludge in Pipelines (see also p. 267)[68, 69]

So that chokes do not occur in pipelines and pumps, certain types of material should be prevented from entering the sludges. Such materials are sometimes removed by screening or may be disintegrated and returned to the sedimentation tanks or added directly to the sludge, first separating metallic particles which are not readily disintegrated. Sludge may be pumped for long distances, but if the grease content is high clogging tends to occur and for this reason it may be preferable first to digest the sludge. Commonly, provision is made for flushing sludge mains regularly with water. Compressed air may be used for clearing chokes but cannot be used where digestion tank gases are utilized and may be adulterated with air. For ordinary municipal sludges containing 3–4% solids, friction losses may be about one and a half times as great as for mains water and for more concentrated sludges may be as much as four times greater.[69] Sludge drawn off from digestion tanks may not flow so freely if gritty material is present in unusually large proportions, which often occurs, for example, following periods of wet weather. Usually the velocities in sludge mains should be at least 3 ft/s (0·9 m/s), but velocities of only about 2 ft/s (0·6 m/s) are permissible for digested sludge free of sand or grit. Long sludge mains must be provided with arrangements (vents) for the release of gases, because sludge usually produces gases during pumping.

BIBLIOGRAPHY

Isaac, P. C. G. (Ed.). *Waste Treatment* (1960) Pergamon, Oxford

McCabe, Bro. J. and Eckenfelder, W. W., Jr (Eds). *Biological Treatment of Sewage and Industrial Wastes*, Vol. 2, *Anaerobic Digestion and Solids–Liquid Separation* (1958) Reinhold, New York

American Society of Civil Engineers Committee, 'Advances in Sludge Disposal 1954–60', *Proc. Am. Soc. civ. Engrs* **58** (SA2) (1962) 13

'Anaerobic Sludge Digestion', *Wat. Pollut. Control Fed. Manual of Practice No. 16* (1968) Washington D.C.

'Sludge Dewatering', *Wat. Pollut. Control Fed. Manual of Practice No. 20* (1969) Washington D.C.

REFERENCES

1. Chmielowski, J., Simpson, J. R. and Isaac, P. C. G. 'Use of Gas Chromatography in Sludge Digestion', *Sewage ind. Wastes* **31** (1959) 1237; also Pohland, F. G. and Bloodgood, D. E., **35** (1963) 11
2. Logsdon, G. S. and Jeffrey, E. A. 'Estimating the gravity dewatering rates of sludge by vacuum filtration', *J. Wat. Pollut. Control Fed.* **38** (1966) 186; also Lengyel, W., *Wass.Abwass.* (1963) 92
3. Gale, R. S. 'Filtration Theory with Special Reference to Sewage Sludges', *Wat. Pollut. Control* **66** (1967) 622
4. Stanbridge, H. H. 'Operation and Performance of the Hogsmill Valley Sewage Treatment Works of the Greater London Council 1958–66', *Wat. Pollut. Control* **67** (1968) 20
5. Pöpel, F. *Sludge Digestion and Disposal* (1964) Forschungs- und Entwicklungsinstitut für Industrie- und Siedlungswassenvirtschaft, Stuttgart
6. Isaac, P. C. G. 'Principles of Waste Treatment', *Effl. & Wat. Treat. J.* **5** (1965) 138, 196
7. Stalmann, V. 'Zur Eindickung von Abwasserschlamm', *Gas- u. WassFach* **108** (1967) 485
8. Voshel, D. 'Sludge Handling at Grand Rapids, Michigan, Wastewater Treatment Plant', *J. Wat. Pollut. Control Fed.* **38** (1966) 1506
9. McCarty, P. L. 'Sludge Concentration—Needs, Accomplishments and Future Goals', *J. Wat. Pollut. Control Fed.* **38** (1966) 493

10. McCarty, P. L. and Vath, C. A. 'Volatile Acid Digestion at High Loading Rates', *Int. J. Air Wat. Pollut.* **6** (1962) 65: see also Kaplovsky, A. J., *Sewage ind. Wastes* **23** (1951) 713
11. Dague, R. R., McKinney, R. E. and Pfeffer, T. T. 'Solids retention in anaerobic treatment systems', *J. Wat. Pollut. Control Fed.* **42** (1970) R29
12. Hindin, E. and Dunstan, G. H. 'Effect of detention time on anaerobic fermentation', *Sewage ind. Wastes* **32** (1960) 930; (with Mueller, L. E. and Lunsford, J. V.) **31** (1959) 669, 697
13. Jeris, J. S. and McCarty, P. L. 'The Biochemistry of Methane Fermentation using C^{14} tracers', *J. Wat. Pollut. Control Fed.* **37** (1965) 178
14. Buswell, A. M. 'Important considerations in sludge digestion. II. Microbiology and theory of anaerobic digestion', *Sewage Wks J.* **19** (1947) 28
15. Burgess, S. G. 'Anaerobic Digestion of Sewage Sludge', *J. Proc. Inst. Sew. Purif.* (1962) Pt 5, 411
16. Schulze, K. L. 'Studies of sludge digestion and methane fermentation. I. Sludge digestion at increased solids concentrations', *Sewage ind. Wastes* **30** (1958) 28
17. Fair, G. M. and Moore, E. W. 'Effect of temperature of incubation upon the course of digestion', *Sewage Wks J.* **4** (1932) 589
18. Imhoff, K. 'A new method of treating sewage', *Survr Munic. Cty Engr* (1909) May 21
19. Müller, W. J. 'Neuere Konstruktionsformen von Emscherbrunnen', *Zentbl. Bauverw.* **47** (1927) 594
20. Pönninger, R. *Mechanische Abwasserreinigung durch Emscherbrunnen* (1962) Verlag der Österreichischen Abwasserrundschau, Vienna
21. Griffiths, J. 'The Practice of Sludge Digestion', *Waste Treatment* (Ed. Isaac, P. C. G.) (1960) 367, Pergamon, Oxford
22. Hewitt, C. H. 'Sludge Digestion at the Works of the Birmingham Tame and Rea District Drainage Board', *J. Proc. Inst. Sew. Purif.* (1954) Pt 1, 39
23. Imhoff, K. 'Über ungeheizte Schlammfaulräume', *Gesundheitsingenieur* **74** (1953) 273
24. Fuhrman, R. E. 'Scum control in sludge digestion', *Sewage ind. Wastes* **26** (1954) 453
25. Griffiths, J. and Withnell, I. 'Experiments with Gas Recirculation in Sludge Digestion Tanks', *J. Proc. Inst. Sew. Purif.* (1956) Pt 2, 125
26. Torpey, W. N. and Lang, M. 'Elutriation as a substitute for secondary digestion', *Sewage ind. Wastes* **24** (1952) 813 (and discussion Senter, A. L.)
27. Swanwick, J. D., Shurben, D. G. and Jackson, S. A. 'A Survey of the Performance of Sewage Sludge Digesters in Great Britain', *Wat. Pollut. Control* **68** (1969) 639: see also Simpson, J. R., *J. Proc. Inst. Sew. Purif.* (1960) Pt 3, 330
28. Langford, L. L. 'Digester volume requirements', *Wat. Sewage Wks* **107** (1960) 453

29. Pfeffer, T. T. 'Increased loadings on digesters with re-cycle of digested solids', *J. Wat. Pollut Control Fed.* **40** (1968) 1920: see also Keefer, C. E., *Sewage ind. Wastes* **31** (1959) 388
30. Stecher, E. 'Betrachtungen zu den Betriebsergebnissen der Münchener Abwasserkläranlage im Betriebsjahr 1933–34', *Bautechnik* **13** (1935) 113, 200
31. Burgess, S. G. and Wood, L. B. 'The Properties and Detection of Sludge Gas', *J. Proc. Inst. Sew. Purif.* (1964) Pt 1, 24
32. Dugdale, J. 'The Provision of Electrical Energy for Sewage Treatment', *Wat. Pollut. Control* **67** (1968) 223
33. Imhoff, K. and Keefer, C. E. 'Sludge gas as fuel for motor vehicles', *Wat. Sewage Wks* **99** (1952) 284
34. Henson, J. B. 'Sludge Heating', *J. Proc. Inst. Sew. Purif.* (1960) Pt 1, 59
35. Rawn, A. M. 'Sludge digestion temperature control with live steam', *Wat. Wks Sewer.* **89** (1942) 310
36. Gunson, C. P. 'Sludge heating and circulation by direct steam injection', *Sewage Wks J.* **19** (1947) 661
37. Rudolfs, W. 'Trends in Digestion of Sludges and Wastes', *J. Proc. Inst. Sew. Purif.* (1953) Pt 3, 185
38. Fischer, A. J. and Greene, R. A. 'Plant scale tests on Thermophilic Digestion', *Sewage Wks J.* **17** (1945) 718
39. Imhoff, K. 'Treibgasgewinnung aus festen Abfallstoffen', *Gesundheits-ingenieur* **68** (1947) 3
40. Reinhold, F. 'Faulgas aus organischen Stoffen', *Gas- u. WassFach* **96** (1955) 176
41. Sanders, F. A. and Bloodgood, D. E. 'The effect of Nitrogen to Carbon Ratio on Anaerobic Decomposition', *J. Wat. Pollut. Control Fed.* **37** (1965) 1741
42. Drew, E. A. and Swanwick, J. D. 'Sludge treatment at Ryemeads—consequences of a recent inhibition of digestion', *Publ. Wks. Munic. Serv. Congr. (1962) Paper No. 6* (see *Wat. Pollut. Abstr.* **36** (1963) 767); see also Cotton, P., *Wat. Pollut. Control* **68** (1969) 627
43. Masselli, J. W., Masselli, N. W. and Burford, M. G. 'Sulfide saturation for better digester performance', *J. Wat. Pollut. Control Fed.* **39** (1967) 1369
44. Imhoff, K. and Imhoff, K. R., Jr. 'Die natürlichen Verfahren der Schlammentwässerung sind immer noch wirtschaftlich', *Gas- u. Wass-Fach,* **105** (1964) 710
45. Kershaw, M. A. and Wood, R. 'Sludge treatment and disposal at Maple Lodge', *J. Proc. Inst. Sew. Purif.* (1966) Pt 1, 75
46. Pearse, L. *et al.* 'Sludge lagoons', *Sewage Wks J.* **20** (1948) 817
47. Jeffrey, E. A. 'Laboratory Study of Dewatering rates for digested sludge in lagoons', *Proc. 14th Ind. Waste Conf., Purdue Univ.* (1959) 359; also Bubiss, N. S., *J. Wat. Pollut. Control Fed.* **34** (1962) 830
48. Hamlin, N. J. and El Hattab, I. 'Factors affecting the dewatering of sludge', *Instn publ. Hlth Engrs J.* **66** (1967) 101
49. Braun, R. 'Problems of Sludge Disposal', *Proc. 2nd Int. Conf. Wat.*

Pollut. Res. Tokyo **2** (1964) 217; also SWANWICK, J. D., *Wat. Pollut. Control* **67** (1968) 374

50. HOLROYD, A. 'The Sheffield Filter Pressing Plant', *J. Proc. Inst. Sew. Purif.* (1964) Pt 2, 159
51. GENTER, A. L. and KENNEDY, R. M. 'Vacuum Filtration, Heat Drying and Incineration of Sewage Sludge in U.S.A.', *J. Proc. Inst. Sew. Purif.* (1957) Pt 4, 361
52. ETTELT, G. A. and KENNEDY, T. J. 'Research and Operational Experience in Sludge Dewatering at Chicago', *J. Wat. Pollut. Control Fed.* **38** (1966) 248
53. GRIFFIN, G. E. and BROWN, J. M. 'Studies on concentrating digested sludge prior to barging', *J. Wat. Pollut. Control Fed.* **38** (1966) 996
54. EYE, J. D. and BASU, T. K. 'The use of flyash in wastewater treatment and sludge conditioning', *J. Wat. Pollut. Control Fed.* **42** (1970) R125
55. SIMPSON, G. D. 'Operation of vacuum filters', *J. Wat. Pollut. Control Fed.* **36** (1964) 1460
56. HURWITZ, E. and DUNDAS, W. A. 'Wet oxidation of sewage sludge', *J. Wat. Pollut. Control Fed.* **32** (1960) 918
57. HARDING, J. C. and GRIFFIN, G. E. 'Sludge Disposal by Wet Air Oxidation at a Five M.G.D. Plant', *J. Wat. Pollut. Control Fed.* **37** (1965) 1134
58. NICHOLSON, R. W., PEDO, D. J. and MARTINEK, J. 'Wet Air Oxidation of Sewage Sludge', *Am. Cy Mag.* **97** (April 1966) 170
59. PORTEUS, I. K. 'Sewage sludge heat treatment at Huddersfield', *Wat. Waste Treat. J.* **7** (1960) 543
60. LEWIN, V. H. 'Disposal of Industrial and Sewage Sludges', *Rep. Effl. & Wat. Treat. Conv.* London (1965)
61. BROOKS, R. B. 'Heat Treatment of Sewage Sludge', *Wat. Pollut. Control* **69** (1970) 92, 221 and **67** (1968) 592
62. MICK, K. L. and LINSLEY, S. E. 'Sewage Solids Incineration Costs', *Wat. Sewage Wks* **105** (1958) R327
63. QUIRK, T. P. 'Economic Aspects of Incineration vs. Incineration Drying', *J. Wat. Pollut. Control Fed.* **36** (1964) 1355
64. HOLROYD, A. 'Recent Progress in the Blackburn Meadows Reconstruction and Extension Scheme at Sheffield', *Wat. Pollut Control* **68** (1969) 357
65. VELZY, C. R., FUHRMAN, R. E. *et al.* 'Dewatering, incineration and use of digested sludge' (Symposium), *Proc. Am. Soc. civ. Engrs* **69** (1943) 41, 323; also SCHROEPFER, G. J., *Sewage Wks J.* **19** (1947) 559
66. KASS, N. I. 'Sludge disposal at sea', *Wat. Wks Sewer.* **88** (1941) 385; also LIEBMAN, H., 391, WEST, P. E., 395, EMERSON, C. A., 401
67. RAWN, A. M. and BOWERMAN, F. R. 'Disposal of digested sludge by dilution', *Sewage ind. Wastes* **26** (1954) 1309
68. RAYNES, B. C. 'Economic transport of digested sludge slurries', *J. Wat. Pollut. Control Fed.* **42** (1970) 1379
69. MERKEL, W. *Die Fliesseigenschaften von Abwasserschlamm* (1933) Oldenbourg, Munich; also MERKEL, W., *Physics* **5** (1934) 355

14

TREATMENT OF INDUSTRIAL WASTES

INTRODUCTION

The general nature of industrial waste (trade waste) discharges was discussed briefly in Chapter 3 (pp. 21, 22, 24 et seq.) and some typical data were set out in Table 3.4 (p. 23). Methods of laboratory examinations of aqueous wastes were reviewed in Chapter 4.

It was seen that whereas many industrial wastes may be compared readily with ordinary municipal sewages, using the results of laboratory examinations as a basis of direct comparisons, there are wastes from some industries which are so distinctly different from municipal sewages that direct comparisons are difficult or even impossible. However, most of the treatment processes discussed in the foregoing Chapters 9–13 are applicable to industrial wastes, excepting those cases where specific properties of the particular waste preclude application, as when, for example, the presence of toxic concentrations of one or more substances prevents the use of biological methods; these chapters, therefore, are part of the general background to the subject of industrial waste treatment which is discussed more specifically in the following pages.

CHARACTERISTICS OF INDUSTRIAL WASTES

(see also Chapter 3, p. 21)

There are few, if any, industries which do not make some use of water, either directly as part of the product, e.g. the beverage industries, for steam raising in boiler houses, as cooling water, as a transporting medium, e.g. in the paper and mining industries, and so on. Mostly water is used as a solvent, either as a medium for chemical reactions or for washing products and containers, or machines, apparatus, factory floors, etc. Much of the water is discharged finally as waste waters, the composition of which differs according to the usage and the nature of the industrial processes operated.[1]

Industrial waste flows are usually related to the amount of raw material processed, to the amount of finished product (see Table 3.4, p. 23) or to the number of people employed in the factory. Table

14.1 gives some data on the basis of the numbers of employees working in the factory concerned.

The flow patterns of industrial waste discharges are normally very different from those of municipal sewage flows. In some cases the flow of industrial waste may be relatively uniform during working

Table 14.1

Industrial waste flow gal/employee d (l/employee d)	*Nature of industry*
Up to 10 (50)	Leather manufacture and boot-making, clothing manufacture, tobacco goods
10–30 (50–150)	Building trade, instrument and optical industries, potteries, wood working, printing trade, engineering works
30–120 (150–500)	Stone and earth, concrete, stoneware, iron and steel foundries, steel fabrication, automobile industry, electrical industry, metal processings, synthetic resins manufacture, bakeries, yeast manufacture
120–250 (500–1 000)	Wrought-iron works, steel presses, stamping, wire-drawing, cold rolling mills, textile industry (weaving and dyeing), ice cream factories, iron ore mining, laundries, meat industry, fruit and vegetable canning, groceries
250–1 200 (1 000–5 000)	Rubber goods, asbestos industry, distilleries and breweries, coal mining, chemicals and cement industries, blast furnace, steel hot rolling mills, dairy industry, gas works
1 200–10 000 (5 000–50 000) or more	Cellulose and paper industries, sugar industries, oil refineries

Waste discharges usually are substantially reduced if re-use of waste water is practised

hours but may be substantially reduced, even to zero, at night or during weekends or shut-downs. In other cases the rates at which wastes are discharged may be more or less variable, even highly variable within short periods of time. Where a more uniform flow of wastes is necessary, for discharges into rivers or into public sewers, holding or balancing tanks of appropriate sizes (see p. 316) should be installed. (See also Figs 3.1, 3.2.)

Water used in factory bathrooms, kitchens, closets, etc., leads to discharges similar in character to domestic sewage, whereas water used for cooling purposes normally produces relatively clean waste water which may be re-used in the factory. Process wastes, however, are usually polluted to some degree, the nature and concentration of the polluting matters depending on the processes operated in the particular factory.

Laboratory examinations of industrial waste waters are generally carried out according to the procedures used for municipal sewage (see Chapter 4) but additional determinations required in particular cases include acidity, contents of fats, greases, toxic substances, phenols, etc. For the most part industrial wastes are characterized by the following (see also pp. 21–24):

1. High contents of suspended solids are found in wastes from abattoirs, canning plants, textile and paper factories. Increased amounts of screenings and settled sludges result.
2. High figures of oxygen demand (B.O.D.) occur in wastes from sugar factories, canning works, breweries, distilleries, abattoirs, tanneries, wool industries; and (C.O.D.) in wastes from washing of waste gases, in sulphite pulp-mill wastes, dye wastes, tannery wastes, etc. The ratios of B.O.D. to C.O.D. may be much different from those found for municipal sewages.
3. Many industrial wastes, particularly from inorganic industries, have a high content of acids or alkalis or toxic substances (p. 306) which might produce damage in sewerage systems or treatment works or nuisance in rivers, etc.

NATURE OF POLLUTING CONSTITUENTS

There are three main classes of substance to be considered in relation to the problems of industrial waste pretreatment and disposal, (1) *non-fermentable or biologically inert* substances; (2) (*readily*) *fermentable* substances and (3) *toxic* (*including radioactive*) substances.

Mining wastes, such as from salt mines or coal mines, provide excellent examples of the first class. Widespread pollution of natural waters has resulted from the discharge of these wastes in many countries where such industries are concentrated, owing in some cases to excessive amounts of soluble salts, in others to insoluble solids, or both. Generally it is not practicable to separate soluble salts such as sodium chloride or potassium sulphate, and discharges carrying relatively large amounts of these may be tolerated only where sufficient natural dilution is available. Such wastes must often be stored in specially constructed storage tanks or basins so that advantage may be taken of natural flood discharges in wet weather or from melting snows, or alternatively so that they may be conveyed by pipelines into streams with sufficient flow or into the ocean itself. Acid wastes and many chemical wastes also belong to this class. Acid and alkaline wastes generally must be treated, either directly, by neutralizing agents (p. 179) or indirectly at the expense of the natural reserves of alkalinity and acidity (buffer capacities) of natural waters or soils, and their chemical contents must be dispersed finally by dilution in waters or soils. Community rubbish dumps—normally used to reclaim areas of swamp lands, etc.—containing especially waste building materials, scrap metal, etc., may also provide considerable reserves of alkaline materials and of carbonic acid. The ultimate fate of soluble chemicals should always be taken carefully into account.

Further examples are provided by mineral sludges from coal mines, metallurgical and chemical industries, and mixed sludges which may be effectively inert because of toxic substances, such as insoluble matters from chrome tanning processes. Water-carried wastes of these types may be treated by ordinary sedimentation processes, by mechanical filtration, or in a few cases simply by screening. Simple setting basins or lagoons in which the settled sludges may remain, or may be removed from time to time after draining and drying (pp. 168, 290), are often well suited for the treatment of such wastes. The use of inert additives or coagulants to assist settling or draining and drying is often economical and may improve the overall removal of polluting matters in suspension or solution.

Fermentable polluting matter is also characteristic of many industrial wastes. Even phenolic compounds such as those from coal carbonization industries may be treated biologically to effect decomposition of the phenols, and sodium sulphide (from xanthate manufacture) has been treated biologically with quantitative conversion to sodium sulphate. Wastes from food industries mostly belong to the fermentable class (e.g. wastes from dairies, distilleries, breweries, yeast and starch factories, abattoirs and sugar refineries), but other industries also discharge organic wastes, including dyehouses, paper mills, flax-retting plants, wool-scouring plants and tanneries. All such wastes may usually be discharged untreated into public sewers or may be treated and disposed of by means similar to those used for ordinary municipal sewage. Of course, many of the details of the processes, and the apparatus used, may differ considerably from municipal plants because the composition of some of these wastes is quite different, and unit loadings and quantities given for municipal plants must be adjusted accordingly. It is possible generally to make many adjustments on the basis of the calculated population equivalents, but this must be done very carefully, taking relative volumes and concentrations of total and organic solids and biological oxygen demands, temperatures and other special factors into account.

Toxic (*including radioactive*) substances occur characteristically in some wastes (pp. 319, 322, 327). The possibility of toxic concentrations of physiologically active substances is of primary importance in trade waste problems, both for treatment works design and operation[2–5] and for final disposal.[6–8] Thus compounds of chromium,[9] for example, may not interfere in any way with sedimentation plant operations, but may be present in the settled effluent in such concentrations that the effluent cannot be disposed of without harm to the river water and likewise to the land. Then it is necessary to treat the waste in the first place to remove practically all the chromium as an insoluble compound which is then harmless to animals and plants. Table 14.2 sets out the limits of concentration of typical substances which may be permitted in wastes without endangering biological treatment processes. Higher concentrations are likely to interfere with such processes. However, in some cases biological activity may be affected only slightly at the concentration levels given, depending on various other factors such as temperature or the presence of other compensating conditions or substances.

The determination of tolerance levels for toxic substances presents considerable difficulty.[10] Observations of toxicity levels for, say, a particular species of fish will not be valid for other species, because specific tolerances vary among species for different substances; also the values for certain species of smaller aquatic animals, such as water fleas of the genus *Daphnia*, usually are entirely different. On the other hand, poisoning of *Daphnia* or other such small animals may be of direct significance for the survival of the fish, because the active life and growth of the water fleas, dependent in turn on the growth of microscopic plants (planktonic algae), may control the whole food chain on which the fish are nourished. Other problems arise in determining toxicity, for example, the interdependence of the tolerance of fish to poisons at low concentrations and the content of dissolved oxygen; if the oxygen content itself is rather low, the fish may succumb to concentrations of the substance in question at levels which otherwise may be tolerated without effect.

Fish and other aquatic animals, with plant life such as algae, fungi, bacteria, etc., are the principal forms of life to be taken into account when considering the toxic effects of the discharge of raw wastes or treated effluents into surface waters, but other groups of organisms are more important for biological processes which may be used for treatment prior to disposal. Various specific bacteria, protozoa, fungi and important predators such as insect larvae or adults, snails, aquatic nematode and other worms, etc., are the main organisms for the balanced ecology of percolating filters, activated-sludge processes, and so on. For anaerobic processes it is bacterial activity which is most important and in these processes the role of metallic elements may sometimes be protective, rather than damaging, because they can combine with ions such as sulphides or cyanides, thus *controlling* the potential toxicity of the environment, rather than causing toxicity.[11] When present in greater proportions, however, they inhibit digestion.

An important difference between the two types of environment, i.e. that of natural waters and that of 'artificial' treatment works using natural processes, lies in the possibility of acclimatization. Even the more complex organisms such as fish, or human beings, can become adapted to special conditions which may be intolerable (in some cases even relatively very toxic) to their ordinary kind. Relative freedom of movement from place to place is characteristic for fish, etc., in streams and lakes, and this freedom carries with it a degree of intolerance or of susceptibility. The factors controlling adaptability are various and complex, mainly related to individual biochemistry and physiology, but the percolating filter bed or activated-sludge tank system is much more adaptable, in general, than unrestrained natural systems. This is illustrated by earlier references to phenol oxidation (p. 97). The effect of cyanides on percolating filters, which has been studied by the staff of the Water Pollution Research Board,[12] provides another illustration. However, these possibilities of acclimatization to potentially toxic concentrations of waste substances often cannot be exploited in treatment systems or in receiving waters unless the more extreme variations in concentration are prevented, for example, by the use of balancing or storage (see p. 316).

Following a detailed presentation of the whole subject,[13] Liebmann has presented a tabulated summary of the tolerance levels for more than 130 toxic substances, which takes account of seven different criteria of toxicity. Table 14.2 presents some data from laboratory studies and practical experience of different workers.

The combined effect of all toxic substances present in a waste water may be determined by composite tests, such as the A–Z test,[14] ranging perhaps, from oxygen consumption tests to provide an assessment of the rate of oxygen demand of the waste water, to simulation tests aimed at detecting inhibiting or toxic effects, e.g. inhibiting oxygen production by green algae.

Most toxic substances owe their toxicity to their interference with biochemical reactions which are vital for the particular organism in

Table 14.2

LIMITING CONCENTRATIONS FOR VARIOUS TOXIC EFFECTS

Toxic substance or chemical unit in waters	*Test organism or other criterion of toxicity*	*Acceptable concentration limits, found by different workers*, mg/l
Acetic acid	Water flea (*Daphnia magna*)	150
(in natural waters)	Mosquito larva (*Culex* sp.)	1 500, 47
	Goldfish	10–100 (pH 6·7)
	Bluegill fingerlings	80–130 (pH 3·75)
Ammonia	Fish life	5–7
Arsenic		
in sodium arsenate	*Daphnia magna*	14
	Sea lettuce (*Ulva lactuca*)	375
in sodium arsenite	Diatoms (*Navicula* sp.)	75
	May-fly larvae (*Caenis* sp.)	2·3
	Drinking water	0·2
Chromium		
Cr in dichromate	*Daphnia magna*	0·4
	Fish (*C. Carassius*)	800
	Aerobic biological treatment	1–2, 10
Cr as chromic ion	Sludge digestion	200 (in sludge)
	Daphnia magna	0·1
	Diatoms (*Scenedesmus* sp.)	5
Copper, as Cu^{++} ion	*Daphnia magna*	0·08
	Fish life	0·2–0·3
	Aerobic biological treatment	1
	Sludge digestion	5–10 (in municipal waste)
Cyanide (as CN^{-} in KCN):		
in soft water	Fish life	0·2–0·4
in hard water		0·1–0·15
Sulphide (as S in Na_2S)	*Daphnia magna*	about 4
	Sludge digestion	about 1 000
Zinc (as Zn ion)	Aerobic biological treatment	2, 5–10
	Sludge digestion	10 (in municipal waste)

question. In systems where unorganized organic matter is undergoing decomposition (fermentation), such as in percolating filters, aeration or digestion tanks, similar possibilities of poisoning arise, but may be lessened because these systems are heterogeneous and accordingly often more adaptable. However, it is common experience that dilution of toxic wastes with relatively large volumes of ordinary (hence non-toxic) municipal or domestic sewage will often reduce the level of toxicity sufficiently to obviate any direct interference with normal purification of the mixed wastes.[15] This is generally what should be done, because it usually is not safe to dispose of potentially harmful waste, untreated, directly to a watercourse above or below ground. It may be too toxic to be decomposed in the form and concentration discharged, but eventually it may become unstable and decompose in some less suitable place as it becomes progressively more diluted.

POPULATION EQUIVALENTS

If the polluting matter present in industrial wastes is characteristically similar to that of municipal sewage, it may be measured adequately by B.O.D. tests and expressed in terms of B.O.D., for example, as the B.O.D.$_5$ or B.O.D.$_{20}$ of the waste, and by measuring the rates of flow the total amount of the B.O.D. (per hour, or per shift, and hence per day) can be estimated.[15] In this way the 'population equivalent' of wastes may be determined. Thus from Table 3.2 (p. 19) the total amount of polluting organic matter in municipal sewage corresponds to B.O.D.$_5$ of 0·12 lb/hd d (55 g/hd d), and hence if the wastes from a factory contain each day an average total of, say, 6 000 lb (2 700 kg) of B.O.D.$_5$, the wastes have a population equivalent of 6 000/0·12 = 50 000 persons. By way of example, commonly accepted values for particular industries are given in Table 14.3. Usually such values are related to recognized units of raw material or finished product and may be applied accordingly to factories both large and small. Large factories may conserve their raw materials better than small and lower values are commonly more appropriate for them; detailed consideration of each case is usually necessary before reliable estimates can be made of the actual population equivalent. In areas with higher living standards where sewage from residential areas may have a B.O.D.$_5$ of 0·15 lb/hd d (68 g/hd d) or even more (see Table 3.2), the population equivalent figures of Table 14.3 have to be altered accordingly, e.g. multiplied by the factor 0·12/0·15.

The usefulness of such figures is readily apparent. An example has been given already (p. 83) showing how such figures simplify river pollution studies and pollution control problems. From the table we see that a factory engaged in the scouring of 30 000 lb (say 14 000 kg, or 40 bales) of wool daily is likely to discharge polluting matter equal to the sewage discharge from a community of about 300 × 100 or, say, 30 000 persons. Treatment by trickling filtration or by activated sludge (after plain sedimentation) may then be designed on a population basis for this case, 30 000/15 = 2 000 yd³ (1 530 m³) of high-rate filters (see example, p. 351). However, such a basis of calculation has its limitations, and from Table 3.4 (p. 23) the expected volume of this waste would be 300 × 200 = 60 000 gal/d (270 m³/d), whereas the expected sewage flow from 30 000 persons would be about 1 200 000 gal/d

Table 14.3

POPULATION EQUIVALENTS TYPICAL FOR PARTICULAR INDUSTRIES

Class of industry	*Unit basis*	*Population equivalent per unit*
Abattoirs	1 head cattle = 2½ hogs = 2½ sheep	20– 100
Bleaching house	100 lb (45 kg) bleached textile	6– 16
Brewery	100 gal (455 l) beer	100– 500
Cellulose manufacture	1 ton (1 016 kg) cellulose	about 500
Simple dairy	1 000 gal (4 550 l) milk	100– 300
Dairy with cheese manufacture	1 000 gal (4 550 l) milk	400–1 000
Dyehouse (including sulphur dyeing)	100 lb (45 kg) dyed textile	about 100
Distillery	1 ton (48 bushels, 1 000 kg) cereals	about 1 500
Flax retting and milling	1 ton (1 000 kg) flax fibre	about 1 000
Laundry	100 lb (45 kg) washed textile	20– 100
Paper milling	1 ton (1 000 kg) paper	100– 300
Starch manufacture	1 ton (48 bushels, 1 000 kg) maize	100– 400
Beet-sugar manufacture	1 ton (1 000 kg) beet	100– 400
Sulphite pulp milling	1 ton (1 000 kg) pulp	1 000–4 000
Tanning	1 ton (1 000 kg) raw hide	1 000–4 000
Wool scouring	100 lb (45 kg) raw wool	70– 140

(5 400 m³/d). The waste, therefore, may be twenty times more concentrated than sewage. Accordingly, sedimentation tanks may be relatively much smaller for the wool-scouring wastes while, on the other hand, recirculation is necessary for trickling filters; thus the population equivalent basis cannot be used for all details of the treatment plant.

Calculations of population equivalents may be extended to particular substances. Thus, if phenol exerts a B.O.D.$_5$ of 1·7 lb/lb (or kg/kg), hence 1 lb (0·45 kg) of phenol has a pollution potential equal to that of the raw sewage of 1·7/0·12 = 14 persons. The population equivalent of phenol, therefore, may be taken as 14 persons/lb (31 persons/kg).

Where the disposal of sludge or solids is involved it may be more appropriate to base calculations of population equivalents on the quantities of sludge, or of dry solids. In wool-scouring wastes, suspended solids may amount to 3 000 p.p.m. (Table 3.4, p. 23) of which 65–75% may be organic. Taking the volume of wastes at 200 gal/100 lb (22 l/kg) (again from Table 3.4) and reckoning 60% of the suspended solids as settleable solids (equal to 1 800 p.p.m.) raw sludge from plain sedimentation would contain $22{\cdot}4 \times 200 \times 10 \times 1\,800 \times 10^{-6} = 80$ lb dry solids/ton (35 kg/tonne) of wool, equal in amount to the settleable solids from (see Table 3.2, p. 19) $80/0{\cdot}12 = 670$ or about 30 persons equivalent population units per 100 lb of wool (66/100 kg). In this way better estimates of the possibilities of sludge treatment or solids disposal may be obtained, especially if the characteristics of the solids can be taken into account (e.g. drainability, organic content, C : N ratio, etc.).

On the other hand, there are many and various industrial wastes characterized by specific properties or constituents and having no parallel for a basic comparison with ordinary domestic or municipal sewage. Acids, toxic compounds, metallic compounds and mineral salts include a variety of substances which may be damaging to sewerage structures or harmful to plant and animal life and, therefore, to biological treatment units and to the life and self-purifying capacity of receiving waters or soils. Then it becomes necessary to ensure that the concentrations of such objectionable substances do not exceed the permitted levels for efficient treatment and satisfactory disposal.

SEWERAGE SYSTEMS IN FACTORIES

In residential areas the sewage from homes and similar premises is taken directly into the public sewerage system usually without any pretreatment and in such a way that ventilation of the public sewerage system is effected through the house vents (Chapter 2). Waste discharges from factories often require arrangements of factory sewerage schemes very different in concept and layout. The wastes produced in particular departments of a factory may make it necessary to install screens, cooling pits, scum or oil traps, balancing tanks, etc. Such pretreatment or storage facilities may be provided only for particular departments, or on the other hand for the whole of the premises, to comply with the requirements of the sewerage or river authority concerned. Generally, separate systems within the factory area are more economical than combined systems, whether the public sewers to which the wastes are to be connected are separate or combined. Segregation of the effluent from a particular department of a factory may make pretreatment easier, e.g. with more concentrated wastes in some cases; on the other hand mixing of different wastes may result in equalization of flow or of composition and thus possibly simplify the pretreatment or even render it unnecessary.

In many cases also special attention is required for ventilation within the factory drainage system. Ventilation systems may need to be separated because some wastes may be dangerous (sulphides, cyanides, explosive gases or vapours, etc.) or may need special attention to avoid dead-ends, or accumulation of carbon dioxide in pits, or because of foul odours. Septicization of organic wastes may also be a possibility (see p. 316); the bacterial reduction of sulphates may need to be guarded against and so on.

Separate drainage is preferred for factory areas. The volume and character of the wastes in dry weather only must be the main consideration when designing sewers and facilities for pretreatment and disposal. If, for example, some of the wastes are highly acidic, it may be economical to lay acid-resistant pipelines large enough for the acid waste stream only, but uneconomical to include provision for harmless surface waters in addition.

Where they are to be disposed of separately (not in admixture with other municipal wastes) any special characteristics of the wastes must also be considered carefully. Acidity or alkalinity, toxic or surface-active compounds must be allowed for and other important differences in nature or composition relative to ordinary sewage taken into account. Cooling water is usually relatively free of pollution and should be kept separate from polluted wastes. Generally, before wastes are discharged, on to land or into natural waters, it is necessary or desirable to separate any settleable solids in primary sedimentation units and in many cases to reduce the B.O.D. of the settled waste by biological treatment just as in ordinary sewage treatment. Chemical treatment is often specially suitable as a preliminary treatment for industrial wastes ahead of primary sedimentation units (p. 175 et seq.).

Other aspects of factory drainage always to be considered include the possibility of the recovery of materials from the wastes, or re-use of the water only. Some treatment is usually required before any recovery or re-use is possible, but this often is economical. In other cases it may be economical for the industry to change or modify the manufacturing process so as to reduce the volume of the wastes or the amount of polluting matter in the wastes and consequently to reduce the costs of disposal, whether into a public sewerage system or into natural receiving waters directly.

Disposal into Public Sewerage Systems

The discharge of industrial wastes into municipal sewers is generally considered the best practice.[15–22] Large municipal sewage treatment plants can generally treat mixed domestic sewage and industrial wastes more efficiently and more economically than any individual establishment treating and discharging its own waste into the nearest receiving water. This applies particularly in smaller industrial undertakings.

If discharged into public sewerage systems, separate treatment of the trade wastes usually is not necessary except as may be required for the protection of the sewerage maintenance staffs of the operating authority against personal dangers, for the protection of the structures of the sewerage system against corrosion or other damage and to obviate disturbance of treatment and disposal works processes.[19, 20] Most public authorities have established regulations, by acts and bye-laws, to ensure such protection. Typical regulations include the following conditions: wastes discharged into public sewers shall have a temperature not exceeding 95°F (35°C) (sometimes 100°F (38°C) or more) and shall not contain excessive quantities of solids or greases which may form deposits or cause choking, or certain toxic substances, explosives or inflammable substances, acids, oils, etc. (Although domestic wastes are mostly not regulated, they are also subject to similar prohibitions.) To comply with these regulations, or sometimes for protection of the drainage systems within factories, some degree of pretreatment is often required[21] and may include coarse or fine screens, neutralization units, grease separation units (grease-traps or save-alls), grit arrestors (to retain sandy or other coarse solids), skimming tanks to separate mineral oils, waste cooling apparatus, etc., possibly together with holding tanks and flow-metering apparatus. Special treatments, perhaps using special apparatus, may be necessary specifically to eliminate particular contaminants. However, very often the combined discharges of all domestic and industrial wastes may so reduce the effect of the objectionable constituents or unfavourable qualities of individual discharges that their effects may then be negligible (see pp. 21–26). So variable are all the factors[22] from one community to another that each discharge of industrial waste into public sewers requires individual consideration of the local circumstances.

Large variations in flow are objectionable regarding both the capacity of sewers and the efficiency of municipal sewage treatment works processes, and for such reasons industrial establishments commonly—especially when the volumes of wastes involved are large compared with those of the mixed municipal sewages—are required to install storage or holding (balancing) tanks for equalization of the waste flows (e.g. over 8 or even 24 h periods). These tanks may also be used to equalize the concentrations of suspended solids, etc. Such balancing is particularly desirable where the water-borne wastes of the community include large proportions of widely differing industrial effluents (see p. 316).

Disposal into Natural Waters (see Chapter 6)

The discharge separately of industrial wastes into receiving waters is quite different from the disposal into public sewerage systems. This may be a matter of economy, it may be because no public sewerage system is available, or it may be preferable for protection of the public sewerage and disposal works. In such cases the treatment and disposal of the wastes must be considered in much the same way as for domestic or municipal sewage treatment and disposal. The special characteristics of the waste, including acidity or alkalinity and toxic and surface-active compounds must be considered,[1] and allowances made for other important differences in composition relative to ordinary sewages. Cooling water is comparatively pure and may be discharged separately (see p. 98). In general it is necessary to separate the settleable solids in primary sedimentation units, and in many cases the B.O.D. may then be removed or reduced by biological treatment as in ordinary sewage treatment. Chemical precipitation is often specially suitable as a preliminary treatment before primary sedimentation (pp. 175 et seq.).

Biological treatment of industrial wastes[23] may not be feasible if the available organic matter and other constituents are not balanced properly for utilization in the metabolism of micro-organisms and associated feeding organisms. Thus sometimes a particular waste has to be mixed with compensating wastes, or nitrogen compounds or phosphates added, to speed, or make possible, biological activity (e.g. phenols, p. 97). Nitrogen and phosphorus must always be present in relatively substantial amounts for cell nutrition, along with traces of other elements—e.g. magnesium, iron, zinc, calcium, cobalt, copper, sulphur—necessary for activation of enzyme systems or for incorporation into specific proteins. Ratios of B.O.D.$_5$: nitrogen : phosphorus of about 100 : 6 : 1 are of the required order[24,25] often expressed otherwise as the amounts required for each unit of B.O.D.$_5$ removed, e.g. about 6 lb nitrogen and 1 lb phosphorus/100 lb of B.O.D.$_5$ removed.

As stated above, even compounds such as phenols can be oxidized biologically, if domestic sewage is mixed with the waste at a rate sufficient to provide the required amount of nutrients for the support of bacterial life. High-rate trickling filters using recirculation may be loaded with phenols up to 1·2 lb/yd^3 d (0·7 kg/m^3 d). In activated-sludge plants the load may be say 0·06 lb/ft^3 aeration tank daily[26] (i.e. about 1·7 lb/yd^3 (1 kg/m^3)). Recent experience with coke oven wastes shows that loadings three times greater, that is about 0·2 lb/ft^3 d (3·2 kg/m^3 d) may be permissible, especially when the activated sludge is fertilized by the addition of phosphates and nitrogen compounds. (Regarding disposal of phenols and effects on water supplies, see p. 97.)

A characteristic of such wastes as those from distilleries, starch factories and dairy wastes, is a marked tendency to undergo acid fermentation. Sometimes advantage may be taken of this by providing for acid fermentation tanks as a primary treatment prior to biological oxidation, whereas in other cases it may be advisable, or even necessary, to control acid fermentation, perhaps by chlorination to stabilize the wastes during treatment by sedimentation or chemical precipitation.

Where the wastes are too concentrated for ordinary methods, treatment may be feasible after dilution with water, or with weak sewage or treated sewage. Dilution may also be obtained by the recirculation of oxidized effluent from the waste treatment works, for example, by recirculating the humus tank effluent to the settled waste before it is applied to trickling filters (p. 208), or by recirculation of highly diluted return sludge in the case of activated-sludge treatment (p. 238). Treatment by intermittent sand filtration may also be suitable for the treatment of industrial wastes polluted by organic matter, this method of treatment being reliable and efficacious and, therefore, very important, even where the waste is unusually variable in character and flow. Recirculation can be used for dilution before treatment by intermittent sand filtration. It should be noted that recirculation may be used, too, as a means of controlling the acid character of some wastes, particularly if primary settling tanks or holding tanks are included in the recirculation cycle (see Figs 12.8, 12.9, pp. 208, 209).

Anaerobic treatment of industrial wastes may be applied successfully to certain types of industrial waste, using a process similar to that used for sludge digestion.[27, 28] When the wastes are highly polluted, mainly by readily fermentable organic substances, anaerobic treatment may be efficient and economical and, therefore, preferable to other methods.[29] In some cases where such treatment is applied, it may be most economical to collect a highly concentrated or thickened waste, thereby saving digestion tank space and reducing costs of digestion. Such thickening might be obtained, for example, by recycling the wastes through the manufacturing processes within the factory—avoiding bacterial contamination and decomposition meanwhile—until further concentration is no longer feasible or economical when balanced against cost of treatment. The capacity of digesters is determined by the quantity and character of the wastes and the digestion period necessary for efficient treatment. The effluents from anaerobic treatment processes commonly contain sulphides or other such malodorous compounds and so may require treatment accordingly, e.g. by regulated chlorination to destroy sulphides. Subsurface irrigation methods are often suitable for effluent treatment whereby malodorous compounds are oxidized simply by aeration within the subsurface distribution drains so that odour difficulties are obviated. Similarly, where secondary treatment on trickling filters or by activated sludge is used, recirculation of the settled oxidized effluent or diluted return sludge may overcome satisfactorily the septic condition of digester effluent.

It should be noted that successful anaerobic biological treatment again requires appropriate proportions of nitrogen and phosphorus, and other elements, for the necessary metabolic processes of the different bacterial or other micro-organisms. It has been found that phosphorus and nitrogen are required in about the ratio phosphorus : nitrogen : total sludge accumulation = 1 : 9 : 65.

The use of balancing tanks, as suggested in the foregoing, can bring many advantages, but sometimes requires caution, particularly where organic matter is present in the works. This can decay septically if retention is extended, and if sulphates are present in moderate proportions only (e.g. cotton mill–dyehouse wastes) intolerable concentrations of sulphides may result. Regular cleaning of apparatus and drainage system may be useful for control in such cases, but detention or settling tanks generally should be avoided when wastes are susceptible to anaerobic decomposition.

Similarly, balancing tanks may be dangerous if alkaline solutions (e.g. from electroplating works carrying high concentrations of cyanides, or sulphur dye wastes) are allowed to mix with acidic wastes (unless the latter are always neutralized adequately beforehand), since poisonous vapours may be generated.

Often, however, even in such circumstances as above, the proper use of balancing-tank systems offers possibilities of minimizing many problems of waste disposal, including costs. High rates of discharge may be controlled so that the size and cost of drainage systems and treatment works (p. 132) are kept down. Similarly, concentrations of organic matter may be kept more uniform for discharge to treatment works, or (perhaps even more important) unduly high concentrations of toxic chemicals may be entirely prevented from poisoning biological treatment works because the concentration is held to low average levels by the balancing system.

Recycling of Waste Water (see also pp. 109, 117)

In many industries recycling of waste water is practised for reasons of economy, especially where costs of supply water are high, or where costly treatment of waste waters is required before discharge, or because of high rates to be paid for trade waste discharges into municipal sewerage systems.

In coal mining, for instance, it has long been the practice to recycle coal-washing waters. Intermediate sedimentation plants are provided, to keep the solids contents of the recycled waters within acceptable limits. In steel mills, recycling of water makes possible a substantial reduction in water requirements, say, from 350–700 ft^3/ton (10–20 m^3/tonne) crude iron to less than 180 ft^3/ton (5 m^3/tonne). Cooling water already used again once or several times may sometimes be used as process water.

Recovery of Valuable Substances from Industrial Wastes[30] (see p. 116)

This not only reduces the load on any treatment or pretreatment works, whether municipal or private, but usually brings additional economic advantages. Generally, recovery is cheaper the higher the content of valuable substances to be recovered, and lower the content of unwanted matter. Therefore, for efficient recovery, different types of waste are best collected separately, and their concentrations built up, e.g. by recycling, before recovery processes are applied.

TREATMENT AND DISPOSAL PROBLEMS OF SOME TYPICAL INDUSTRIAL WASTES

In the following pages some typical industrial wastes are discussed only in sufficient detail to indicate the variety of problems which arise, and the many variations of available technique applied to their solution. Whereas domestic and municipal wastes treatment and disposal methods are now fairly well developed, the processes of industrial wastes treatment and disposal are still in the stage of development because of the rapid changes in industrial technology.

It is evident from the rapidly growing and specialized literature that industry itself is playing an increasing part in the co-operative study of industrial waste disposal problems along with local communal and regional water pollution control authorities. Individual industries have established central technical groups to handle the industry's special problems, and even particular factory units have set up their own special groups to study and control their water usage and wastage, etc.

Non-fermentable (inorganic or otherwise biologically inert) types of waste can give rise to harm or damage including obstruction or silting of waterways, chemical spoilage (salinization, increased corrosion, etc.), direct poisoning of water supplies or aquatic life, indirect effects such as the clogging of fish gills by suspended matter or by increasing turbidity (reducing light penetration underwater), etc.

Coal and potash mining and tin dredging provide examples of these problems. *Potash wastes*[31] contain only dissolved salts in significant amounts and cannot be treated by any of the usual methods. They require adequate dilution in surface waters, or dispersal into the ocean, or alternatively directly into suitable strata below surface (controlled to prevent excessive salinities in rivers or local groundwaters).

Wastes from the mining of coal[32–35] vary in character according to the nature of the associated geological strata and the method of mining employed—pit, shaft, or open-cast working. In many cases fine coal and other earthy matter are carried in suspension and some treatment by sedimentation is necessary. The solids are largely granular (see p. 150) and settling beds, basins or lagoons may be sufficient for treatment. Where finer solids must be removed flocculation may be required too, as is common with wastes from brown-coal mining, which are slightly acidic and for which lime treatment is effective. Sometimes coal-mining wastes are highly saline and need considerable dilution for satisfactory disposal into streams.

Wastes from the iron and metallurgical industries may cause problems of siltation or metal hydroxide precipitation, steel works and the steel industries[36] particularly. Mill scale is formed in large tonnages in rolling mills, etc., and must commonly be removed by pickling or electrolytic descaling procedures. Mill scale is relatively dense and hence settles readily and drains freely, but it is not readily assimilated into soils, and hence tends to be an accumulating waste.

The pickling wastes include overflow waters from rinsing baths and spent pickling liquors, and carry high concentrations of metallic salts, such as ferrous sulphate, copper and zinc, nitrates, sulphates and chlorides, along with mineral acids. Neutralization is usually necessary and the metals may be largely precipitated as hydroxides by lime treatment and so removed before discharge. Alternatively, products such as ferrous sulphate may be recovered[37] and marketed, or used as a coagulant for sewage treatment.[38, 39]

Wastes from gas scrubbers, smelting furnaces, etc., may be dangerous on account of sulphur dioxide, cyanides, or phenolic bodies.[40] Re-use of the wastes by recycling results in saving in water consumption and substantially alleviates the disposal problem.[41]

Besides salts of toxic metals, copper, zinc, chromates, etc., the wastes from electroplating works[42, 43] often contain large quantities of cyanides. Copper may be removed, for example, by means of metallic iron, and chromium by reduction and precipitation. Cyanides may be precipitated as complex iron compounds, or distilled off as hydrogen cyanide, or may be completely destroyed by chlorination in alkaline solution.[44, 45] Ozone is also recommended.

The water-borne wastes from the many different kinds of *chemical manufacture*[46, 47] are very different. The means of disposal must depend upon the particular nature of the industry concerned; for example, in synthetic ammonia manufacture it may be the content of copper (from the catalyst) in the waste which determines the method of treatment before disposal, but in the manufacture of D.D.T. insecticide, or of T.N.T. explosives, it may be determined by both high acidity and the content of toxic substances. Usually some chemical treatment is required before wastes from chemical factories can be discharged. Base-exchange resins may provide methods for removing undesirable elements or compounds. Scrap iron, wire or turnings may be used to remove copper from solution, or cheap industrial chemicals may be used for precipitation of metallic oxides or other reactions. Then simple sedimentation, skimming or centrifuging may effect sufficient removal of undesirable constituents to allow discharge of the treated wastes into natural waters. Treatment by activated-sludge processes, by land filtration, by ponding or in simple settling beds may be applicable to other cases.[48] Sometimes, however, it is necessary to volatilize certain constituents using high stacks for dispersal, in special cases even to evaporate the whole of the waste.

Other inorganic industries are the common salt industry, and those based on gypsum and cement manufacture. The latter may be like the mining industries in discharging wastes of an almost entirely mineral character, but the discharges may be alkaline enough to be toxic. Enamelling works and the ceramic industries are generally characterized also by strongly mineralized wastes.

Sedimentation with or without coagulation is the main process used for treating such wastes. Screening is usually not applicable, but the specialized 'classifying' systems used in the mining industry may sometimes help as a preliminary treatment. Flotation may also provide a means of separating and recovering a particular mineral species and may be good economy in special cases. Reclamation of useful material should always be investigated wherever any possibilities exist.[49]

Wastes from organic industries, except for those referred to above, are mostly fermentable to some extent, so that biological treatment processes may be applicable. In many cases, however, the lack of variety in chemical make-up and often the lack of important nutrients (nitrogen, phosphorus, trace elements, or vitamins) may limit or even prevent a sufficiently vigorous biological activity.

Wastes from coke ovens, gas works, etc., pose many complex problems.[50–51] These wastes contain some of the volatile constituents distilled from the coal during carbonization or from producer-gas manufacture, principally ammonium compounds and free ammonia, phenols, cresols and analogous monohydric and polyhydric aromatic alcohols, sulphides, cyanides, thiocyanates and emulsified tars. The cyanides, sulphides, ammonia and phenolic compounds are dangerous in rivers, while the thiocyanates cause objectionable discoloration and so it is usually necessary to treat the wastes before disposal. Phenols may be oxidized or recovered by special methods, and sludge, tars and oils removed by physical methods (sedimentation, skimming); ammonia is usually recovered by distillation and cyanides and sulphides volatilized. Land filtration may be effective but usually the best solution is discharge to municipal sewerage systems. If the proportion of pretreated waste does not exceed about 0·5% of the sewage flow the ordinary processes of municipal treatment works, including especially the biological processes, effect a high degree of purification even of the toxic constituents.[52] Polyhydric phenols and thiocyanates are oxidized with some difficulty and usually incompletely but the sulphides and monohydric phenols are oxidized almost completely.

Wastes from oil refineries[53] must be freed of mineral oil and acids. They can be treated biologically by the activated-sludge process if pretreatment with the aid of iron salts is provided and deficient nutrients such as phosphates are added to the wastes.[54, 55]

In *viscose rayon*[56] factories the spinning baths must be recharged regularly or continuously and these wastes may be evaporated to reclaim the Glauber's salt. The viscose wastes contain sulphide compounds but may be treated biologically on trickling filters, by which means the sulphur and sulphides are oxidized to sulphates and the alkalinity neutralized at the same time.[57]

Sulphite pulp wastes have been studied extensively in many countries.[58–61] Wastes are produced from cellulose factories in relatively large quantities, manufacture of 1 ton of cellulose yielding commonly about 2 000 gal of water-borne wastes (1 tonne yields 9 000 l) containing about 12% solids (1–2 ton (1 000–2 000 kg) dry solids), i.e. $2\frac{1}{4}$ ton (2 300 kg) of raw material (wood) yield $1\frac{1}{4}$ ton (1 300 kg) of waste material. Nine-tenths of the solids are organic. These concentrated wastes are commonly diluted by secondary wastes which contain much less solid, but in any case the wastes finally discharged have a high B.O.D. and carry considerable suspended matter. The total volume may be reduced if the washing-out processes are arranged for continuous counter-current flow and the quantity of waste solids may also then be less.

Although many experimental studies have been undertaken biological treatment methods have not yet been applied successfully and in most cases these wastes are disposed of by discharge into natural waters, usually into rivers. Purification and dispersal are made easier if the volume discharged is regulated according to stream flow conditions, usually requiring storage basins of capacity equal to three months' accumulation, or more, of the wastes. These are gradually decomposed by the natural processes of self-purification, usually in two stages; about one-third of the organic matter is decomposed first by the activity of ordinary bacterial types and their commensals, after which further decomposition depends upon filamentous and other higher fungi which grow only to die and cause a secondary objectionable pollution of the stream. The whole process requires extensive zones of purification along the course of the stream for its completion.

Partial treatment has sometimes been effected by several stages of chemical precipitation, using lime as a coagulant, recovering valuable by-products such as vanillin, followed by incineration of the concentrated residues. In one case production of fodder yeast from the wastes by a special fermentation process has been operated successfully, and in another the production of alcohol by fermentation has likewise been successful, but in both instances circumstances were specially favourable. B.O.D. reduction is far from complete even in the most successful of the treatment plants using such methods. In some localities stream pollution by sulphite pulp wastes has been ameliorated by artificial aeration of the mixed waters in the streams; under present conditions this is comparatively expensive.

Evaporation followed by incineration (with waste heat recovery in boilers) affords virtually complete treatment but is difficult and expensive, the capital costs being very high. Chemicals can be recovered from the residues, but markets are not certain and generally are not large enough. Pulping plants have been constructed using magnesium, ammonium and sodium lyes in place of calcium, whereby water consumption may be greatly reduced and the cost of evaporation and incineration decreased accordingly.

Strawboard mills and fibre mills operate processes which differ in details, but the wastes are also relatively highly concentrated. Sedimentation is usually efficient, removing, for example, about 60% of the total suspended solids content of strawboard mill wastes.[61] Anaerobic treatment has also proved successful;[62] thus digestion of fibre mill wastes in two stages at 100°F (38°C) has been found[63] to remove 85–90% of the B.O.D. within 24 h. In these wastes about 50% of the B.O.D. is due to dissolved solids. Biological oxidation[64] is usually not successful unless nitrogen is added to the waste.

Flax-retting wastes are treated using lime followed by trickling filters or land filtration. In England the idea has been developed of retting by aerobic bacterial activity,[65] while Nolte[66] has proposed activated-sludge treatment fertilized by addition of phosphates. Biological treatment requires the addition of nitrogen.

Paper mills[67] also produce large volumes of wastes and re-use by recirculation is practised as far as possible. Then suspended solids, especially fibres, must be separated efficiently and finely divided solids, such as kaolin, are precipitated and removed as well. Mechanical filtration and flotation tanks—using alum and bone glue as additives—are commonly employed, often using chlorine to control septicity and the growth of moulds, also chemical precipitation, for example, using ferrous sulphate and aeration and sedimentation methods. Biological treatment, after mixing with domestic sewage and settling, is also successful.

Dyehouse wastes[68, 69] may carry high proportions of polluting matter, especially when sulphur dyeing is carried out. Large sedimentation tanks also serve as balancing tanks to mix the different types of waste produced in the factory. Chlorination may be helpful in decomposing some of the dyestuffs and oxidizing specific chemicals which may be present. Chemical precipitation may be applied successfully, especially using ferric sulphate or chlorinated copperas and aeration in the presence of iron turnings has also been used successfully.[70] Biological methods may be used,[71] the addition of iron compounds and admixture with municipal sewage being helpful to good purification. Often the discoloration of streams is objectionable and some dyehouse wastes are toxic and usually the chemical oxygen demand of raw wastes is high so that a high degree of purification is required before the wastes may be discharged into natural waters. Treatment by lagooning or land filtration is also applicable but large areas are required.[72]

Cannery wastes[73, 74] arising from fruit and vegetable processing are normally moderately concentrated, sometimes rather weaker than ordinary sewage, but generally tending to become acid during storage and often highly odorous and objectionable. Coarse suspended matter consists largely of vegetable material, seed pods, fruit skins, potato peel, etc., and is removed readily by screening methods; it may be composted for fertilizer production or utilized for producing animal foods or alcohol. In many cases brines from pickling and canning add considerable quantities of sugars and mineral salts to the wastes; sugars, citrates and tartrates also occur in wastes from fruit packing. Chemical treatments for neutralization and precipitation are commonly applied; ponding is used in many cases, sometimes using nitrates to prevent septicization during early stages of natural purification.[73, 75, 76]

The *sugar industry* is usually on a large scale, whether sugar-beet or sugar-cane is the source. The processes consist in the extraction of the raw sugar from the beets or canes by diffusion or by pressing, followed by concentration and crystallization resulting in the production of pure sucrose together with more or less concentrated residues such as molasses. Sugar industry wastes are normally very strong, including both inorganic and organic suspended matter and particularly dissolved organic matter consisting mainly of carbohydrates.

Washing water and water used for transport in beet-sugar factories are contaminated mainly by relatively inert vegetable matter and inorganic dirt, and may be settled in plain sedimentation tanks and re-used, chlorination being applied to prevent septicization. Process water from precipitators, presses and other concentration plant units, and from by-products sections, may be treated by ponding or chemical precipitation and final irrigation, but the process which is inherently most suitable is that of anaerobic treatment in two stages; in such a process, acid fermentation for about one day at 105°F (40°C), followed by neutralization and subsequent alkaline digestion for a further two days or more, yield an effluent which may be aerated and disposed of by land treatment or recirculated and used again.[77, 78]

In cane-sugar manufacture,[79] the bagasse, or waste cane after extraction of the sugar, has been used widely for the production of acoustic and general purpose wall-board. Such waste vegetable matter (from sugar-beet also) can be useful in various ways;

e.g. as cattle food or as fuel for steam production in the sugar factory.[80]

Distillery and yeast factory wastes vary considerably according to the raw materials used and from stage to stage of processing. Where corn or potatoes are the sources of carbohydrates the residues are utilized for fodder production, but if molasses is the basic raw material then no unused material is recovered.[81, 82]

Wastes are highly concentrated, the B.O.D. being very high, but the biological treatment methods used for sewage treatment are applicable, the temperature of the wastes also being favourable. Treatment along with municipal sewage is preferable, otherwise primary treatment by anaerobic digestion, as suggested by Buswell, effects considerable decrease in strength. Secondary treatment by high-rate trickling filtration may then follow. Usually, two-stage digestion is used, digesting for only three days, using loadings up to, say, 0·10–0·14 lb organic solids/ft^3 d (1·6–2·2 kg/m^3 d) or even more.[83] Gas production is high (about 8 ft^3/lb (0·5 m^3/kg) organic solids) and the gas contains about 73% methane, 25% carbon dioxide and 2–3% hydrogen sulphide. Digested sludge is produced in small amounts only. Normally about 85% of the B.O.D. is removed by such digestion (septic tank treatment) and this may be sufficient, but secondary treatment (biological oxidation) sometimes is also necessary.

Brewery wastes are of average strength.[84] Some constituents may be reclaimed for re-use or special uses. The wastes may be treated readily along with ordinary sewage, or using similar methods.[85, 86]

Organic wastes arise from *other fermentation industries*, which include starch factories, penicillin and streptomycin manufacture,[87] vinegar, yeast and yeast foods, and are generally relatively strong and suitable for primary anaerobic treatment.[88] B.O.D. removals of up to about 80% result, but secondary treatment is usually necessary, pre-aeration and intermittent sand filtration being suitable. Trickling filters or activated-sludge plants and sometimes land filtration are applicable in many cases.[89]

Tannery wastes[90, 91] are normally very strong and may contain decomposable organic matter together with toxic compounds, alkaline sulphides and acids, from different sections of the factory. Anthrax bacilli may be present in significant numbers. Usually the various wastes should be mixed together before treatment. Chemical precipitation methods, using carbon dioxide and ferric chloride, or other coagulants, are commonly applied. Secondary treatment methods commonly used include lagooning, but where practicable the pretreated waste is discharged into municipal sewers and can be treated successfully with the sewage by ordinary biological methods.[92] Sludges from primary sedimentation (pretreatment) plants are normally dried mechanically and burnt, or digested at high temperatures before disposal.

Wool-scouring[93, 94] factories produce large volumes of waste, which are usually very strong and include considerable amounts of earthy matters scoured from the wool.[95] The organic matter is derived mainly from the suint and the woollen fibres and hence includes large quantities of grease (lanoline, etc.) together with grass, burrs and other seeds, dried excreta and faecal matter, so that the wastes are highly contaminated and decompose rapidly. During scouring some of the hairs (wool) are broken down and lost and these appear in the wastes. The total amount of solids discharged in the waste may exceed the weight of scoured wool. Greases are valuable as by-products and in many cases are recovered, using centrifugal methods (high-speed centrifuges). Preliminary treatment by sedimentation is efficient. Simple skimming or aeration and skimming may also be used. Acid treatment (acid cracking) is also applied, usually for secondary grease removal[96, 97] and calcium hypochlorite has been used.[97] After the grease content has been reduced by these treatments, ordinary methods may be applied. The wastes are suitable for anaerobic treatment.[98]

Most of the various *meat industries*[99–102] produce characteristic wastes. These may come from abattoirs, meat or poultry packing or bacon factories, etc., and are usually highly concentrated and decompose very rapidly. They should be collected and pretreated within the factory, discharged into the sewers and thence carried to the treatment works as rapidly as possible so that treatment can be applied before septic conditions develop; it is then cheaper and more successful.

Generally these wastes contain nutrients in optimum proportions and the elemental composition is well balanced. They are very suitable, therefore, for biological treatment, both aerobic and anaerobic, but allowance must be made for the very high concentrations of organic matter when designing units for aerobic treatment. In many cases anaerobic treatment may be applied with marked success.

The content of organic matter is nearly always very high, including both dissolved and suspended solids, also large amounts of earthy matter may be carried in suspension and high salinity is often characteristic. The suspended matter usually includes considerable proportions of animal fats and oils, the proportion depending on the nature of the by-product and recovery processes operating. For abattoirs and associated by-product departments the content of fats and oils (grease) amounts usually to about 2 lb (1 kg) grease/head of cattle killed, or more generally about 0·1% by animal weight. The odour, colour and appearance of such wastes, owing mainly to nitrogenous compounds and to blood, are usually unmistakable and very objectionable. The B.O.D. is usually very high and the content of nitrogen is also relatively high.

Pretreatment should always include screening and skimming to retain coarse solids and as much grease as possible. After such pretreatment the wastes are easily treated in municipal plants and hence can be received into public sewers.[101] If sufficient capacity is available for the additional B.O.D. loadings and for sludge treatment, ordinary municipal plants using plain sedimentation followed by biological treatment (filtration or activated sludge) can treat up to about 10% by volume of such concentrated wastes without modification; higher proportions may be permissible where two-stage biological treatment or dilution by recirculation are provided. Either activated-sludge or trickling-filtration plants may be adapted readily using the various techniques[99–103] as described in Chapters 12 and 13 (aerobic and anaerobic processes).

The dissolved air flotation technique has been applied successfully to meat industry wastes for the removal of a large proportion of the suspended matter, including grease.[104] Dried sludges may be hot pressed to recover grease as a by-product commanding a good price in a consistent market.

Septic tank treatment can be successful, especially for smaller works, but full treatment always requires final oxidation for which subsurface irrigation methods are usually suitable (p. 376). Effluents from anaerobic processes are not suitable for discharge into public sewers or into natural waters but should first receive secondary treatment by aerobic processes. Again for small works, simple treatments by chemical precipitation methods, including ferric chloride or chlorine treatment, may be suitable for primary treatment.

The slaughterhouse and meat-packing industries generally provide many good examples of how good house-management within factories can be a major factor in waste handling and treatment economy. First, the economic benefits of waste recovery (by-product recovery), are greatly improved by locating the first recovery units as close as possible to the point of origin of the wastes to be reclaimed. The value of blood may be recovered by intercepting the highest possible proportion of blood with the lowest possible volume of diluting water (floor washings) at the killing floor. Likewise wastes such as fatty tissues and gelatin-rich material such as animal hoofs may be collected directly in their original form. Then the water-borne wastes, for example, from the cooking of the fatty material to recover high-grade tallows and from evaporators for protein recovery, may be treated in the plants before they are discharged into the common factory sewers; and so on. Careful usage of water at all points leads to overall savings in pretreatment or final treatment stages of the individual or combined wastes.

The *dairy industry* includes dairies, milk depots and a variety of associated manufacturers, such as cheese or butter factories, evaporated-milk factories (producing condensed and dried milk, baby foods, etc.) and ice-cream factories. In all these cases significant quantities of whole milk are wasted as a result of spillage and the necessity for thorough cleansing of all vessels and pipelines used for collecting and handling the raw milk. Careful design and operation can result in considerable savings in whole milk wastage and hence greatly reduce the strength of the wastes.[105]

According to the various processes, the essential constituents of the milk are utilized more or less fully in different factories so that the composition of the wastes varies considerably. They may contain mineral matter in suspension (salinity usually being low), but the grease (butter-fat content) is mostly emulsified and may not be removed even partly by skimming. All the wastes are highly fermentable, tending to ferment rapidly with formation of acids such as lactic acid, butyric acid, etc.

Treatment by sedimentation is usually of little benefit, but aerobic oxidation by activated sludge or by trickling filtration, for example, by alternating double filtration, may be applied successfully and is commonly used.[106–108] Land treatment may also be successful, preferably using spray irrigation at low rates.[109–111] Anaerobic treatment is not usually successful because of the marked tendency to acid fermentation which can be overcome only by careful seeding, for example, with relatively large proportions of sewage sludge.

Radioactive wastes[112, 113] arise from the use of radioactivity in industry and in hospitals and research institutions, where the use of radioisotopes is increasing steadily. From these sources will come increasing amounts of waste radioactivity[114] (some waste activity will also appear along with domestic and municipal sewage accordingly). The amounts discharged to sewers or to streams are now subject to regulatory controls in most countries, with considerable uniformity in the limits specified. So far, these sources do not appear to have reached significant levels, but there are possibilities of concentration in sludges, etc. Plant and animal life can also result in high degrees of concentration of radioactivity and these possibilities may need to be taken into account where radioactivity occurs in sewage or other waste waters.

On the other hand, atomic reactor installations, e.g. atomic power plants, usually discharge large quantities of waste containing sufficient radioactive material to be dangerous or toxic to higher forms of life. Some radioactivity may be absorbed and dissipated incidentally during treatment by the ordinary methods of sewage treatment and it is also possible to use sludges or muds for absorption. Similarly sufficient radioactive material may be settled out in large storage reservoirs followed by gradual dissipation in the bottom muds. In other cases the radioactivity may be sufficiently dispersed simply by dilution in receiving waters or in the oceans.[115]

Ion-exchange resins of suitable specificity may provide means for concentrating radioactivity. The final disposal of material such as sludges or ion-exchange resins into which radioactivity has been concentrated, or indeed of restricted volumes of waste carrying initially very high levels of radioactivity, requires very special techniques. They may be transferred to elaborately protected and enormously strong underground tanks where they may be stored indefinitely or at least long enough for their radioactivity to be dissipated sufficiently—which usually requires arrangements for the transfer of the radiant energy (commonly as heat) to the surroundings—or they may be stored in containers reckoned to be indestructible and dropped into known areas of great depth under the ocean.

It should be realized that pollution by radioactivity, no matter whether the radioactive substances be organic or inorganic, is not capable of self-purification in the ordinary sense, but is subject only to decay at the particular rate (the residual concentration remaining dangerous in some cases only for periods of seconds, hours or days, in others centuries or longer) according to the particular isotope in question.

COSTS OF INDUSTRIAL WASTE DISPOSAL

It will now have been seen that the cost of waste disposal is generally about the same for industrial wastes as for municipal sewage. This is especially so if the relative costs be compared on the basis of population equivalents (p. 309), allowing for volume and concentration, including any of the many special aspects referred to—such as nutrient requirements or inhibitory properties—which might necessitate special sequences of treatment processes, special preliminary treatments, special arrangements of balancing tanks, or other provisions not required for ordinary municipal water-borne waste.

Disposal by discharge into other sewerage systems, public or private, is not normally attained without costs which might not be much different from the costs for disposal separately (i.e. without first mixing into other waste flows). However, the dilution resulting from admixture with other wastes of a different character, such as ordinary municipal sewage, or wastes from a different type of industry may reduce the concentration of toxic substances to levels below those at which they are inhibitory to biological processes, or alternatively may supply additional nutrients or other substances essential for adequate treatment; in which cases the disposal of the mixed wastes could become cheaper *pro rata.*

Public sewerage authorities normally make specific charges for accepting industrial wastes into their sewers, usually designed to cover the cost of collecting and piping the wastes to the disposal works and finally disposing of them, including the cost of any treatment required for the wastes in question.[116–121] These charges are made up from one or more of the following (often parts 1 and 2 or part 1 only):

1. A charge made on a volume basis, proportional to the average total cost per volume unit of collecting and piping the wastes.
2. A composite charge made on a proportional basis
   (a) for preliminary and primary treatments (screening, disintegrating, settling, etc., plus solids disposal) usually including any primary chemical treatment costs, digestion costs, etc.
   (b) for oxidation (normally biological) of non-settleable organic solids, again on a proportional basis related to operating costs for mixing, aerating, tertiary treatments, etc.
3. A penalty charge (not very frequently) aimed at discouraging toxic discharges.

These charges must be added to the costs of installation (interest only), maintenance and operation, of any preliminary treatment required before the discharge is accepted into the sewers.

BIBLIOGRAPHY

BESSELIEVRE, E. B. *The Treatment of Industrial Wastes* (1969) McGraw-Hill, New York

COLLINS, J. C. *Radioactive Wastes, Their Treatment and Disposal* (1960) Spon

ECKENFELDER, W. W. JR. *Theory and Design of Biological Oxidation Systems for Organic Wastes* (1965) University of Texas, Austin

ECKENFELDER, W. W. JR. *Industrial Water Pollution Control* (1966) McGraw-Hill, New York

ECKENFELDER, W. W. JR and O'CONNOR, D. J. *Biological Waste Treatment* (1961) Pergamon, Oxford

GURNHAM, C. F. *Industrial Waste Water Control* (1965) Acad., New York

ISAAC, P. C. G. (Ed.) *The Treatment of Trade Waste Waters and the Prevention of River Pollution* (1957) University of Durham—Contract Record, London

ISAAC, P. C. G. (Ed.) *Waste Treatment* (1960) Pergamon, Oxford

ISAAC, P. C. G., WHEATLAND, A. B., LUMB, C. and LITTLE, A. H. *Textile Effluent Treatment and Disposal*, Symposium Shirley Institute (1965) St. Ann's, Altrincham, England

McCABE, BRO. J. and ECKENFELDER, W. W. JR (Eds) *Biological Treatment of Sewage and Industrial Wastes*, Vol. 1 (1956) and 2 (1958) Reinhold, New York

MEINCK, F. J., STOOF, H. and KOHLSCHUTTER, H. *Industrie-Abwasser*, 4th Edn (1968) Fischer, Stuttgart

NEMEROW, N. L. *Theories and Practices of Industrial Waste Treatment* (1963) Addison-Wesley, Reading, Mass.

SADDINGTON, K. and TEMPLETON, W. L. *Disposal of Radioactive Waste* (1958) Newnes, London

WHISTANCE, D. J. and MANTLE, E. C. *Effluent Treatment in the Copper and Copper-Alloy Industry* (1965) British Non-ferrous Metals Research Association, London

International Union of Pure and Applied Chemistry, *Re-use of Water in Industry* (1963) Butterworths, London

Min. of Housing and Local Govt, Gt Brit. *Pollution of Water by Tipped Refuse* (1961) H.M.S.O., London

Proc. Ind. Waste Conf., Purdue Univ. Annually 1946 to date

Annual Reviews of Literature. Wat. Pollut. Control Fed.

REFERENCES

1. ISAAC, P. C. G. and SIMPSON, J. R. 'The Technical Approach to the Disposal of Trade Effluents', *Effl. & Wat. Treat. J.* **1** (1961) 16, 68, 180; see also *Effl. & Wat. Treat. J.*, Feb. (1970) 72
2. BARTH, E. F., ENGLISH, J. N., SALOTTO, B. V., JACKSON, B. N. and ETTINGER, M. B. 'Field Survey of Four Municipal Wastewater Treatment Plants Receiving Metallic Wastes', *J. Wat. Pollut. Control Fed.* **37** (1965) 1101; see also p. 86
3. McDERMOTT, G. N., POST, M. A., JACKSON, B. N. and ETTINGER M. B. 'Nickel in Relation to Activated Sludge and Anaerobic Digestion Processes', *J. Wat. Pollut. Control Fed.* **37** (1965) 163
4. BAILEY, D. A., DORRELL, J. J. and ROBINSON, K. S. 'The Influence of Trivalent Chromium on the Biological Treatment of Domestic Sewage', *Wat. Pollut. Control* **69** (1970) 100
5. *Rep. Wat. Pollut. Res. Bd., Lond.* (1961) 58, (1962) 56, (1964) 71,

H.M.S.O., London; also O'NEILL, J., *J. Proc. Inst. Sew. Purif.* (1957) Pt 2, 150

6. DOWDEN, B. F. and BENNETT, H. J. 'Toxicity of Selected Chemicals to Certain Animals', *J. Wat. Pollut. Control Fed.* **37** (1965) 1308
7. PICKERING, Q. H. and HENDERSON, C. 'Acute Toxicity of Some Important Petrochemicals to Fish', *J. Wat. Pollut. Control Fed.* **38** (1966) 1419
8. IRUKAYAMA, K. 'The Pollution of Minamatta Bay and Minamatta Disease', *Proc. 3rd Int. Conf. Wat. Poll. Res., Munich* **3** (1966) 153
9. INGOLS, R. S., FETNER, R. H. and ESCHENBRENNER, A. B. 'Chromate Toxicity', *Wat. Sewage Wks* **14** (1964) 548
10. ABRAM, F. S. H. 'The Definition and Measurement of Fish Toxicity Thresholds', *Proc. 3rd Int. Conf. Wat. Poll. Res., Munich* **1** (1966) 75; also ALABASTER, J. S. and ABRAM, F. S. H., *Proc. Conf., Tokyo*, **1** (1964) 41
11. MASSELLI, J. W., MASSELLI, N. W. and BURFORD, M. G. 'Sulfide Saturation for Better Digestion Performance', *J. Wat. Pollut. Control Fed.* **39** (1967) 1369
12. *Rep. Wat. Pollut. Res. Bd., Lond.* (1952) 36; also Rep. (1963) 19 et seq.
13. LIEBMANN, H. *Handbuch der Frischwasser- u. Abwasser-biologie*, Vol. 2 (1960) Oldenbourg, Munich, Table 85, 974
14. KNÖPP, H. 'Der A–Z Test, ein neues Verfahren zur toxikologischen Prüfung von Abwässern', *Dt. Gewässerk. Mitt.* **5** (1961) 66
15. WOODRUFF, P. H. 'An Industrial Waste Sampling Programme', *J. Wat. Pollut. Control Fed.* **37** (1965) 1223
16. BYRD, J. F. 'Municipal Waste Ordinances—The Views of Industry', *J. Wat. Pollut. Control Fed.* **37** (1965) 1635
17. GAILLARD, J. R. 'The Acceptance by Local Authorities of Industrial Effluents into the Public Sewers', *J. Proc. Inst. Sew. Purif.* (1957) Pt 1, 21
18. SWETS, D. H., RANNEY, C. C., METCALF, C. C. and PURDY, R. W. 'Combined Treatment at Kalamazoo', *J. Wat. Pollut. Control Fed.* **39** (1967) 204; also LEARY, R. D. and ERNEST, L. A. **39** 1223
19. MÜLLER, W. J. *Treatment of mixed domestic sewage and industrial waste waters in Germany* (1966) O.E.C.D., Paris
20. JACOBITZ, K.-H. *Die Ableitung industrieller Abwasser in öffentliche Entwässerungen und die gemeinsame Behandlung mit häuslichem Abwasser*, Dr.-thesis, Technische Hochschule Darmstadt, 1965
21. HURLEY, J., MCNICHOLAS, J. and JONES, C. B. O. 'Symposium on the Pretreatment of Trade Effluents', *J. Proc. Inst. Sew. Purif.* (1940) 184
22. MÜLLER, W. J. 'Déversement et traitement en commun des eaux usées domestiques et industrielles', *Tribune de CEBEDEAU* (1966) 399
23. ECKENFELDER, W. W. JR. 'Theory of Biological Treatment of Trade Wastes', *J. Wat. Pollut. Control Fed.* **37** (1967) 240
24. HELMERS, E. N., FRAME, J. D., GREENBERG, A. E. and SAWYER, C. N. 'Nutritional requirements in the biological stabilization of industrial wastes. III. Treatment with supplementary nutrients', *Sewage ind. Wastes* **24** (1952) 496

25. KILGORE, H. D. JR and SAWYER, C. N. 'Nutritional requirements in the biological stabilization of industrial wastes. IV. Treatment on high-rate filters', *Sewage ind. Wastes* **25** (1953) 596
26. BICZYSKO, J. and SUSCHKA, J. 'Investigations on phenolic wastes treatment in an oxidation ditch', *Proc. 3rd Int. Conf. Wat. Pollut. Res., Munich* **2** (1966) 285
27. CILLIE, G. G., HENZEN, M. R., STANDER, G. J. and BAILLIE, R. D. 'Anaerobic Digestion IV. The Application of the Process in Waste Purification,' *Wat. Res.* **3** (1969) 623
28. MCCARTY, P. L. 'Anaerobic waste treatment fundamentals, 4. Process Design', *Publ. Wks, N.Y.* **95** (1964) No. 6, x, xi, xii
29. PETTET, A. E. J., TOMLINSON, T. G. and HEMENS, J. 'The treatment of strong organic wastes by anaerobic digestion', *Instn Publ. Hlth Engrs J.* **58** (1959) 170
30. RICKLES, R. N. 'Waste Recovery and Pollution Abatement', *Chem. Engng Albany* **72** (1965) 133
31. SEIFERT, F. 'Die Abwasserfrage in der Kali-Industrie', *Ber. Abwassertech. Verein* **3** (1952) 158; see also LÜSSEN, H., *vom Wass.* **29** (1962) 9
32. *Rep. Glamorgan River Board 1962–3* and *1963–4*; also ADAMSON, G. F. S., *J. Proc. Inst. Sew. Purif.* (1965) Pt 4, 370, and BRALEY, S. A., *Ind. Wastes* **5** (1960) 89
33. HARPER, W. G. 'Coal preparation effluents and their treatment', *J. Inst. Publ. Hlth Engrs* **59** (1960) 119
34. MÜLLER-NEUHAUS, G. *Die Abwasser der Bergbaubetriebe* (1961) ZfGW-Verlag, Frankfurt/M.
35. HUSMANN, W. 'Der heutige Stand der Abwasserreinigung der Kohlenindustrie', *Münch. Beitr.* **11** (1964) 67
36. HOWELL, G. A. 'Re-use of Water in the Steel Industry', *Publ. Wks, N.Y.* **94** (1963) 114, 168, 170
37. MOLYNEUX, F. 'Waste Acid Recovery', *Chem. Process Engng* **45** (1964) 485
38. HICKS, R. C. 'The use of iron pickle in sewage sludge treatment', *Sewage Wks J.* **21** (1949) 591
39. GROEN, M. A. 'The use of steel pickling liquors for sewage sludge conditioning', *Sewage Wks J.* **21** (1949) 1037
40. BURKERT, E. 'Abwasserprobleme eines gemischten Hütterwerkes und ihre technische Lösung', *vom Wass.* **24** (1957) 315
41. POEHLMANN, H. 'Von der Umstellung der Wasserwirtschaft au. Eisenhüttenwerken', *Gas- u. WassFach* **95** (1954) 37
42. TANNER, M. 'The Design and Operation of a Plating Waste Treatment Plant', *Wat. Pollut. Control* **67** (1968) 401
43. STONE, E. H. F., MACKENZIE, K. C. E. and ROBINSON, J. B. C. 'The Kynock-Effluent Treatment Plant', *J. Proc. Inst. Sew. Purif.* (1964) Pt 4, 423 and *22nd Ind. Waste Conf. Purdue Univ. Series 129* (1967) 848: also TALLMADGE, J. A. and BARBOLINI, R. R. 'Rinsing Effectiveness—Cube and Other Shapes', *J. Wat. Pollut. Control Fed.* **38** (1966) 1461

44. Ridenour, G. M. 'Laboratory studies on effect of alkali-chlorinated cyanide, case hardening, copper and zinc plating wastes on aerobic and anaerobic sewage treatment processes', *Sewage Wks J.* **20** (1948) 1059
45. Corcoran, A. N. 'Treatment of cyanide wastes from the electroplating industry', *Sewage ind. Wastes* **22** (1950) 228
46. Hewson, J. L. 'The problems of the British Chemical Industry regarding the Discharge of Effluents and the prevention of pollution', *Tribune de CEBEDEAU* **16** (1963) 406
47. Bess, F. D. and Conway, R. A. 'Aerated Stabilization of Synthetic Organic Chemical Wastes', *J. Wat. Pollut. Control Fed.* **38** (1966) 939
48. Michalson, A. W. and Burhans, C. W. Jr. 'Chemical Waste Disposal by Ion Exchange', *Ind. Wat. Wastes* **7** (1962) 11
49. Prüss, M. 'Die Emscher-Kläranlage bei Essen-Karnap', *Gesundheitsingenieur* **52** (1929) 615
50. Noble, T. G., Jackman, M. I. and Badger, E. H. M. 'Biological Treatment of Gas Liquor in Aeration Tanks and Packed Towers', *Effl. & Wat. Treat. Jnl* **4** (1964) 415, 532; also Blackburn, W. H. and Kershaw, M., *Res. Com. Gas Council* (1963) G.C. 96
51. *Rep. Br. Coke Res. Ass.*, 19th, 20th and 21st Annl Rep. (1962, 1963, 1964); also Cooper, R. L., *Gas J.* **304** (1960) 221
52. Institute of Sewage Purification, 'Memorandum on gasworks effluents and their disposal by discharge to public sewers', *J. Proc. Inst. Sew. Purif.* (1956) Pt 1, 5
53. Davis, R. W., Biehl, J. A. and Smith, R. M. 'Pollution Control and Waste Treatment at an inland refinery', *19th Indt. Waste Conf., Purdue Univ.* (1964) 126
54. American Petroleum Inst. *Industry Manual on Disposal of Refinery Wastes—Liquid Wastes* (1969) API Division of Refineries; also von der Emde, W., *Gas- u. Wass Fach* **104** (1963) 94
55. McKinney, R. E. 'Biological treatment systems for refinery wastes' *J. Wat. Pollut. Control Fed.* **39** (1967) 346
56. Neas, G. M. 'Treatment of Viscose Rayon Waste', *Proc. 14th Ind. Waste Conf., Purdue Univ. Series 104* (1959) 450; see also Wincor, W., *Wass. Abwass.* **2** (1957) 254
57. Hughes, J. W. 'Industrial waste treatment at a viscose rayon factory', *Survr munic. Cty Engr* **110** (1951) 781
58. Faulkender, C. R., Byrd, J. F. and Martin, D. W. 'Green Bay Wisconsin—joint treatment of pulpmill and municipal wastes', *J. Wat. Pollut. Control Fed.* **42** (1970) 361
59. Gehm, H. W. and Gellman, I. 'Practice, Research and Development in Biological Oxidation of Pulp and Paper Mill Effluents', *J. Wat. Pollut. Control Fed.* **37** (1965) 1392
60. Vasseur, E. 'Progress in Sulfite Pulp Pollution Abatement in Sweden', *J. Wat. Pollut. Control Fed.* **38** (1966) 27
61. Coogan, F. J. and Stovall, J. H. 'Incineration of Sludge from Kraft Pulp Mill Effluents', *TAPPI* **48** (1965) 94A

62. JUNG, H. 'Schnellfaulung zur Behandlung stark organisch verschmutzter Abwasser', *vom Wass.* **17** (1949) 38
63. BUSWELL, A. M. and SOLLO, F. W. JR. 'Methane fermentation of a fibreboard waste', *Sewage Wks J.* **20** (1948) 687
64. QUIRK, T. P., OLSON, R. C. and RICHARDSON, G. 'Bio-oxidation of concentrated board machine effluents', *J. Wat. Pollut. Control Fed.* **38** (1966) 69
65. GIBSON, M. 'The large-scale development of the aerated retting process in Great Britain', *J. Soc. chem. Ind. Lond.* **67** (1948) 337
66. NOLTE, E. 'Die Reinigung von Flachsrostabwässern mittels des Belebtschlammverfahrens', *Beitr. WasserChem.* **4** (1949) 3
67. LIEBMANN, H. 'Abwasserbeseitigung aus Papier- und Pappefabriken', *Münch. Beitr.* **11** (1964) 340
68. SMITH, A. L. 'Waste Disposal by Textile Plants', *J. Wat. Pollut. Control Fed.* **37** (1965) 1607; also JONES, L. L. JR. p. 1693
69. LITTLE, A. H. 'The Treatment and Control of Bleaching and Dyeing Wastes', *Wat. Pollut. Control* **68** (1969) 178
70. JUNG, H. and SCHRÖDER, W. 'Das Niersverfahren zur Reinigung textiler Abwässer', *Text. Prax.* **8** (1950) 534
71. GREENE, G. 'Experimental Work on the Treatment of the British Celanese Ltd. Trade Waste', *J. Proc. Inst. Sew. Purif.* (1957) Pt 2, 116; also PETRU, A., Pt 4, 321
72. MCCARTHY, J. A. 'Characteristics and treatment of wool dyeing wastes', *Sewage ind. Wastes* **22** (1950) 77
73. DOUGHERTY, M. H. 'Activated-Sludge Treatment of Citrus Waste', *J. Wat. Pollut. Control Fed.* **36** (1964) 72
74. SPARHAM, V. R. 'Treatment of steam peeler effluent from carrot preparation', *Br. Fd J.* **65** (1963) 130
75. TEMPLETON, C. W. 'Cannery Waste treatment and disposal', *Sewage ind. Wastes* **23** (1951) 1540
76. LULEY, H. G. 'Spray irrigation of vegetable and fruit processing wastes', *J. Wat. Pollut. Control Fed.* **34** (1962) 1117
77. NOLTE, E. 'Drei Jahrzehnte Erfahrungen mit Zuckerfabrikabwässern', *Beitr. Wass.- Abwass.- u. FischChem.* **1** (1946) 24
78. OFFHAUS, K. 'Abwasserbeseitigung der Zuckerfabriken', *Münch. Beitr.* **11** (1964) 223
79. BHASKARAN, T. R. and CHAKRABARTY, R. N. 'Pilot plant for treatment of cane-sugar waste', *J. Wat. Pollut. Control Fed.* **38** (1966) 1160
80. KELLER, A. G. and HUCKABAY, H. K. 'Pollution abatement in the sugar industry of Louisiana', *J. Wat. Pollut. Control Fed.* **32** (1960) 755
81. BORUFF, C. S. and BLAINE, R. K. 'Grain distillery feeds and wastes', *Sewage ind. Wastes* **25** (1953) 1179
82. BHASKARAN, T. R. 'Utilization of Materials derived from Treatment of Wastes from Molasses Distilleries', *Proc. 2nd Int. Conf. Wat. Pollut. Res., Tokyo* **2** (1964) 85

83. SEN, B. P. and BHASKARAN, T. R. 'Anaerobic digestion of liquid molasses distillery wastes', *J. Wat. Pollut. Control Fed.* **34** (1962) 1015
84. BUCKSTEEG, W. and KORNATZKI, K. H. 'Untersuchungen an Brauerei-Abwässern', *Wass., Luft und Betr.* **11** (1967) 26
85. LINES, G. 'Liquid Wastes from the Fermentation Industries', *Wat. Pollut. Control* **67** (1968) 655
86. TIDSWELL, M. A. *Sewage disposal research at Burton-on-Trent*, Efflt. and Wat. Treat. Convention, London (1962); see also *J. Proc. Inst. Sew. Purif.* (1960) Pt 2, 139
87. HILGART, A. A. 'Design and operation of a treatment plant for penicillin and streptomycin wastes', *Sewage ind. Wastes* **22** (1950) 207
88. GEHM, H. W. and MORGAN, P. F. 'A novel high-rate anaerobic waste treatment system', *Sewage Wks J.* **21** (1949) 851
89. GREENFIELD, R. E., CORNELL, G. N. and HATFIELD, W. D. 'Cornstarch processes', *Ind. Wastes Symp., Ind. Engng Chem.* **39** (1947) 583
90. BIANUCCI, G. and DE STEFANI, G. 'Tannery Wastes', *Effl. & Wat. Treat. Jnl* **5** (1965) 407
91. SPROUL, O. J., ATKINS, P. F. JR and WOODARD, F. E. 'Investigations on Physical and Chemical Treatment Methods for Cattleskin Tannery Wastes', *J. Wat. Pollut. Control Fed.* **38** (1966) 508
92. FRENDRUP, W. 'Von der Gemeinsamen Reinigung von Gerberieabwässern und häuslichen Abwässern', *Leder* **17** (1966) 79
93. MENDIA, L. 'Aspetti Tecnichi del Problema degli Scarichi Industriali', *Ingegn. sanit.* **10** (1962) 4
94. PETRU, A. 'Combined Treatment of Wool-scouring Wastes and Municipal Wastes', *J. Proc. Inst. Sew. Purif.* (1964) Pt 5, 497
95. EVERS, D. 'Carpet Manufacturing Effluents and Their Treatment', *J. Proc. Inst. Sew. Purif.* (1966) Pt 5, 464
96. HILLIER, W. H. 'The Recovery, Processing and Marketing of By-products at the Esholt Works of the Bradford Corporation', *J. Proc. Inst. Sew. Purif.* (1947) Pt 1, 65
97. FRANKLIN, J. S., BOWES, E. L. and COLVILLE, J. F. 'Treatment of wool scour effluent with $CaCl_2$ using rotary vacuum filtration', *Trans. Instn chem. Engrs* **42** (1964) 3
98. SINGLETON, M. T. 'Experiments on anaerobic digestion of wool scouring wastes', *Sewage Wks J.* **21** (1949) 286 (cf. *Chem. Tr. J.* **154** (1964) 78)
99. GRIFFITHS, J., BOARD, R. G., GIBBONS, J., RILEY, C. T., POUTIN, R. A., BAXTER, S. H., MUERS, M. M. and LINES, G. T. 'Symposium on Effluent for the Food Industries', *Wat. Pollut. Control* **67** (1968) 605–672
100. STEFFEN, A. J. and BEDKER, M. 'Operation of full-scale anaerobic contact treatment plant for meat packing wastes', *Proc. 16th Ind. Waste Conf., Purdue Univ.* (1961) 423; see also *Sew. ind. Wastes* **22** (1950) 807
101. WEBER, W. *Der Einfluss von Schlachthofabwässern auf die Gestaltung städtischer Entwässerungsnetze und Sammelkläranlagen* (1964) R. Oldenbourg, München
102. LIEBMANN, H. *Die Reinigung von Abwässern aus Schlachthöfen und Krankenhäusern* (1961) R. Oldenbourg, München

103. Silvester, D. K. 'The treatment of slaughterhouse waste by anaerobic digestion', *Instn Publ. Hlth Engrs J.* **61** (1962) 266
104. Hirlinger, K. A. and Gross, C. E. 'Packinghouse waste trickling filter efficiency following air flotation', *Sewage ind. Wastes* **29** (1957) 165
105. Muers, M. M. 'Waste from the Processing of Milk and Milk Products', *Wat. Pollut. Control* **67** (1968) 644; see also Bloodgood, D. E., *Sewage Wks J.* **20** (1948) 695
106. Imhoff, K. 'Tropfkörper mit Wasserrücklauf für Molkereiabwasser', *Gesundheitsingenieur* **64** (1941) 367
107. Southgate, B. A. 'Waste disposal in Britain', in Symposium on Liquid Industrial Wastes, *Ind. Engng Chem.* **44** (1952) 524
108. Wieselberger, F. 'Die Reinigung von Molkereiabwässern mit Hilfe von Wechseltropfkörpern', *Ber. Abwasser Tech. Verein.* **11** (1960) 264
109. Nolte, E. 'Biologische Reinigung von Molkereiabwässern', *Beitr. Wass.- Abwass.- u. FischChem.* **3** (1948) 9
110. McKee, F. J. 'Dairy waste disposal by spray irrigation', *Sewage ind. Wastes* **29** (1957) 157
111. Lawton, G. W., Breska, G., Engelbert, L. E., Rohlich, G. A. and Porges, N. 'Spray irrigation of dairy wastes', *Sewage ind. Wastes* **31** (1959) 923
112. Belter, W. G. and Regan, W. H. 'Radioactive Waste Research and Development Activities—A Report of Progress', *J. Wat. Pollut. Control Fed.* **37** (1965) 316
113. Straub, C. P. 'Radioactivity Removal by Water and Waste Treatment Processes', *Publ. Wks, N.Y.* **96** (1965) 104; also Burns, R. H. and Glueckauf, E., *Proc. 2nd Inter. Conf. on Peaceful Uses of Atomic Energy*, Paper No. 308 (1958) UNO, Geneva
114. Lewin, V. H. 'Radioactivity and Sewage Treatment', *J. Proc. Inst. Sew. Purif.* (1965) Pt 2, 166
115. Whipple, R. T. P. 'Consideration on the siting of outfalls for the sea disposal of radioactive effluent in tidal waters', *Int. J. Air Wat. Pollut.* **7** (1963) 889
116. Symons, J. M. 'Rate formulae for industrial wastes', *Wat. Sewage Wks* **101** (1954) 540
117. Griffiths, J. and Kirkbright, A. A. 'Charges for Treatment of Trade Effluent', *J. Proc. Inst. Sew. Purif.* (1959) Pt 4, 505
118. Asher, S. J. A. 'Trade Effluent Control in Birmingham', *J. Proc. Inst. Sew. Purif.* (1962) Pt 6, 497
119. Bubbis, N. S. 'Industrial waste control in Metropolitan Winnipeg', *J. Wat. Pollut. Control Fed.* **35** (1963) 1403
120. Jackson, C. J. 'Trade Wastes—Charge Considerations', *Effl. & Wat. Treat. Jnl* **5** (1965) 557
121. Collom, C. C. and Hicks, R. C. 'Trade Wastes By-laws and Charges in Auckland', *J. Proc. Inst. Sew. Purif.* (1965) Pt 2, 106; also Waterfall, C. E. (1964) Pt 4, 520

15
GENERAL TREATMENT WORKS DESIGN DETAILS

In the foregoing chapters the problems and technology of waste treatment and of disposal of water-borne wastes, including reclamation and re-use, have been discussed and data given regarding pollutional loadings and the purification expected. The data are based on experimental studies or on working examples, and may be used for practical designs, but many simple details must also be considered carefully so that the works when constructed may achieve the expected efficiencies and operate successfully. For example, the flow must be properly subdivided before treatment in parallel units, otherwise none of the units can operate as designed, some being underloaded, others overloaded; again, the efficiency of a trickling filter in practice may be seriously reduced and some of its benefits (such as the virtually complete removal of many pathogenic organisms) actually lost if even a small fraction of the settled waste escapes treatment by leakage or blowing or splashing from the distributor discharge directly into the effluent drains. In the same way, maintenance and operation of treatment works require the careful consideration of the designing engineer, who should prepare adequate instructions to ensure proper maintenance and operation of the works and this should be done before designs are finally authorized for construction.

Special aspects of the treatment of stormwaters or wet-weather flows in sewerage systems are first considered.

TREATMENT OF WET-WEATHER FLOWS

In most, even separate sewerage systems (p. 7), the sewage flow is increased during rain or snowstorms, owing to the access of stormwater either by design or by infiltration through defective joints, fractured pipes, etc., or from surface waters incidentally or through unauthorized connections. In areas where surface drainage is properly provided wet weather conditions should produce only a relatively small increase in the sewage flow in sewers of separate sewerage systems, say up to three times average dry-weather flow. The peak wet-weather flow is usually conducted to the works for full treatment.

This requires some allowances in the design of treatment units (e.g. see p. 340, also below). Surface run-off water is collected by the stormwater drainage system and discharged into the natural receiving waters, sometimes after treatment in stormwater sedimentation tanks operated only during short periods after storms.

In partly separate and combined sewerage systems, however, the wet-weather flow is markedly different, the maximum normally exceeding the average dry-weather flow in sewers by 50–100 times or more. The total flow cannot be treated at the treatment works and in such cases it is customary to provide overflows to discharge excess flow from the system directly into natural waters which at that time might also carry additional flow. The dilution ratio at which an overflow might start to operate ranges from 4 : 1 (4 parts of stormwater to 1 part of sewage) to 10 : 1, but smaller ratios down to 2 : 1 have been used (see p. 13). The greater the dilution ratio the less frequently is overflow water discharged, and the more is the receiving water protected from pollution.

Stormwater from settled areas is polluted more or less depending on local conditions.[1–7] In addition the first run of stormwater in combined sewers will produce a flushing effect and carry downstream the pollutional matter such as sewer slimes and sewer deposits removed from the sewerage system. This wet-weather flow may be polluted as much as normal sewage or even more. It is important that this first flush is conducted to the treatment works and only wet-weather flow following this flushing period ($\frac{1}{2}$–1 h, or more in larger systems) should be released through overflows into the natural waters. For this purpose sewage treatment works might be designed to treat increased sewage flows in wet weather periods, but provide storage especially for the first flush of stormwater flow, so that it may be given full treatment when flows have subsided. The composition of stormwater run-offs should be assessed carefully, considering the nature of the area as regards climate, topography, kind of residential or industrial area, traffic, population density, etc. Some data have been published, but may not be applicable, in which case it may be necessary to undertake studies of comparable areas nearby. A consideration of the possibilities of pollution by wet-weather sewage overflows will permit a proper assessment of treatment requirements during and following rain.

The works must be designed to ensure proper treatment of dry-weather flows, but at the same time the various units must be considered in relation to the maximum rate of waste flow into the works and the dilution ratio of the various overflows upstream of the raw waste inlet. Primary treatment must be effective enough for dry-weather flows, but must provide sufficient capacity for peak wet-weather flows to ensure sedimentation of coarse suspended matter and effective retention of floating matter, to prevent choking of screens, pumps, etc. This capacity is principally determined by surface areas, provided tanks are large enough to obviate excessive turbulence. Storage capacity may be provided by secondary weir arrangements (Fig. 15.1) or by other such arrangements whereby

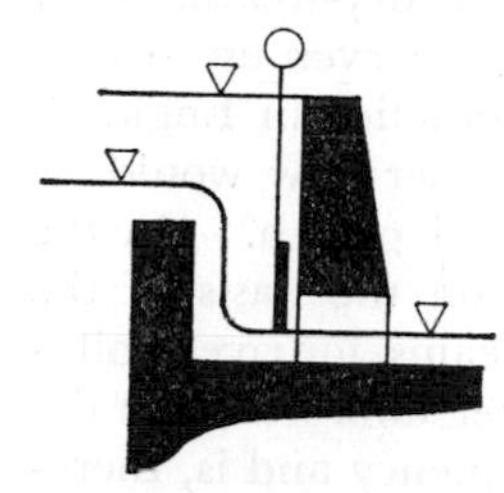

Fig. 15.1. *Secondary weir for sedimentation of wet-weather flows*

overflow rates (surface loading rates) may be partly controlled. For most cases, rates should be kept below about 1 500–3 000 gal/ft^2 d (3–6 m^3/m^{2}h). Efficient grit chambers are required for pretreatment before sedimentation.

Treatment for the mixed waters overflowed from the sewers may be provided by constructing special stormwater tanks (Fig. 15.2) into which the overflow is discharged. Excess flows are thereby

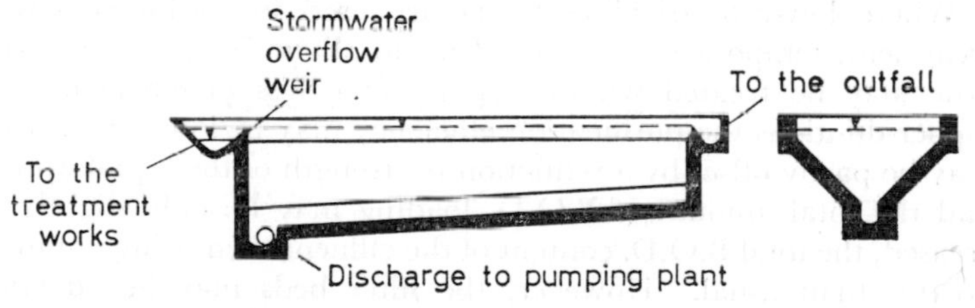

Fig. 15.2. *Stormwater storage and sedimentation tank*

given partial treatment by sedimentation. The entire content of these tanks, including any sludge which settles out, is pumped back into the sewers after cessation of the rainfall and treated along with the ordinary sewage flow. Thus the highly polluted run-off at the beginning of the rainfall receives efficient primary treatment at least, and in storms of short duration, even of high intensity, all or most of the overflow may be retained in the stormwater tanks and later treated properly before final disposal, the tanks acting as retardation basins.[6]

Stormwater tanks may be designed simply as sedimentation tanks (Chapter 10) and when located at the treatment works may be of the same type as the dry-weather primary sedimentation tanks so that they may be used as stand-by tanks during periods of repair or manual cleaning, or other temporary shut-downs. Fig. 15.2 shows a special design of self-cleansing stormwater tank, which may be especially suitable for small municipal sewerage installations. The tanks may be designed on the basis of equivalent dry-weather flow capacity, for example, to equal at least 3 h, 6 h, or even up to 12 h of ordinary dry-weather flow, as is common practice in England. Then a wet-weather flow of six times dry-weather flow would be detained for a settling period of respectively $\frac{1}{2}$, 1 or 2 h. On the other hand, the design might be calculated on the basis of the desirable frequency of overflow into natural streams, for river pollution control, and then the basis for stormwater tank design is the maximum run-off from storms of a certain frequency and is, therefore, dependent on local rainfall records and independent of contributing population and sewage flow.

The amount of sludge settled from stormwater treatment may be such as to increase the average total quantity of sludge (from all treatment) by up to 20% or more according to all the circumstances, and the proportions of inorganic and organic matter may also be altered significantly. Sludge treatment and disposal works must be enlarged accordingly.

Where low-rate trickling filters are used for ordinary sewage treatment, temporary increases of up to about 50% of the sewage flow may be treated without appreciable loss of efficiency. At higher dilutions the purification efficiency may be lowered, but this may be partly offset by a reduction in strength of the liquid applied and the total amount of B.O.D. loading may be only slightly increased, the total B.O.D. content of the effluent then being not much greater than usual. However, the filter beds may be adversely affected by scouring during periods of higher and more dilute flows and may take some days to recover. With high-rate filters, recirculation ratios may be reduced during wet weather so as to maintain much the same conditions of loading. In either case it is common practice to design trickling-filter plants so that flows up to about three times average dry-weather flow, at least, are given full treatment.

On the other hand, activated-sludge plants are only able to treat increased flows during wet weather successfully if they have been suitably designed. Usually the excess wet-weather flow is stored in retention basins and returned gradually to the raw sewage inlet.

TREATMENT WORKS DETAILS

Reinforced concrete is generally the most suitable construction material for channels and tanks containing water-borne wastes and sludges, but cast iron or steel may be suitable for open structures and for pipelines and special fittings. Earthenware also is suitable for pipes and channels. Generally, earthen works are used only for temporary construction, using either clay puddle or concrete linings for watertightness. If pipe material is well chosen and pipes are laid properly to grade, etc., the use of piping may prove suitable, especially for raw or settled sewage passing from one process to another.

Commonly choice must be made between shallow or deep construction. Treatment works may often be located on low-lying areas near watercourses where the groundwater table is not far below surface and may rise considerably during periods of high water-level, so uplift pressures and foundation conditions often require special attention. Shallow basins, therefore, are not necessarily cheaper than deep basins, because floors and walls may need to be made specially heavy. Openings left in floors and walls may be effective in preventing serious increases in uplift pressures, but leakage under normal conditions may often be undesirable. Well-points may be used for groundwater table control by pumping, especially for use in emergencies, e.g., when tanks must be emptied for repairs or to clear choking, etc. It should be feasible to empty all structures when required, if necessary by pumping, without much trouble or inconvenience. Deep tanks are usually most cheaply constructed if they are circular in plan and if about half of the effective volume stands above ground-level. Overall costs may be lowest where suitable land is available—taking soil, subsoil and underlying strata and groundwater conditions properly into account, considering the relation of contours to the arrangement of the various units, even though pumping of the sewage may be involved.

For practical reasons it is usually advisable to provide at least two units for parallel operation of each process of treatment so that any one unit may be taken out of operation at least temporarily, if necessary, without interrupting the course of treatment (even though some overloading of the remaining unit may then result).

A liberal water supply to the treatment works is essential but back-siphonage should be obviated, for example, by supplying the works from an elevated tank which is 'disconnected' by air-break from the main supply, while drinking water and amenities buildings are directly supplied. The works supply should be reticulated to all units and outlets fitted for hose connections should be available within easy reach of all sections of the works.

Within the sewage treatment works sewage should preferably be conveyed by open channels. Upstream of the grit chambers the velocity should be kept above 2 ft/s (0·6 m/s), and elsewhere, to avoid sludge deposits, above 1·3 ft/s (0·4 m/s). It is most important that the sewage flow be subdivided accurately and this may be done, for example, by providing special means (see Figs 15.3 and 15.4). Straight lengths of channel may be subdivided as shown in Fig. 15.3, with which arrangement the flow may be divided uniformly, whatever the variations in the depth of flow. Fig. 15.4 shows an alternative arrangement (plan only) which depends on uniform weir lengths to control the rate of overflow into the various channels. Careful design is necessary to ensure accurate subdivision under varying conditions. In large plants it may be necessary to subdivide the

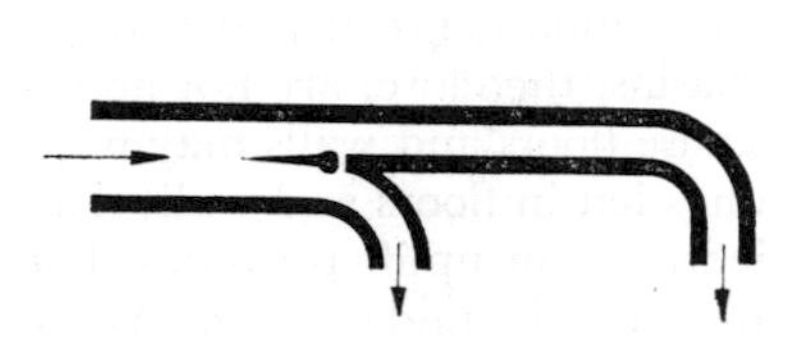

Fig. 15.3. *Adjustable vane*

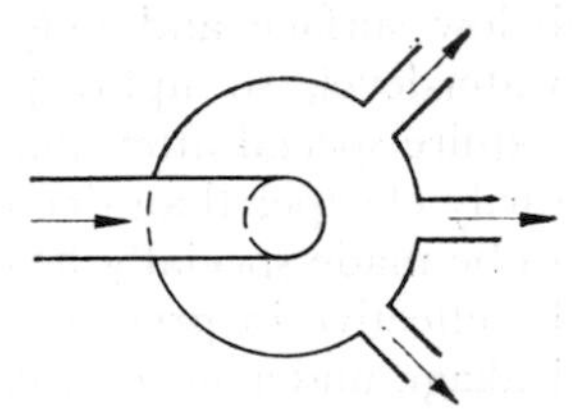

Fig. 15.4. *Circular chamber*

flow for many units in parallel and channels may be widened to provide for a number of partitions. Velocities may then be too low, but aeration can be used to maintain sufficient turbulence to prevent deposition of solids, provided the air is introduced in such a way that accurate subdivision of the flow is always obtained. Very accurate subdivision of the inflow is especially necessary in two-storey tanks, not only where one sludge compartment is used for the sludge from two or more sedimentation compartments (see p. 266) but also wherever there are two or more tanks in parallel, because sedimentation efficiencies and sludge quantities and digestion periods all depend on inflow subdivision.

Complicated machinery and apparatus should generally be avoided, especially in small towns where skilled attendance and service may be difficult to obtain. Mechanical apparatus may increase the cost and complication of the works, which may, therefore, be neglected, the resulting purification efficiency then being lower than simpler works could achieve. Generally local circumstances must be considered and sometimes may favour manual work and sometimes mechanical work. Manual control is more flexible and usually more reliable, but there is a growing need to eliminate dirty or disgusting manual work as much as possible, even in small works. In any case it is important to simplify works such as sewage treatment works to reduce unpleasant work to a minimum so that proper maintenance and operation are not hindered by aesthetic objections, and to make the work easy so that it may always be done properly. Suitable accommodation for a sufficient number of operators should be provided in all cases.

Cleanliness is one of the most important requirements in the maintenance and operation of waste-treatment works. This should be facilitated and encouraged by good design and efficient layout, and by attractive landscape treatment within the grounds. Sufficient space for areas of green grass, gardens, shrubs and trees should be provided not only for these and other aesthetic reasons, but also to encourage bird-life, which provides some control of flies and other insects.

Metering installations may be more or less complicated according to local circumstances. The amount of waste flow should be measured by a suitable indicating, integrating and recording apparatus, while other measurements to be made include temperatures, sludge quantities, sludge-gas production, water and power consumption. In larger plants additional meters should be installed to measure all significant quantities. Ratios of recirculation, excess and return sludge should also be measured. In small plants, particularly, records of sludge quantities and sludge-gas production provide a basis for supervision which may be more reliable and much cheaper than a suitable number of chemical investigations, but in larger plants laboratory investigations of reliable samples, additional to complete records of measurements, provide data as the basis for more economical operation and more efficient purification. The designing engineer should provide for such factors.

It is to be remembered that the purpose of treatment works is to ensure always at least a certain degree of purification, and this must be kept always in mind in deciding all details of design, maintenance and operation.

REFERENCES

1. WEIBEL, S. R., ANDERSON, R. J. and WOODWARD, R. L. 'Urban Land Run-off as a factor in Stream Pollution', *J. Wat. Pollut. Control Fed.* **36** (1964) 914
2. LUMB, C. 'The Storm Sewage Pollution Problem', *J. Proc. Inst. Sew. Purif.* (1964) Pt 2, 168; also *Rep. Wat. Pollut. Res. Bd. Lond.* 1955 and

1960 to date, and *Tech. Paper No. 11* (*Pollution of the Thames Estuary*) (1964) H.M.S.O., London

3. Burm, R. J. and Vaughan, R. D. 'Bacteriological comparison between combined and separate sewer discharges in south-eastern Michigan', *J. Wat. Pollut. Control Fed.* **38** (1966) 400
4. Weller, L. W. and Nelson, M. K. 'Diversion and Treatment of Extraneous Flows in Sanitary Sewers', *J. Wat. Pollut. Control Fed.* **37** (1965) 343
5. Dunbar, D. D. and Henry, J. G. F. 'Pollution Control Measure for Stormwaters and Combined Sewer Overflows', *J. Wat. Pollut. Control Fed.* **38** (1966) 9
6. Sharpe, D. E. and Shackleton, D. S. 'Balancing reservoirs. Their use in surface water drainage systems', *J. Instn munic. Engrs* **86** (1959) 80
7. Jenkins, S. H. 'Purification of Sewage at some works of the Birmingham, Tame and Rea District Drainage Board', *J. Proc. Inst. Sew. Purif.* (1961) Pt 5, 435

16

EXAMPLES OF CALCULATIONS FOR TREATMENT WORKS DESIGN

IT IS ASSUMED that the aim of treatment works is to ensure always at least a certain degree of purification required before final disposal of the wastes. This basis must be kept in mind in deciding finally all details of design and construction, and of maintenance and operation.

Design begins with the basic data which have been discussed in Part A. There may be aspects requiring special consideration and allowances, such as pumping station capacity or possible septicization of the wastes, otherwise the amounts of waste flow and the composition of the wastes must first be determined or estimated and the available methods of disposal considered accordingly. Having selected the means of disposal, the requisite degree of purification is then determined and methods of treatment chosen, having regard to all the circumstances of the particular case. The considerations involved, including aspects of re-use and reclamation, are discussed in Part B; in Part C the more important processes and techniques of waste-water treatment have been reviewed.

Sewerage and drainage systems are designed usually for peak inflows from place to place, calculated according to the time of concentration and the rate of discharge of wastes, together with groundwater and stormwater infiltration. However, these sewerage or drainage allowances cannot be used directly as the basis for treatment works design. Detention periods, and biological processes which require certain periods for ripening and adsorption and continuous activation must be allowed for. Average rates of flow and composition accordingly are more important for considerations of treatment works efficiency; the average daily flow, the average daytime flow or the average flow during working hours, therefore, provide reasonable bases from which the effective load on the treatment works can be estimated and hence may be used as a design basis. Of course the design must also consider the peak rates of inflow which may be expected occasionally, as during wet-weather flow, to ensure adequate purification under such circumstances. For municipal sewage treatment works, the average mean daily dry-weather flow (average D.W.F.) and corresponding average composition are usually taken as the design basis. For industrial waste treatment it is often necessary to consider in considerable detail the patterns of variation in rate of flow and composition before a proper basis for design can be established. However, in many cases it is feasible to use simple average values, particularly where holding or settling tanks, or mixing pits, etc., are large enough to counterbalance variations in flow and composition (cf. Chapter 3).

MUNICIPAL SEWAGE TREATMENT

The following examples of treatment works design are based on average quantities and qualities of municipal sewage generally to be found in communities with average or high living standards, as set out in Chapter 3 and Tables 3.1, 3.2 and 3.3. Since the examples have been chosen to amplify and illustrate the processes which have been discussed, the imperial system of units only has been used in the following sections, both for clarity and ease of reading. Treatment processes considered are:

Example 1 Primary treatment by Imhoff tank (pp. 166, 265).

Example 2 Primary treatment, with mechanical sludge collection and separate sludge digestion (pp. 165, 270).

Example 3 Primary treatment followed by low-rate trickling filtration (p. 196 et seq.).

Example 4 Primary treatment followed by high-rate trickling filtration (p. 196 et seq.).

Example 5 Primary treatment followed by activated-sludge treatment, including sludge digestion and utilization of sludge gas (p. 217 et seq., 279 et seq.).

Example 6 Primary treatment followed by intermittent sand filtration (land filtration) (p. 192).

Other examples of sewage disposal and treatment design are included in the text, pp. 70–77, 83, 157, 382.

Example 1. Primary Treatment by Imhoff Tank

Given a town of 10 000 population, with average living standard, without industry, combined sewerage system. Sewage D.W.F., 30 gal/hd d, peak W.W.F. to treatment works six times D.W.F.

Average daily flow $= 30 \times 10\,000 = 300\,000$ gal/d $= 12\,500$ gal/h

Peak W.W.F. $= 6 \times 12\,500 = 75\,000$ gal/h

Stages in the treatment works flow diagram are:

Raw Sewage → Screening → Grit Removal → Imhoff Tank → Treated Effluent to outfall.

(a) *Screening* Suppose 2-in bar screens are used to withhold only very coarse solids. They remove 0·1 ft^3/hd yearly, i.e. 0·1 × 10 000 = 1 000 ft^3 screenings/year, about 3 ft^3 to be burned or buried daily.

(b) *Grit chambers* Say, Essen (horizontal) type grit chambers are used, designed to remove particles of sand above 0·15 mm size under conditions of peak D.W.F. Average D.W.F., 12 500 gal/h; peak D.W.F., say, (24/10) × D.W.F. (p. 17) = 2·4 × 12 500 = 30 000 gal/h. From Table 10.2 (p. 151), settling rate for quartz particles of 0·15 mm diameter is about 2 000 in/h, hence permissible surface loading rate (p. 150) is 2 000 × 12·5 = 25 000 gal/ft^2 d, which for a flow rate of 30 000 gal/h requires (30 000 × 24)/25 000 = 29 ft^2. Peak D.W.F. of 30 000 gal/h = 1·34 ft^3/s; velocity of 1 ft/s (p. 154) taking depth of 8 in requires width of $1{\cdot}34/\frac{2}{3}$ = 2 ft, hence length is 29/2 = say, 15 ft. A second chamber would provide for peak W.W.F. of 75 000 gal/h = $3\frac{1}{3}$ ft^3/s, total area 58 ft^2 allows a settling rate of (75 000 × 24)/(58 × 12·5) = 2 480 in/h, corresponding to particles of size about 0·17 mm. If depth of flow increases to 9 in in each chamber, velocity is $2 \times \frac{3}{4} \times 3\frac{1}{3}/2 \simeq 1{\cdot}1$ ft/s. Expected quantity of grit from combined residential area (p. 156), say, 10 yd^3/1 000 persons yearly = (10 × 10)/52 = 1·9 yd^3, say, 50 ft^3 weekly, requiring storage below invert of 25 ft^3 in each chamber, or, say, 1 ft deep below invert. In wet weather, more frequent cleaning may be necessary.

(c) *Imhoff tank*

(i) Settling compartment, allow 2 h minimum capacity average D.W.F. at 12 500 gal/h = 25 000 gal, but ½ h for peak W.W.F. of 75 000 gal/h requires 75 000/2 = 37 500 gal: then, say, two Imhoff tanks with settling capacity each of 18 000 gal above slots.

(ii) Sludge compartment, from Table 13.4 (p. 277) allow 2 ft^3/person below slots, i.e. 10 000 × 2 ft^3, provide 10 000 ft^3 below slots in each tank.

(d) *Sludge disposal* The alternatives are (i) drying beds requiring for 10 000 persons at $1\frac{1}{3}$ ft^2/hd (p. 289), a total of 13 000 ft^2, say, two sets of four beds each 1 650 ft^2 drainage returned to Imhoff tank influent, or (ii) sludge lagoons, providing capacity for 2 year's digested sludge output. From Table 13.1 (p. 257), output is 0·09 gal/hd d; 0·09 × 10 000 × 365 × 2 = 657 000 gal or 105 000 ft^3, area for 2 ft depth is 50 000 ft^2; for 5 ft depth 20 000 ft^2, say ½ acre.

Example 2. Primary Treatment with Mechanical Sludge Collection and Separate Sludge Digestion

Consider a town of 50 000 population, with high living standard, separate sewerage; sewage flow about 45 gal/hd d, plus some industry discharging wastes during the working period of 8 h as follows:

Type of waste	*Actual population or population equivalent*	*Waste flow per equivalent population unit* gal/d	*Total flow* gal/d
Residential	50 000 actual	45	2 250 000
Mixed industry (6%)	3 000 equivalent	30	90 000
Abattoirs	12 000 equivalent	6–7	(say) 80 000
Wire factory	3 000 equivalent	40	120 000
Total population equivalent	68 000	(37)	2 540 000

Suspended solids contribution per equivalent here is taken to be similar to normal contribution per head, therefore the population equivalents are also suitable for sludge treatment design.

Particular volumes of waste:

Mean D.W.F.

domestic sewage 2 250 000 gal/d — 94 000 gal/h
industrial wastes 290 000 gal/8 h — 36 000 gal/h
daytime mean flow = 130 000 gal/h

Peak D.W.F.

domestic sewage $24/14 \times 94\,000$ — 162 000 gal/h
industrial wastes $1{\cdot}5 \times 36\,000$ — 54 000 gal/h
peak hourly D.W.F. = 216 000 gal/h

Peak W.W.F.

peak D.W.F. — 216 000 gal/h
plus $0{\cdot}5 \times$ D.W.F. — 65 000 gal/h
peak W.W.F. = 281 000 gal/h = $12{\cdot}5\ \mathrm{ft^3/s}$

Treatment works flow sheet:

Raw wastes → Screening → Grit Removal → Sedimentation (Sludge to digestion tanks) → Effluent to disposal system (digested sludge to drying beds, finally disposed of as fertilizer).

(a) *Screening and shredding* of the whole waste flow is carried out using a disintegrating machine of 'comminutor' type (p. 145). Total volume of residual screenings negligible (bury).

(b) *Grit removal* Select horizontal type (Fig. 10.6) design, say, to retain sand of nominal 0·1 mm at peak hourly D.W.F. of 216 000 gal/h = 9·6 ft^3/s. Corresponding settling rate is 950 in/h, whence permissible loading rate is 950 × 12·5 gal/ft^2 d (Table 10.2 and equation p. 150) and required surface area is (216 000 × 24)/(950 × 12·5) = 436 ft^2: velocity of 1 ft/s (p. 154) for 9·6 ft^3/s requires 9·6 ft^2, say 7 ft wide, flow 16 in deep, required length, then 436/7 = 62 ft 3 in, say 62 ft; sand storage, allowing, say 5 yd^3/1 000 population units annually (p. 156), should be, for fortnightly cleaning, (68 × 27 × 5)/26 ≏ 350 ft^3, or for chambers 7 ft wide and 62 ft long, a depth below invert of, say, 10 in. Provide two chambers 3 ft 6 in wide for daytime D.W.F., and a third chamber of the same size for W.W.F., allowing more frequent cleaning as necessary from time to time. Say, flow level is then 15 in deep approximately for three chambers under peak W.W.F. conditions, then velocity for 12·5 ft^3/s is 12·5/(10·5 × 1·25) ≏ 0·95 ft/s; and loading rate for 10·5 × 62 = 651 ft^2 is 10 400 gal/ft^2 d, corresponding to a settling rate of 830 in/h and a minimum size of about 0·09 mm, which are satisfactory values: design accordingly. (See also Example 5.)

(c) *Sedimentation tank* The average peak hour's D.W.F. of 216 000 gal/h requires minimum 1·5 h detention, say, 324 000 gal capacity, while 2 h average D.W.F. of 130 000 gal/h requires 260 000 gal, and ¾ h for peak W.W.F. of 281 000 gal/h requires 210 000 gal. Provide then 320 000 gal capacity, using mechanical sludge removal and 20% to provide additionally for scrapers and mechanization (p. 166), or in all, say, 400 000 gal; taking depth of mechanical tanks at 7–9 ft, say 8 ft, a capacity of 400 000 gal corresponds to surface area of 400 000/(6·25 × 8) = 8 000 ft^2; peak D.W.F. of 216 000 gal/h corresponds to a surface loading rate of (216 000 × 24)/8 000 = 650 gal/ft^2 d, and a minimum settling rate of 650/12·5 ≏ 50 in/h, which are satisfactory (p. 166): provide two tanks each 200 000 gal capacity.

(d) *Sludge pumping* From Table 13.1 (p. 257) say up to ½ gal/hd d (allowing for mechanical desludging), or in all up to 0·5 × 68 000, say 30 000 gal sludge, is to be pumped to the separate sludge digestion plant daily, or during 8 h at daytime, say, 30 000/8 = 3 700 gal/h; pumping capacity required, say, 70 gal/min.

(e) *Digestion plant* Using tanks heated, say, to 90°F (32°C), allow 0·8 ft^3/population unit or 68 000 × 0·8 = 54 000 ft^3 (Table 13.4, p. 277) in this case, say, two heated tanks each 30 000 ft^3 (equipped for sludge gas collection and heating with sludge gas) and a third unheated tank for sludge storage purposes and for separation of supernatant liquor, also 30 000 gal.

(f) *Sludge disposal* The required sludge drying area for a dry, warm climate with short winters is about $1\frac{1}{3}$ ft²/population unit, say 68 000 × $1\frac{1}{3}$ ≏ 100 000 ft². Quantity of digested sludge (Table 13.1, p. 257) about 0·09 × 68 000 ≏ 6 100 gal/d wet digested sludge, yielding about 0·03 gal/hd d × 68 000 or about 4 500 yd³ annually of dried sludge bulk.

Example 3. Primary Treatment Followed by Low-Rate Trickling Filtration

Designed for 90–95% overall purification—taking a small town with average living standard as in Example 1, population 10 000 inhabitants, no industry, a further requirement for wet weather conditions is that up to three times D.W.F. shall be treated by trickling filters.

Treatment flow sheet:

Screening → Grit Removal → Imhoff Tank Treatment → Low-rate filtration → Humus Removal → Effluent Disposal (Wet sludge disposed of by lagooning).

(a) *Screening*—as Example 1

(b) *Grit chambers*—as Example 1

(c) *Imhoff tank*

(i) Settling compartment as for Example 1.

(ii) Sludge compartment, enlarged for treatment of trickling filter humus sludge which results in an increase in digested sludge (Table 13.1, p. 257), 0·11 gal/hd d as against 0·09 gal/hd d for Example 1. Table 13.4 (p. 277) suggests digestion tank capacity requirement as 3·0 ft³/hd, then Imhoff tank sludge compartments capacity should be 10 000 × 3 = 30 000 ft³, 15 000 ft³ in each below slots.

(d) *Low-rate trickling filters* Being a small town allow, say, 8 lb B.O.D.$_5$ daily/1 000 ft³ (p. 207); taking 0·12 lb B.O.D.$_5$/hd d of which 0·04 lb is settleable (p. 19), allowing 90% removal of suspended solids by sedimentation 0·12 – (0·04 × 0·9) = 0·084 lb/hd, total 0·084 × 10 000 = 840 lb B.O.D.$_5$ loading on trickling filters, requiring (840/8) × 1 000 = 105 000 ft³ (3 900 yd³). This corresponds to about $2\frac{1}{2}$ persons/yd³. For 6 ft depth, required area is about 17 500 ft², 2 000 yd² (about 0·4 acre) and loading is 300 000/3 900 = 77 gal/yd² (p. 206), say then four filter beds, each about 75 ft diameter. Peak hourly D.W.F. average, is 24/10 × average hourly D.W.F., i.e. 12 500 × (24/10) = 30 000 gal/h (500 gal/min); 3 times average D.W.F. = 3 × 12 500 = 37 500 gal/h (625 gal/min): design dosing tanks and rotary distributor accordingly for, say, 200 gal/min (to cover emergencies) in each of four beds; total 800 gal/min.

(e) *Humus tanks* Reckon 1·5 h detention for peak W.W.F. from filters of (37 500 × 1·5), say 56 000 gal; provide two Dortmund type hopper-bottomed tanks, vertical velocity not to exceed 75 in/h (p. 200); minimum area, then, (37 500 × 24)/(75 × 12·5) (p. 150), not less than 960 ft^2, design accordingly; pump required for humus sludge, say 0·04 × 10 000 = 400 gal sludge/d (Table 13.1, p. 257), provide, say 600 gal/h = 10 gal/min.

(f) *Sludge disposal* Use lagoons as in Example 1, but total digested sludge volume increased to 0·11 gal/hd d; total annual volume = 0·11 × 10 000 × 365 gal, say 65 000 ft^3; the area required for 2 ft depth is 33 000 ft^2 (drying beds for nine fillings annually, 8 in deep would require about 11 000 ft^2).

Example 4. Primary Treatment Followed by High-Rate Trickling Filtration (pp. 198, 206)

Designed for 85–90% overall purification taking a city of 50 000 population with high living standard, with industry, as in Example 2, total equivalent 68 000 population units, separate sewerage system. As Example 2:

Mean D.W.F. = 130 000 gal/h (daytime mean flow)
Peak hourly D.W.F. = 216 000 gal/h
Peak W.W.F. = 281 000 gal/h = 4 700 gal/min = 12·5 ft^3/s

Treatment works flow sheet (see Fig. 16.1):
Raw waste water → Screening → Grit Removal → Sedimentation → High-rate filtration → Humus Removal → Effluent.
Sludges to separate sludge-digestion tanks, digested sludge to drying beds.

(a) *Screening*—as Example 2
(b) *Grit removal*—as Example 2
(c) *Primary sedimentation*—as Example 2

(d) *High-rate filters* The principal distinction from low-rate filters is a distribution rate high enough to maintain the flushing of excess filter slimes (p. 198); taking a moderate loading of 45 lb $B.O.D._5$/1 000 ft^3 d (p. 206), use 9 ft deep beds because the sewage is strong, 9 × 43 6 × 45 = say 18 000 lb/acre; assuming 0·15 lb $B.O.D._5$/hd d containing 0·05 lb of $B.O.D._5$ in settleable solids, 95% removal of settleable solids by sedimentation gives a loading on filters, for 68 000 population units, of 68 000 × (0·15 − 0·05 × 0·95) = 7 000 lb $B.O.D._5$/d requiring, therefore, 7 000/18 000 = 0·39 acre. An average D.W.F. of 130 000 gal/h = 3·12 × 10^6 gal/d for 0·39 acre corresponds to a loading rate of only 8 × 10^6 gal/d per acre which is grossly inadequate for flushing, the normal requirement being greater than 10 × 10^6 gal/d per acre (pp. 198, 207) preferably 15 × 10^6 gal/d per acre or more; hence reckon on doubling the flow by recirculation ratio of 2 : 1 (p. 208) thus providing a hydraulic loading rate of about $8 \times 2 = 16 \times 10^6$ gal/d per acre or 4 300 gal/min combined rate of discharge on to filters in dry weather; in wet weather reduce circulation to maintain total rate not exceeding 281 000 gal/h = 4 700 gal/min, design filters accordingly 9 ft deep, total effective area 0·39 acre; say three filters each 85 ft diameter; distributor

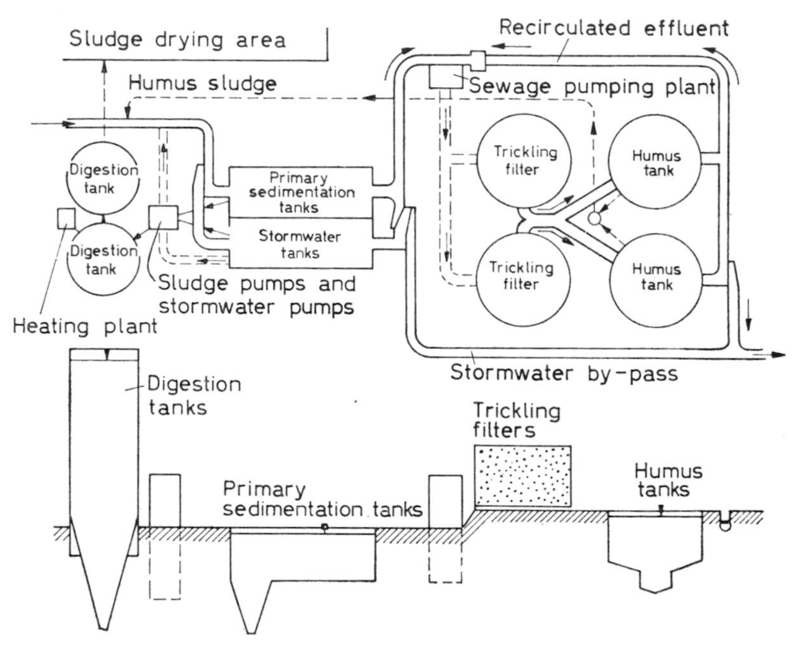

Fig. 16.1. *High-rate filtration plant design* (*Example 4*)

capacity each filter 1 600 gal/min. Ventilation openings at least 0·5% of 0·39 acre (p. 202), say about 85 ft², make about 30 ft² each filter. Being flushing filters, media of size about 3 in would be suitable; specification might exclude material sized less than 2 in.

(e) *Humus tanks* The recirculation cycle includes humus tanks; allow for $1\frac{1}{2}$ h detention at peak W.W.F. of 4 700 gal/min = 281 000 gal/h requiring, say, 420 000 gal capacity; this corresponds to $420\,000/(130\,000 \times 2) = 1{\cdot}6$ h average detention under normal dry weather operation; reckoning rate of overflow not to exceed 84 in rise/h (p. 200), surface area should be not less than $(4\,700 \times 1\,440)/(84 \times 12{\cdot}5)$ (p. 150), or 6 500 ft²; say three Dortmund tanks each 140 000 gal capacity and 2 200 ft^2 of surface; expected volume of humus sludge increased compared with low-rate filtration, from Table 13.1 (p. 257) 0·07 gal/hd d; 68 000 $\times$ 0·07 = 4 760 gal, say 5 000 gal to be pumped from humus tanks to digestion plant daily.

(f) *Sludge treatment* Digestion tank capacity required for mixed primary and humus sludges is considerably larger than Example 2, from Table 13.4 (p. 277) say 1·7 ft^3/population unit for heated tanks, total capacity required 68 000 $\times$ 1·7 = 116 000 ft^3, say two heated tanks each 58 000 ft^3 and a third unheated tank for storage purposes and separation of supernatant liquor, also 58 000 ft^3. Raw sludges (thickened primary with humus sludge), say $\frac{1}{3}$ gal/hd d totalling 23 000 gal/d. Sludge gas production reckoning 1·3 ft^3/population unit daily (see p. 279) $\simeq$ 90 000 ft^3/d.

(g) *Sludge disposal* The volume of wet digested sludge is considerably greater compared with primary treatment only (higher moisture content) 0·12 $\times$ 68 000 = 8 200 gal/d, when dry amounting to a bulk of about 68 000 $\times$ 0·04 = 2 700 gal or 430 ft^3/d available for sale as fertilizer of high humus content; drying bed area assuming average warm, dry climate, allowing nine fillings annually each 8 in deep (p. 289), totalling 6 ft, average total annual volume of wet digested sludge 8 200 $\times$ 365 gal/year = say 480 000 ft^3 of sludge annually, requires 480 000/6 = 80 000 ft^2 or about 1·2 ft^2/population equivalent.

Example 5. Primary Treatment Followed by Activated-Sludge Treatment including Sludge Digestion and Utilization of Sludge Gas

Take a town of population 50 000 with high living standard, with industry, separate sewerage system as in Examples 2 and 4, designed for 95% purification. Flow data estimated as Example 2; D.W.F. to be treated by activated sludge, daytime mean flow (dry weather) 130 000 gal/h, peak hourly D.W.F. 216 000 gal/h; excess to be treated only by sedimentation; peak W.W.F. 12·5 ft^3/s or 280 000 gal/h.

Treatment flow sheet:

Raw Sewage → Screening → Grit Removal → Primary Sedimentation (together with excess activated sludge) → Aeration with activated sludge → Final sedimentation (with separation of activated sludge) → Effluent, with digestion of mixed sludge (Fig. 16.2).

(a) *Screening*—as Example 2

(b) *Grit removal* Select an aerated grit chamber (p. 155) capacity 2 min peak W.W.F., 2 $\times$ 12·5 $\times$ 60 = 1 500 ft^3. Make 10 ft deep, 6 ft wide and 25 ft long, surface area 150 ft², required air flow $1{\cdot}5 \times 1\,500/60 \simeq 40$ ft³/min. Cross section 60 ft² and longitudinal flow velocity at peak W.W.F. amounts to $(12{\cdot}5/60) \times 60 = 12{\cdot}5$ ft/min $\simeq 0{\cdot}2$ ft/s, which is satisfactory. The grit removed from the sump estimated at $(50\,000 \times 5)/1\,000 = 250$ yd³ annually (p. 156). Particle size and cleanliness under various conditions of flow controlled by adjustment of air flow. Provide for aeration similarly as for Fig. 10.7 allowing up to 50 ft³/min compressed air supply to grit chambers. If overflow to downstream channels or piping is not completely free, some provision should be made for skimming off any accumulated scum (pass skimmings on to sludge seeding or mixing wells for digestion).

(c) *Primary sedimentation* Plant is to provide also for separation of excess (surplus) activated sludge (p. 239), say 3% additional flow,

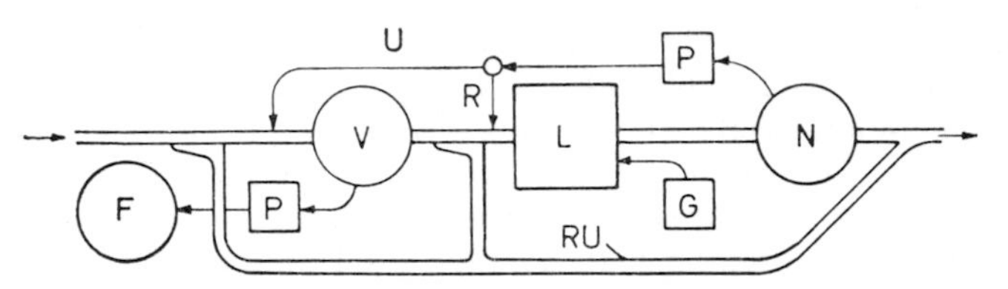

Fig. 16.2. *Layout of activated-sludge plant*

V = *primary sedimentation tanks*
L = *aeration tanks*
N = *humus tanks*
F = *digestion tanks*
P = *pumping plant*
G = *air-compressor house*
R = *returned activated sludge*
U = *excess activated sludge*
RU = *stormwater by-pass*

design generally as for Example 2; reckon peripheral speed of sludge scrapers at about 10 ft/min (see p. 165).

(d) *Sludge pumping* Sludge volume increased allow, say twice the normal allowance (see Example 2) 68 000 gal/d; sludge pump say 70 000 gal operating 12 h total daily; or $70\,000/12 =$ say 6 000 gal/h or 100 gal/min discharging to sludge thickening tanks.

(e) *Aeration tanks* Selecting normal diffused air system, take loadings equal to a nominal 7 h aeration period (cf. Table 12.3, p. 233) for mean daytime hourly flow tank volume $= 130\,000 \times 7 = 910\,000$ gal. This corresponds to $910\,000/216\,000 = 4{\cdot}2$ h for peak D.W.F. Taking 7 000 lb B.O.D.$_5$ in primary tank effluent (Example 4(d)) we have $(7\,000/910) \times 6{\cdot}23 = 48$ lb B.O.D.$_5$ input daily/1 000 ft³. Furthermore, taking the suspended solids residual in the primary tank effluent (92% settling efficiency) at 0·09 lb/hd d, and the mixed liquor suspended solids (MLSS) concentration at

3 000 p.p.m. (minimum) we have a Gould Sludge Age of $(910\ 000 \times 10 \times 3\ 000 \times 10^{-6})/(0{\cdot}09 \times 68\ 000) = 4{\cdot}5$ d minimum. The B.O.D.$_5$ loading of $7\ 000/(910\ 000 \times 10 \times 3\ 000 \times 10^{-6}) = 0{\cdot}26$ lb/lb MLSS. These values are generally satisfactory but suggest that the MLSS level probably should be maintained at a level of 4 000 p.p.m. rather than 3 000. Design tanks, say 16 ft deep with diffuser plates 13 ft below surface, 17 ft wide and hence $910\ 000/(16 \times 17 \times 6{\cdot}24) = 540$ ft total length; say then four tanks each 135 ft long. 95% overall purification, taking B.O.D.$_5$ at 0·15 lb/population unit daily (Table 3.2, p. 19) corresponds to an amount of $0{\cdot}15 \times 0{\cdot}05 \times 68\ 000 = 510$ lb remaining in the final effluent. Allowance must be made for the return of digestion tank supernatant, say equivalent to 2% of raw sewage B.O.D.$_5$ (p. 276) $= 0{\cdot}02 \times 68\ 000 \times 0{\cdot}15 = 204$ lb B.O.D.$_5$/d; total removal by aeration of $7\ 000 + 204 - 510 =$ say 6 300 lb B.O.D.$_5$ requiring 500 ft^3 air/lb of B.O.D.$_5$ removed (p. 227) or $6\ 300 \times 500 = 3\ 150\ 000$ ft^3 of air supply daily, equal to 2 200 ft^3/min; provide say 3 300 ft^3/min to allow for peak demands, aeration of channels, aerated grit chamber, etc.; pressure of air supply, say 14 ft water, and allowing 75% efficiency, then power requirement is $(3\ 300 \times 14 \times 62{\cdot}3)/(33\ 000 \times 0{\cdot}75) = 116$ hp or 85 kW or $116/67 = 1{\cdot}7$ hp/per 1 000 population units (p. 242).

(f) *Final sedimentation tanks* Suppose hopper-bottomed tanks are used designed for 2 h detention peak D.W.F. $216\ 000 \times 2 = 432\ 000$ gal with space for sludge collection, say then 500 000 gal or four tanks each 125 000 gal; reckoning rate of upflow 72 in/h maximum (p. 238), surface area required for 216 000 gal/h not less than $(216\ 000 \times 24)/(72 \times 12{\cdot}5) = 5\ 800$ ft^2.

(g) *Returned activated sludge* Reckon 60% of mean daytime D.W.F. or $130\ 000 \times 0{\cdot}60 = 78\ 000$ gal/h or 1 300 gal/min; also allow for excess activated sludge to be pumped to primary sedimentation tank inlet; from Table 13.1, 1·1 gal/hd d, i.e. 75 000 gal/d/1 440 = 52 gal/min, allowing 12 h pumping daily make 100 gal/min; total pumping capacity 1 300 + 100, say 1 500 gal/min.

(h) *Sludge treatment* Reckon digestion of mixed raw and excess activated sludge, selecting temperature of 90°F (32°C) from Table 13.4 (p. 277) required capacity is $1\frac{1}{2}$–$2\frac{1}{2}$, say 1·8 ft^3/population unit, hence $68\ 000 \times 1{\cdot}8 \simeq 120\ 000$ ft^3, say two primary digestion tanks, heated to 90°F (32°C) each 60 000 ft^3 and one secondary tank, unheated, also 60 000 ft^3 for storage purposes and for separation of supernatant liquor, totalling 180 000 ft^3; equip two first-stage (primary) tanks with fixed covers for gas collection, allowing that separation of liquor will not be made from these tanks, and equip the secondary tank for gas collection under floating covers; thickening for sludge from primary sedimentation (two thickening tanks each 34 000 gal) will yield about 68 000 × 0·43 = say 30 000 gal thickened mixed sludge to be digested daily, corresponding digestion periods about 25 d first stage and more in second stage, which is sufficient (p. 264, Fig. 13.3, Table 13.3).

(k) *Sludge gas collection* Gas yield may range from 1·2–1·5 ft³/hd d (p. 279), reckon 1·5 ft³/hd d or say 100 000 ft³ d; taking power yield as about 1 hp for 18 ft³ sludge gas/h (p. 284), power yield will be 100 000/(18 × 24) = 230 hp. Power requirement for plant is, say 120 hp leaving about 110 hp available for heating the digestion tanks (i.e. (100 000 × 110)/230 = about 50 000 ft³ gas); this should be ample for heating digestion tanks from 50–90°F (10–32°C) (p. 284); gas holder capacity, provide 25% of daily gas production plus volume of daily raw sludge addition (30 000 gal or 4 800 ft³) hence (110 000/4) + 4,800 ≏ 30 000 ft³ total (see p. 280).

(l) *Sludge disposal* The digested sludge volume from Table 13.1 (p. 257) is 68 000 × 0·17 = say 12 000 gal/d; dried sludge bulk (68 000 × 0·06)/6·23 = 650 ft³ or say, 25 yd³ d; for sludge drying, allowing 2¼ ft²/population unit (p. 289) drying bed area should be 68 000 × 2¼ = 153 000 ft² or about 3½ acres.

Example 6. Primary Treatment Followed by Intermittent Sand Filtration (Land Filtration)

Take a town of 10 000 population with average living standard without industry, as for Examples 1 and 3, assume that areas of suitable sandy soil (p. 192) are available, and land filtration, therefore, is selected to achieve about 95% overall purification.

Treatment flow sheet as follows:

Raw Sewage → Screens → Grit Chambers → Imhoff Tanks → Intermittent Sand Filters → Outfall:

Flow data as for Example 1.

Average D.W.F. 30 × 10 000 = 300 000 gal/d
= 12 500 gal/h average

Average peak hourly D.W.F. = 24/10 × 12 500 gal/h
= 30 000 gal/h

Peak W.W.F. (6 × D.W.F.) = 75 000 gal/h = 3·34 ft³/s

(a) *Screens*—as Example 1
(b) *Grit chambers*—as Example 1
(c) *Imhoff tanks*—as Example 1
(d) *Sludge disposal*—as Example 1
(e) *Intermittent sand filtration* Allow 1 600 persons/acre, hence

10 000/1 600 = 6·3 acres, corresponding sewage load 300 000/6·3 = 48 000 gal/d per acre, average rate of drainage 48 000/($6\frac{1}{4}$ × 43 560 × 24) = about 0·007 ft/h (Table 12.2, p. 194): provide eight beds, each 0·8 acre, normally using four beds for daytime flow and two beds for night flow, all beds in rotation; thus in effect two beds are available for emergencies in wet weather; taking daytime flow during 8 h at (300 000/12) × 8 = 200 000 gal, filling 2 h on to four beds gives 50 000 gal/bed or for 0·8 acre (= say 35 000 ft^2) approximately 3 in depth which is about right for such sewage: *storage basin* required capacity equal to the volume of charge per bed, required therefore 50 000 gal capacity: *influent and effluent* channels; each bed should be charged in 10 min (from storage basin) hence dosing (inlet) channels (each bed) require capacity of 50 000/10 = 5 000 gal/min or 13·3 ft^3/s; allowing 2 h for drainage, effluent pipe capacity (each bed) should be 50 000/($6\frac{1}{4}$ × 7 200) = about 1·1 ft^3/s.

For very small works see Part D (Examples, p. 382).

PART D

DISPOSAL AND TREATMENT OF WASTES FROM RESIDENCES, INSTITUTIONS AND SMALL COMMUNITIES

17

DISPOSAL OF WASTES FROM ISOLATED RESIDENCES AND INSTITUTIONS

GENERAL CONSIDERATIONS

The problems of disposal of wastes considered in Parts A to C were such as may be expected in connection with communal sewerage schemes or large factories, involving so many individual sources of waste, or such large amounts, that the sewerage and drainage systems and treatment and disposal works must be relatively substantial, otherwise the nuisances which arise become widespread and the living conditions of whole communities may be adversely affected.

However, there are in all countries many buildings not connected either to public sewers or to private systems of comparable dimensions to the sewerage and drainage systems of moderate sized towns or villages. This may be because the buildings are isolated from the more densely settled metropolitan and urban areas so that their connection to public sewerage systems would be extremely costly compared with independent arrangements; in some cases it is rather a matter of standards of living, particularly local standards of sanitation and communal economy. Whatever the case, except in the most primitive types of civilization, there must always be some water-borne wastes—from kitchen, bathroom, laundry or water closet,[1] or from factory, farmyard, dairy, etc.—and it is the disposal of these wastes from independent or isolated buildings with which this part of the book is concerned.

In any community many such buildings must exist including not only residences, small or large, but institutions such as hospitals, schools or camps, or isolated factories.[2–5] Two types of installation may be used: (1) where body wastes are collected and disposed of by dry-closets or otherwise as described later and sullage wastes must be dealt with as water-borne wastes and disposed of separately; and (2) where water-closets are installed, so that the drainage systems, although individual, are similar to those connected to public sewers.

In most civilized countries these independent arrangements are usually subject to regulations imposed by local (district) authorities, or by the supervising (national) government public health authorities, or at least are subject to their concurrence or specific approval, but the laws, regulations or standards which are applicable differ widely among the many countries. Despite differences in regulations the techniques used have approximately the same basis in all countries and the special techniques of wastes engineering required for small sewerage schemes are outlined in the following pages, at least so far as they may differ from ordinary sewage disposal systems discussed in the earlier chapters.

In considering any particular case the first question to investigate should be the possibilities of linking up the drainage of the particular building (or projected building) with the public sewerage of the community. Individual waste disposal is always relatively more costly, allowing for the greater attention and maintenance required, and it is often more economical and more satisfactory generally to pay higher costs for extended pipelines, perhaps with individual pumping arrangements, than to pay for independent individual disposal schemes. This may apply particularly to cases where trends of suburban growth have carried housing or other development beyond ordinary drainage limits or beyond economical limits of existing sewerage.

The magnitude of the individual disposal scheme depends first on the water consumption—excluding agricultural or horticultural use—so that to this extent the standards of water supply and waste disposal are related. Buildings with a piped water supply usually require more water per person (or equivalent) than those without piped water; the wastes also are generally different, and they are, therefore, considered separately. There may be two possibilities in each case, either when the use of water is confined to kitchens (with bathing or laundering) only, or when water-closets are also provided. The provision of the latter is unusual where there is no piped water supply,[6, 7] because the expected water consumption with water-closets, which will be much the same as where unrestricted public water supply and public sewerage is available, can be met practically only with a piped, albeit private, water supply.

WASTES FROM BUILDINGS WITHOUT PIPED WATER

Isolated buildings, especially small residences, may use roof water, water from wells or bores, or from open watercourses, delivered or collected in buckets or jars from which it is poured for use: in such cases water consumption is practically restricted to a minimum and water-closets are not used.

Body wastes then are disposed of by various different methods, according to the circumstances, including pail- pan- or earth-closet (dry-closet), cesspit, or chemical closet systems.

Pail-, pan- or earth-closet systems[2] are used widely in some countries, especially in Asia and Africa, also in Australia. They consist in their most developed form in municipally controlled two-pan systems whereby pans are exchanged at least weekly—in many Eastern communities even daily. The used pans are collected (using tight lids during transport in special wagons) and the contents disposed of by burial in trenches 1–2 ft (0·3–0·6 m) deep in suitable well-drained soils. The empty pans are then cleansed and disinfected to some extent ready for use again, the second pan being substituted for use meanwhile, and so on in rotation. Pans may be up to about 5–6 gal (23–27 l) capacity, the closets (where used) being designed accordingly. Some are designed so that urine and faeces are collected separately, as far as is practicable, otherwise in some countries it is common practice to add specially graded earth, sawdust or other such more or less inert drying agent from time to time to the pan in use. The *night soil*—as the contents of such pans are named—may also be disposed of where possible by dumping into existing sewerage systems, directly into sludge-digestion works or may be composted suitably (see p. 115), usually with waste vegetable material to offset the relatively low carbon : nitrogen ratio. Digestion of night soil requires special adjustment of the carbon : nitrogen ratio, otherwise the rapid formation of ammoniacal compounds may result in alkalinities too highly buffered for the maintenance of active decomposition.

In rural areas especially, cesspit or cesspool methods of disposal of body wastes are used where suitable ground is available. In some cases cesspits are made watertight, for example, according to regulations in Great Britain, and then consist of closed tanks from which the accumulated wastes must be removed periodically, say, each six months, and disposed of on to agricultural or horticultural areas, or elsewhere. Where cesspits are not required to be watertight, the bottom and walls may be made pervious so that the liquid portion of the waste seeps away, the solids then accumulating in the pit at a rate of about 10 gal/hd (45 l/hd) yearly. A watertight cesspit requires a capacity of about 15 gal/hd (70 l/hd) per month, thus if it is to be emptied twice yearly, i.e. each six months, its volume must be at least 90 gal/hd (410 l/hd) served. Odour may be reduced or controlled by the regulated addition of granulated peat, which accordingly requires provision of additional capacity.

Cesspits should be located at least 60 ft (18 m) from residences, public buildings, watercourses, dams or wells and not on the up-stream side of wells supplying drinking water.

Chemical closets consist usually of comparatively small watertight tanks or pits into which the body wastes are carried or deposited directly, with intermittent charges of the chemical, some elementary manually operated mechanism (operated for example by the raising of a lid) ensuring intimate admixture of chemical, faeces and urine. Mostly caustic soda is used, added in flake (solid) form with a small quantity of water (say once or twice daily), the mixture thus being 'liquefied' and practically sterilized. The treated liquid waste over-flows from the tank to some absorption area or in some cases to a collection pit from which it is removed regularly as required.

In all cases, using dry-closets, cesspits or chemical closets, additional arrangements are necessary for the disposal of *sullage wastes* (pp. 6, 366) from kitchen, washroom and laundry. With primitive water supply, the total volume of sullage seldom exceeds 5 gal/hd d (23 l/hd d) and is commonly less, and may be distributed crudely by throwing on to gardens or grassed areas and so dispersed and absorbed, or may be drained to an absorption pit from which it overflows into or is absorbed by the surrounding soil. Such absorption pits may be lined with brick or other coursed structures, laid dry (without joint filling), using coarse gravel or rubble beds about 1 ft (0·3 m) deep below the lining, or may be simple unlined excavations filled with coarse rubble, the pits being sunk in either case at least down to a porous soil horizon, using dimensions based on percolation tests (p. 379). Where absorption is not readily obtained, pits may be connected finally to absorption trench systems similar to the soil absorption systems described on p. 376, but usually smaller because of the smaller volume of waste waters to be absorbed. Any such pits or trenches should be located so as to avoid contamination of watercourses or underground water supplies.

WASTES FROM ISOLATED BUILDINGS WITH PIPED WATER SUPPLIES

In most isolated buildings supplied with piped water from public utilities or private supplies the water-borne wastes include those from kitchen, bathroom, laundry and water-closet, and the average of such wastes may not differ greatly from ordinary domestic sewage discharged from sewered areas except that the actual composition and quantities depend on the living habits of the particular group of persons concerned and the facilities available. The flow amounts usually to at least 25 gal/hd d (115 l/hd d) or more, and in countries such as the U.S.A., where water consumption is relatively high, may may be about 40–50 gal/hd d (180–230 l/hd d).

The problem usually is to dispose of this sewage individually either into natural surface waters or into the ground, requiring first some treatment designed to withhold objectionable solid matter and to render the wastes less dangerous as carriers of water-borne disease. In most cases the treatment to be provided and the method of disposal are subject to prescriptions and regulations of the local authorities. The individual scheme may be provided for a single residence, in which case it is designed for a minimum of five persons or more, or it may be for a larger building or group of buildings with a served population of up to say 300 persons or equivalent. Where a still higher number of persons is concerned it is more appropriate to develop a scheme of treatment and disposal as outlined in Parts B and C. The principles of disposal and treatment are the same in either case but the units used for larger schemes require modification before they may be applied successfully in small sewage treatment works.

It is essential that roof and surface water be excluded absolutely, otherwise the processes of treatment and the disposal works may be seriously disturbed or choked by excessive stormwater flows, and satisfactory operation may not often be obtained subsequently.

The methods used must be developed according to local conditions and circumstances as economically as possible. Where surface-water bodies with a sufficiently large capacity for dilution and self-purification—having regard to all other potential and actual sources of pollution (pp. 57, 81 et seq.)—are available nearby, the sewage may be disposed of directly into such waters after primary treatment or septic tank treatment only; otherwise *full treatment* by trickling filtration or other biological treatment process may be necessary before discharge. Where such water bodies are not readily available—this being the more usual case—disposal must be on to the land (which commonly involves the groundwaters), and in this case also primary or septic tank treatment may sometimes be sufficient. In some instances biological treatment, e.g. by trickling filtration, may be necessary in addition, particularly for disposal on to land which is not the property of the persons or institution served, also where the area and capacity of the available land are relatively limited. Usually primary treatment, in settling tanks or septic tanks, is sufficient pretreatment, permitting subsequent disposal on to land using the methods of broad irrigation, absorption pits, or subsurface irrigation or filtration. Broad irrigation is seldom used, however, because of its disadvantages of odour and health risks. Absorption pits may be applied successfully in suitable cases. Subsurface irrigation is used most commonly because it is inherently the most suitable method. These methods, as applied to small schemes, are described in Chapter 18.

DISPOSAL OF SULLAGE

The content of polluting matter in sullage water is only about one-third of that in domestic sewage, whereas the volume may amount to as much as two-thirds of ordinary domestic sewage flow, so that sullage is about one-half as strong as normal domestic sewage. In Great Britain the sullage volume amounts to about 20 gal/hd d (90 l/hd d) containing about half as much total solids content and B.O.D. as the values given in Table 3.3 (p. 20), but where water consumption is higher, as in the U.S.A., the volume may amount to two or three times more or, say, 40–50 gal/hd d (180–230 l/hd d). Sullage disposal and treatment do not differ in principle from what has been discussed earlier in Parts A, B and C and centralized sullage treatment plants, for camps or other institutions or for large groups of residences, may be considered as having been dealt with already. It should be noted that the relative proportion of greasy matter, principally animal fats, is greater, and special attention to grease removal is often necessary.

Sullage disposal from small isolated institutions or individual residences commonly presents difficult problems because usually such cases arise only where permanent natural streams or lakes do not exist (otherwise full sewerage services and ample dilution by natural waters are usually available). These problems may be intensified where settlement is of suburban rather than merely rural density, because usually the sullage must be disposed of on to the land, and great difficulty arises particularly if the land available to each house is only small in area, and the ground relatively non-absorbent. Where ample area is available, sullage may be disposed of simply by distributing it over the surface of broken-up ground. The sullage from about 150–300 persons may require an acre (0·4 ha) of such land.

DISPOSAL OF WATER-CLOSET WASTES

In some special cases the wastes to be disposed of may consist entirely of water-closet wastes,[6, 7] for example, from public buildings in picnicking areas and may be very concentrated with a relatively high content of nitrogen. Such wastes are readily treated in septic tanks or other systems as described in Chapter 18.

DRAINAGE SYSTEMS

For isolated buildings and institutions it is practically always preferable to separate the stormwater drainage and waste disposal systems completely. Stormwater from roofs, yards, etc., is collected or directed and discharged into natural watercourses, or where this may be difficult, or where the additional irrigation is desirable, may be spread over land or underground. Water-borne wastes are then also collected within buildings or stables, etc., separately—in the same way as for discharge to separate municipal sewerage systems and treated appropriately before disposal (see Chapter 18).

Where there is a possibility that buildings or institutions now isolated will eventually be connected to a nearby (public) sewerage system, or will be connected into a new system, it is common practice to ensure that any plumbing and drainage conforms to the requirements of the authorities concerned.

BIBLIOGRAPHY

EHLERS, V. M. and STEEL, E. W. *Municipal and Rural Sanitation*, 6th Edn (1965) McGraw-Hill, New York

SALVATO, J. A., JR *Environmental Sanitation* (1958) 188 et seq., Wiley, New York

WAGNER, E. G. and LANOIX, J. N. *Excreta Disposal for Rural Areas and Small Communities* (1958) World Hlth Orgn, Geneva

Min. of Housing and Local Govt, Gt Brit. *Memorandum on the principles of design for small domestic sewage treatment works* (1953) H.M.S.O., London

REFERENCES

1. WATSON, K. S., FARRELL, R. P. and ANDERSON, J. S. 'The contribution from the individual house to the sewer system', *J. Wat. Pollut. Control Fed.* **39** (1967) 2039
2. U.S. Dept of Health Education and Welfare, *Publ. Hlth Rept No. 2461*, 'Individual Sewage Disposal Systems' (1953) Washington, D.C.
3. HILL, F. G. and ACKERS, G. L. 'Small Domestic Sewage Treatment Works', *Wat. Sanit. Engr* **3** (1952–3) 373

4. Thomas, H. A., Jr, Coulter, J. B., Bendixen, T. W. and Edwards, A. B. 'Technology and economics of household sewage disposal systems' (Report of Committee, Natl Res. Council), *J. Wat. Pollut. Control Fed.* **32** (1960) 113
5. Truesdale, G. A., Birkbeck, A. E. and Downing, A. L. 'The Treatment of Sewage from Small Communities', *J. Proc. Inst. Sew. Purif.* (1966) Pt 1, 34
6. Williams, R. K. and Wells, C. G. 'Some Notes on Aqua-Privies', *J. Proc. Inst. Sew. Purif.* (1959) Pt 3, 308
7. Duncan, D. L. 'Individual Household Recirculating Waste Disposal System for Rural Alaska', *J. Wat. Pollut. Control Fed.* **36** (1964) 1468

18

TREATMENT AND DISPOSAL OF WASTE WATER FROM ISOLATED BUILDINGS OR SMALL COMMUNITIES

THE DESIGN of small sewage treatment and disposal works serving isolated buildings or small groups of buildings or small communities generally is based on the same principles as larger treatment plants. However, it must be considered that small plants are relatively expensive in construction and operation per head of population served and so have to be constructed in the simplest way, for example, by using prefabricated structures or 'package units', perhaps employing local unskilled labour or local materials. The design of these plants should also allow for the simplest possible operation. Maintenance likewise should be simplified by eliminating all unnecessary working parts or materials which might corrode, ensuring ready access to facilitate clearing of chokages, pumping or removal of accumulated sludges, etc. For smaller works particularly, regular attention might be impracticable or too expensive, and such difficulties could lead to complete neglect.

In this chapter some of the special features of relatively small treatment and disposal works are discussed, including those classified as treatment works in which specific (usually limited) treatment processes are applied, and those often regarded simply as disposal works in which largely unspecified processes occur in a natural but unregulated way.

Wherever feasible the treatment works and disposal areas should be located down-wind, according to prevailing winds, from dwellings and public roads, and far enough way from these and from any local water supply source. According to experience, minimum satisfactory distances to obviate nuisance or risk of dangerous pollution are approximately as set out in Table 18.1.

It is assumed generally that one house will be occupied by at least five persons. Equivalent population units in schools and hospitals may be taken respectively as ten pupils (daytime attendance) and one bed, while for factories (excluding industrial wastes which usually should be treated and disposed of separately) the unit is about three employees. Other cases may be reckoned accordingly.

Details of some types of small treatment and disposal works suitable for these cases are described below, but it must be remembered that practice in different countries varies widely, not necessarily because of different conditions but often because of tradition or the particular requirements of local health or other authorities. The data and values recommended below, therefore, are given as a

Table 18.1

LOCATION OF TREATMENT WORKS

Number of persons served	*Minimum distance from works area to nearest residence, public way, or water supply source,* ft (m)
Up to 25	50 (15)
26– 50	100 (30)
51– 75	150 (45)
76–100	200 (60)
101–300	300 (90)

guide, and in some cases may be lower, in others higher, than required to satisfy local regulations. However, they are average figures, and should be applied carefully and adjusted for unusual conditions if necessary.

TREATMENT IN SEPTIC TANKS OR BY PLAIN SEDIMENTATION[1-4]

Septic tanks are the most useful of the units for small treatment works and are suitable for all schemes in the range of small works (5–300 population units). Satisfactory septic tank treatment depends on the processes of alkaline fermentation (p. 260); the technical requirements, therefore, are similar to those for satisfactory sludge digestion, including, for example, adequate provision of seeding sludge and avoidance of sudden large changes in temperature.

In most countries single-chambered tanks are used but in some two or more tanks in series are preferred, especially for the smallest schemes. Single-chambered tanks should not be used for schemes serving less than about 30 units of population because the efficiency of very small tanks is commonly too low. Where applicable single-chambered tanks are designed simply as rectangular tanks (inlets and outlets as given below) with a length not less than three times the width and a depth of about 6 ft (2 m). The capacity up to mean water-level should be of the order of the average daily flow (but reckoning not less than 25 gal/hd d (115 l/hd d)), but in any case single-chambered tanks should not have a volume below water-level of less than 750 gal (3 400 l).

Smaller tanks, serving 30 persons or less, should consist of two or more tanks in series (Figs 18.1 and 18.2) with a total capacity of about one and a half times the average daily flow but, in any case, not less than 500 gal (2 300 l) below water-level. Sludge then is collected mainly in the first chamber, which should have about half the total capacity (even where three chambers are provided). The transfer of the liquid between the chambers might be through circular or square ports or through vertical slots.

Scum forms inevitably on the water surfaces in septic tanks and

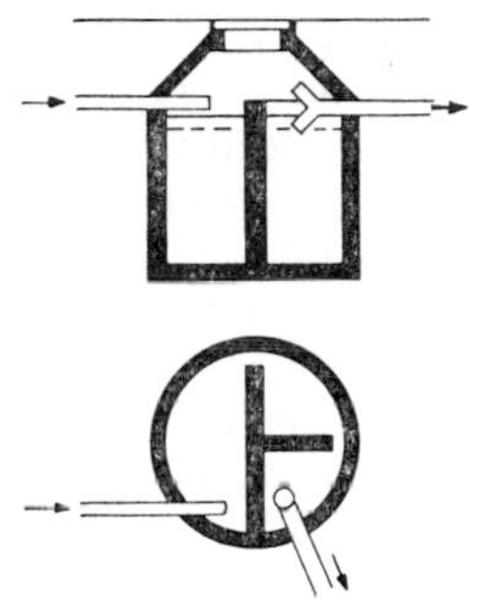

Fig. 18.1. *Circular septic tank*

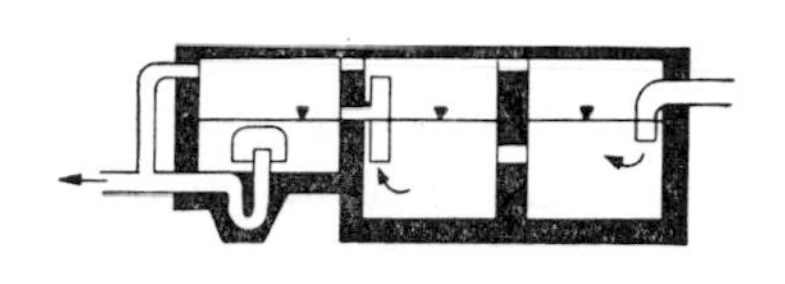

Fig. 18.2. *Two-chambered rectangular septic tank with siphonic dosing tank for intermittent effluent discharge, for example, to sub-surface irrigation system*

hence both inlets and outlets should be taken below the surface. Square junctions or Tees may be used, mounted vertically so that the sockets and spigots of the Tees are located respectively about 1 ft (0·3 m) above and below water-level. In the case of inlets, the branch of the tee should be installed so that the invert of the raw sewage pipeline at its junction to the tank is about 2 in (50 mm) above the outlet level (that is, the invert of the branch of the outlet tee). In larger tanks, however, the outlet arrangements should consist of a weir across the full width of the tank using a 2 ft (0·6 m) deep scumboard (to retain the scum) located 6 in (0·15 m) back from the weir, also spanning the full width of the tank, mounted to extend 1 ft (0·3 m) above and below overflow level.

The accumulated sludge is automatically digested in the septic tank, as in Imhoff tanks or separate digestion tanks, by the same anaerobic processes and normally accumulates at the rate of about 0·08 gal/hd d (0·36 l/hd d) (cf. Table 13.1, p. 257). This accumulation should be removed regularly, at least every 6–12 months, but some digested seed sludge must be retained, important especially for large tanks serving more than thirty persons. The sludge removed may be run into lagoons, or preferably into trenches which may then be filled and covered with earth, or run on to ploughed land and ploughed in after partial drying.

Septic tanks may also be used to digest the dissolved and colloidal solids of the sewage in addition to the settleable solids, by providing a detention period in the septic tanks of about 20 d or more (p. 247).[5] The capacity of the septic tank then should equal about 22 times the average daily flow (allowing for sludge storage), extending the period between sludge removals to a year or longer. Such digestion commonly effects purification comparable to that by trickling filtration of tank effluent and has the advantage that attendance and maintenance work are reduced to a minimum; however, capital costs may be high. The effluent may be odorous (compared to a filtration-plant effluent) and may exert a significant immediate oxygen demand and some nuisance may result accordingly in some cases, on discharge directly into streams or ponds or on to land; simple aeration of the effluent may then be applicable to obviate the nuisance (p. 216).

Two-storey or Imhoff tanks (p. 265) are also suitable for small treatment works serving not less than about 30 persons, but are not suitable for smaller schemes. Within the range 30–300 persons or equivalent such Imhoff tanks may be designed on allowances of about one-third average daily flow (but reckoning at least 25 gal/hd d (115 l/hd d)) for the sedimentation compartment, sludge compartment at least two-thirds daily flow below the slots and about one-quarter of the daily flow above the slots. By comparison with septic tanks, therefore, the required total capacity of the Imhoff tank is about 25% larger, but this may be justified by the advantage of a fresher effluent more suitable for direct discharge into surface waters or more suitable for application to trickling filters. Outlet designs may be similar to those of the settling tanks described below.

The operation of such small Imhoff tanks may be simplified by introducing the influent in such a way, as shown in Fig. 18.3*a*, that it flows first by a short path through part of the sludge compartment (cf. the Travis tank, p. 167). Thus, as shown, the sewage is introduced first into the sludge compartment above the slots so that scum-forming solids are separated within the sludge compartment. An arrangement whereby the raw waste falls into the sludge compartment (falling, say, 2–3 in (50–75 mm) or more) tends to keep the scum layer to a minimum. Foaming may not occur so readily with this arrangement because surface-active compounds formed during digestion of the sludge solids are continuously removed by the raw waste flow into the sedimentation compartment.

Another type of Imhoff tank is shown in Fig. 18.3*b*, wherein a small inlet compartment is provided for the separation of scum also ahead of the sedimentation compartment. Such scum must be removed from time to time. It may be disposed of by mixing with topsoil and burying at shallow depth. Usually scum will digest readily, but because of its strong tendency to separate and rise to the surface, it may require special seeding arrangements to ensure full digestion (see p. 261).

Some sludge also must be removed from the Imhoff tank sludge

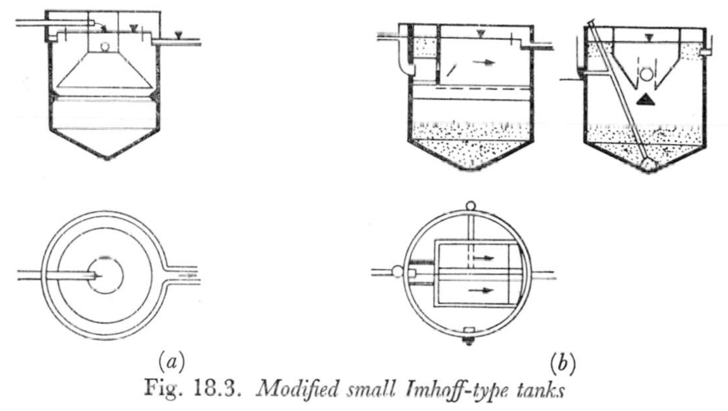

(*a*) (*b*)
Fig. 18.3. *Modified small Imhoff-type tanks*

compartment at say, regular intervals of about one month, removing only that which has accumulated in the meantime so that the tank is never depleted unduly of seeding sludge (p. 269). About half the sludge compartment should be kept full of sludge to make sure that alkaline digestion is maintained.

Settling tanks or plain sedimentation basins (p. 158) may be used as primary treatment units in small sewage treatment works and are designed to separate and to retain from the raw wastes only the settleable solids. These solids can be retained only temporarily as sludge which must be removed before it becomes septic (pp. 160–164). Settling tanks for small treatment works may consist simply in a rectangular chamber about 5 ft (1·5 m) deep below overflow level, length about four or five times the width, designed usually to provide a detention period of at least 6–12 h (average D.W.F. basis, but reckoning at least 25 gal/hd d (115 l/hd d)). Because desludging of such tanks is required frequently, say once weekly or at least fortnightly (in warmer climates twice weekly at least or even daily), this type of tank is not normally effective or economical for small treatment works serving less than about 150 persons or equivalent, and then only where sufficient regular attendance can be assured. Where applicable, the capacity of the tanks should amount preferably to about one-half of the average daily flow, a suitable convenient arrangement consisting of two independent chambers operating normally in parallel, each including a volume equal to about one-quarter average daily flow, whereby either tank may be emptied for cleaning without interruption of treatment.

The floor of such tanks should slope back from outlet to inlet end, draining to a sump near the inlet from which the sludge may be pumped. The most satisfactory form of inlet is a submerged orifice discharging into the centre of the inlet end wall about 2–3 ft (0·6–0·9 m) below surface level, while the outlet should consist of a weir extending over the full width. Scum is retained by a scumboard placed 6 in (0·15 m) back from the weir and extending 1 ft (0·3 m) below weir level. Provision must be made for separate disposal of the sludge removed from the settling tanks (for estimate of quantities, see Table 13.1, p. 257).

The purification effected by plain sedimentation in settling tanks may be less than that effected by septic tanks, but generally the effluent from plain sedimentation is fresher and less malodorous; it also exerts a much lower immediate chemical oxygen demand so that usually it is more suitable—for works serving 150 or more persons—where the effluent is discharged directly into natural waters.

SECONDARY TREATMENT

In some cases, especially for larger groups of homes or for institutions, etc., the amount of effluent to be disposed of may be large enough to require a higher degree of purification than can be achieved readily by primary treatment alone. This may be true especially where there is direct discharge of effluents into natural watercourses. Usually a higher degree of purification is obtained by secondary (biological) treatment of the wastes.

Trickling filtration can be used for biological treatment of effluents from septic tanks, Imhoff tanks or plain sedimentation tanks where required or necessary, but only if sufficient head is available or is provided by pumping. Small trickling filters should be designed on the basis of the principles applied to larger installations, described in Chapter 12 (pp. 196 et seq.). Beds may be either rectangular or circular and various devices are used for distribution of the settled wastes over the surface of the media, including:

1. A series of fixed channels arranged so that the tank effluent overflows from the sides, which may be notched. Such channels may be fed from tipping troughs. Total head requirement is about 1 ft (0·3 m). Distribution may be satisfactory for beds up to about 150 ft² (14 m²), but such distributors are not recommended for larger beds.
2. A rotary distributor consisting of two arms carried on a central standard and extending to the outer margin of the bed surface. Total head required is about 2 ft (0·6 m) for efficient operation, attention and maintenance being required more frequently than for the channel type of distributor.

The depth of the filter must also be taken into account in estimating the overall total head required for trickling filtration.

Settled sewage (tank effluent) may be applied to small trickling filters at a rate of about one and a half persons, or up to about 40 gal/yd³ filter medium daily (2 persons or 0·24 m³/m³ d) for schemes serving up to about 30 persons; for schemes serving 30–300 persons or equivalent the rate may be increased to two persons, or up to 60 gal/yd³ (0·36 m³/m³) of medium daily. Where sewage is weaker and the daily flow *per capita* is much higher, as in the U.S.A., rates of application might also be based on 1½–2 persons/yd³ (2–2½/m³) but correspondingly high volumetric rates of loading may be allowed.

Filter beds should if possible be 6 ft (2 m) deep, in any case not less than 4 ft 6 in (1·4 m), the deeper filter being more efficient and reliable. Efficiencies depend also on the careful grading of the filter medium, 3 in (75 mm) nominal size, uniform over the full depth, being suitable for small treatment works. It is essential for proper treatment to provide sufficient ventilation through the filter, requirements being ample area for the passage of air into the false floor on which the medium rests, out from the medium and out from the outlet channels, also with ample space under the floor; for this purpose glazed half-pipes or special tiles may be used. Suitable ventilation openings must also be provided to ensure unimpeded air supply to top and bottom of the filter.

The provision of a secondary sedimentation tank (humus tank) to remove humus solids (p. 197) from the trickling-filter effluent is not generally recommended owing to the frequent cleaning required to maintain efficiency and avoid septic conditions; humus tanks are, therefore, omitted unless regular attendance and maintenance can be assured. For schemes serving 150 persons or more where regular attention is assured, humus tanks may be designed on the same basis as the settling tanks mentioned above—parallel chambers each of capacity one-quarter average daily flow—reckoning on withdrawal of humus sludge as frequently as circumstances demand to ensure efficient operation, which may be twice or thrice weekly, or more frequently. The humus sludge may be pumped back to the primary or septic tank inlet, but more commonly disposal into lagoons, soakage pits or trenches or on to agricultural land is preferred.

Broad irrigation provides another method for the biological treatment of tank effluents. The same methods are applied in small schemes as have been described in Chapter 12 (p. 190) for ordinary municipal schemes, the tank effluent being distributed through open channels and drains on to the irrigation area. The area should be at least as far from dwellings, public roads and water supplies as the distances set out in Table 18.1 (p. 370), the area required depending on the nature and porosity of the soil. For light loams the required area may be taken as 10 yd^2/person or equivalent (8 m^2/person), but for heavy soils up to 80 yd^2 (65 m^2) may be necessary. The area is subdivided into a suitable number of plots which are used in rotation. They should be forked over, when dry after resting, at least once each year.

However, the use of broad irrigation for tank effluents, even effluents from plain sedimentation tanks, is not recommended because odour nuisance cannot be avoided.

Subsurface irrigation[7] sometimes provides a satisfactory method for the biological treatment and disposal of sewage from small treatment works, and (except where ample surface water is available) is recommended as the best method for small sewerage schemes. It may be used for effluents from septic tanks, Imhoff tanks or settling tanks, but septic tanks are preferred for the primary treatment in this case. However, it should be noted that subsurface irrigation cannot be applied properly if the soil and subsoil are not porous, if the groundwater table rises to less than about 6 ft (2 m) below the ground surface or if there is a danger of pollution of water supplies. The method is not applicable where impervious clay soil occurs or where the soil is waterlogged or unaerated.

Subsurface irrigation systems consist of buried drains into which the waste is fed in such a way that the liquid soaks away or i absorbed and the organic polluting constituents decompose biologically in much the same way as in land filtration system (p. 192). Dosing, therefore, should be intermittent, which may b arranged by tipping trays or buckets, or siphonic dosing tanks (s Fig. 18.2) according to the head (or fall) available, reckoning dosi rates at about 1 gal/yd (5 l/m) of drain. Drains consist normally 4 in (100 mm) diameter agricultural pipes laid in trenches excava not more than 2 ft (0·6 m) deep and about 12–18 in (0·3–0·45 m) wide at bottom, the drains being laid on gravel or rubble at depths not less than 12 in (0·3 m) or more than 18 in (0·45 m) below surface, except in well-drained sandy soil where depths of 2 ft (0·6 m) may be allowed. The drains should be surrounded with rubble and covered as for sullage absorption trenches (p. 364), also using a membrane above the joints to prevent the entry of soil, etc., with the drains. Each individual drain should not exceed 70 ft (20 m) in length and drains should not be laid less than 7 ft (2 m) apart.

The required length of drain depends on the amount of the effluent volume to be disposed of and on the nature of the soil and varies accordingly from about 12 ft (3·5 m) to about 60 ft (18 m) per

Table 18.2

CAPACITY OF SUBSURFACE IRRIGATION DRAINS AND AREAS

Percolation tests (observed time to fall 1 in (25 mm)) min	*Corresponding required length of drain*		*Corresponding irrigation area required for* 25 gal/hd d (115 l/hd d) *sewage flow*	
	ft/*population unit*	m/*population unit*	yd²/hd	m²/hd
5	12	4	10	8
10	18	6	15	13
20	25	8	20	17
30	35	10	30	25
60	60	18	50	42

person or equivalent. The gradient to which drains are laid depends also on the soil and varies with its porosity from about 1 in 350 in heavy soil to 1 in 130 in sand. Trial holes for *percolation tests*[1–4, 8] (as described on p. 379) should always be investigated before drains are laid out, the observed time to absorb 1 in (25 mm) of water being used as a basis according to Table 18.2.

For schemes serving larger equivalent populations, the subsurface irrigation area should be subdivided into two sections used in rotation, thus resting each section of the area in turn. Aeration vents should be provided at the ends of all drains.

Subsurface filtration may provide a satisfactory method of biological treatment of tank effluent, especially in certain cases where subsurface irrigation cannot be used, for example, because of the nature of the soil. The process is that of intermittent sand filtration (p. 192) using specially prepared sand-beds of effective sand size between 0·3 and 0·5 mm (0·012 and 0·020 in) about 2–3 ft (0·6–0·9 m) deep. Such filters are prepared by excavating appropriate areas, allowing about 50 ft^2/hd (5 m^2/hd) and laying a base of gravel about 6 in (0·15 m) deep for under-drainage, agricultural pipe drains being placed near floor level to collect the effluent for disposal. Another layer of gravel about 6 in (0·15 m) deep is placed above the sand filter bed to provide satisfactory distribution of the tank effluent to be treated. This may be introduced also by a system of agricultural drains placed in the top of the gravel layer. Dosing may be arranged as for subsurface irrigation systems (p. 376). A suitable cover of ordinary soil is replaced above the filter to prevent nuisance from odours or sewage flies, etc. In very small works such filters may be constructed in trenches 3 or 4 ft (about 1 m) wide, the length required being then about 15–20 ft/person (4·5–6 m).

Effluents from the under-drains of such filters must be so collected as to permit satisfactory ventilation within the beds. They are usually of suitable quality for disposal into natural waters, or by surface irrigation methods if the right areas are available.

Intermittent sand filtration or *surface filtration* (p. 192) is another possibility for secondary treatment of small sewage flows, but generally requires more attention than subsurface methods if odours are to be avoided.

Recently, simplified *activated-sludge systems* have been developed which may be economical enough even for relatively small works. They are of two different forms, viz., the oxidation ditch system developed in Holland (see p. 224) and the extended aeration system (see p. 231) from which small 'package' units have been developed. Oxidation ditch (ring-ditch) systems require usually overall areas of about 1–2 acres/100 persons (0·4–0·8 ha/100 persons). Package plants based on the extended aeration process are also available with capacities ranging from say 100 to 150 persons (or even less) up to perhaps 2 000 to 5 000 or equivalent. When such plants are equipped with reliable pumps, etc., they need little attention for normal operation. The overall area required is relatively small, say one-eighth acre (0·05 ha) or even less for $\sim$100 persons. Foaming may have to be controlled.

Oxidation (sewage) ponds (see p. 242) also may be used to provide biological treatment of sewage from small communities. Usually low-rate ponds are preferable in order to reduce odours and to keep maintenance to a minimum.

EFFLUENT DISPOSAL

Well-purified effluents from small treatment works may be discharged directly to watercourses, but in many cases watercourses are not available and other means of disposal are required. Effluents may be disposed of finally below ground, with the aim of utilizing the soil's capacity for natural or self-purification (see p. 104). The main technique used is that of absorption into natural soils from covered (subsurface) trenches. Absorption (soak) wells and transpiration areas provide other means.

Absorption trenches[1–4, 8] are prepared by excavating trenches about 18 in (0·45 m) wide and 2 ft 6 in (0·75 m) deep graded according to the soil to about 1 in 130 or flatter, filled to about 18 in (0·45 m) depth (1 ft (0·3 m) below surface) with rubble, on which is then laid a line of 3 or 4 in (75 or 100 mm) agricultural piping. Provided the groundwater table is lower than about 6 ft (2 m) below surface, such trenches provide the most satisfactory means of disposal of sullage from residences, no regular attention being required. A grease-trap of suitable design should be provided upstream of the disposal system and so relieve the system of considerable burden (provided the trap is kept clear of accumulated solids and grease is removed regularly, say, at least weekly or perhaps fortnightly). The required length of trenches, per person contributing sullage, depends on the nature of the soil and subsoil, on groundwater conditions, and on climate, and may be estimated from the data of percolation tests, (see below, also Table 18.3). In average circumstances, where at least 2 ft (0·6 m) of sandy loam is available, an average household of 5–6 persons need perhaps 30–40 ft (9–12 m) of such trenching.

The capacity of the ground is best examined by *percolation tests*.[1–4, 8] A recommended technique is as follows:

> At a spot representative of the soil within the area available for the installation of the absorption trench, excavate a hole 1 ft (0·3 m) square to the proposed depth of the absorption trench. Fill the hole to a depth of 6 in (0·15 m) or more with water and allow to soak away. Fill the hole again to a depth of 6 in (0·15 m) and note the time which it takes the water surface to fall 1 in (25 mm).

From this figure the recommended dosage rate of sullage follows according to Table 18.3.

More than 60 min for a fall of 1 in (25 mm) indicates that the ground or other conditions are unsuitable for absorption trench disposal.

Example Two percolation test holes yielded values of 1 in (25 mm) fall respectively in 8 min and 12 min, a third was then made, value 9 min; a rate of 1·3 gal/ft^2 d (64 l/m^2 d) may be assumed, for a trench 1 ft 6 in (0·45 m) wide, say, 2 gal/d per foot of trench

(30 l/m d); the requirement for 20 gal/hd d (90 l/hd d) is then 10 ft (3 m) per person and for a household of five persons 50 ft (15 m) total length of trench is needed: allow say two more persons, 70 ft (20 m) in all, provide two trenches 35 ft (10 m) long, say, 10 ft (3 m) apart, located symmetrically to suit contours in an area 50 × 20 ft (15 × 6 m).

The length of absorption trench should not be less than 20 ft (6 m) for a residence, whatever the nature of the ground or the data from percolation tests.

Such trenches may also be used for sullage disposal from institutions, etc., and may be improved by arrangements for *intermittent*

Table 18.3

ABSORPTION TRENCH DOSAGE RATES

Percolation test (time for fall of 1 in (25 mm)) min	*Allowable dosage*		*Allowable dosage of* 18 in (0·45 m) *wide trench*		*Length of* 18 in (0·45 m) *trench required/person contributing sullage**	
	gal/ft² *trench bottom* d	l/m² *trench bottom* d	gal/ft d	l/m d	ft	m
5	2·0	100	3·0	45	5– 7	1·5– 2
10	1·4	70	2·1	31	7–10	2– 3
20	1·0	50	1·5	22	10–14	3– 4
30	0·7	30	1·1	16	14–20	4– 6
60	0·4	20	0·6	9	15–35	4·5–10

* Allow 15–20 gal/hd d (70–90 l/hd d)

dosing alternately into duplicated or multiple trench systems, much as for intermittent land filtration. In these cases some regular daily or at least weekly attention to dosing apparatus or distribution pipes is essential and while this may be feasible for collective systems or institutions, it is not for individual residences or other small buildings where regular attention and maintenance are unreliable and cannot be ensured.

The absorption area may be assisted by planting moisture-absorbing trees, shrubs or other vegetation adjacent to the trench, but care should be taken lest roots interfere unduly with the trench, by clogging the openings in rubble or pipes. It is also wise to protect trenches against gradual blockage by soil from above, by placing rubble around and above the agricultural pipes, and covering this with a supporting membrane of roofing felt or other resistant material before replacing the topsoil. Care should be taken to exclude roof water and as far as practicable to divert surface drainage.

The gradient to which trench bottoms are excavated and pipes are laid is most important because on this depends the uniform distribution of the sullage over the whole area, according to the rate of absorption. In heavy soil the gradient may be as flat as 1 in 400, but in porous, sandy soils this should be increased to as much as about 1 in 130 to ensure that sullage will reach the further end of the trench system. Multiple trenches in parallel should be spaced at least 5–10 ft (1·5–3 m) apart, again depending on the nature of the ground.

In warmer climates, wherever the use of absorption trenches is not practicable, artificially prepared *transpiration areas*[8] offer another technique which may be feasible. Such areas may be constructed in various ways. In one method natural soil is broken up to a depth of 9 in (0·23 m) or more and is either left fallow or planted with grasses, annuals or ornamental shrubs characterized by relatively high rates of transpiration. In another method the area is excavated at least 15 in (0·38 m) deep and the bottom levelled off precisely and consolidated. On this is laid a network of unglazed pipes connected from a distribution box and a 3 in (75 mm) bed of soil mixed with sand or ashes spread and consolidated, and on this layer a further 3 in (75 mm) of the topsoil which is then either turfed or planted with vegetation.

Plants suitable for sullage transpiration areas include cereals, some vegetables, perennial rye, prairie grass and clovers. Shade trees should be avoided. The cereals wheat, oats and barley and winter-growing vegetables absorb considerable amounts of water during winter, but in very wet conditions extremely low rates of transpiration are to be expected. Probably the most effective all-year-round disposal is achieved by growing trees, such as willows or similar deciduous shrubs, in conjunction with perennial pasture species such as perennial rye, prairie grass, red or white clover. Rates of transpiration vary widely, but are exemplified by the rates of about 300 000 gal/acre (3 400 000 l/ha) in six months by beech forests (age about 100 years) i.e. about 2 000 gal/acre d (22 000 l/ha d) on average, and about 1 000 gal/acre d (11 000 l/ha d) in summer for grass. In the growing season one plant of maize will absorb (transpire) about 3 gal (14 l) per half-year, or at most about 1 gal (5 l) every five days during the period of most active growth.

These figures correspond roughly to transpiration of sullage from about 30–200 persons/acre (75–500/ha), depending primarily on the volume of sullage per head daily. Normally a transpiration area of about 1 000 ft² (100 m²) is sufficient for a household (sullage only) corresponding to a loading of 200 persons/acre (500/ha).

Where neither absorption nor transpiration is practicable, recourse must be made to discharge into watertight receptacles which may be carried away, or from which the wastes may be pumped, for disposal elsewhere. These arrangements may be so expensive that collective arrangements to serve groups of buildings are cheaper, and then it may be feasible to dispose of the wastes into a stream or on to land some distance away by a common system, using a single treatment works.

Absorption trenches or transpiration areas may be suitable also for the disposal of mixed, domestic wastes (sullage plus water-closet wastes) if they are first given primary, or even secondary treatment. However, allowance must be made for the additional volume of the combined discharges.

Absorption or soak wells[8] provide another means of disposal of tank effluents. They do not provide secondary (biological) treatment unless the groundwater table is at least 6 ft (2 m) deeper than the bottom of the well. Unless this is so they should not be used, and in no circumstance should they be used where danger of contamination of water supplies is possible. They may be of use in suitable situations where sand or gravel beds exist at moderate or even relatively shallow depths below surface, but not if the rock structures are deeply fissured. Their capacity to dispose of waste waters must always be tested by trials over a period of a week or more, but in any case the method is uncertain and should be adopted cautiously if at all.

Before discharge into soak wells, sewage should be treated at least by septic tank treatment or better by trickling filters also. Wells should be lined with stone, brickwork or concrete blocks laid dry and open at the bottom but made watertight at levels above the absorbent strata. The well may be filled with coarse rubble increasing in size from top to bottom, ending the filling about 1 ft (0·3 m) or so below influent level to allow for storage. The wells should be covered and provided with ample ventilation.

Disinfection of the effluent before disposal may be necessary in special cases, e.g. if effluents from hospitals are to be disposed of.

EXAMPLES OF DESIGN OF SMALL TREATMENT WORKS

Take the case of a small scheme serving a group of 16 residences with full domestic sewerage; roof and surface water must be excluded; assume an average daily flow of 25 gal/hd (115 l/hd).

An average of 5 persons/household = 80 persons served.
Average daily flow = 80 × 25 = 2 000 gal (9 000 l)
Take the peak flow to be (2 000/24) × (24/10) = 200 gal/h say 3·5 gal/min (0·25 l/s).

Example 1

Septic tank, reckon required capacity = daily flow = 2 000 gal (9 000 l)

Arrange for one tank, capacity 2 000 gal = 320 ft³ (9 m³)
Let tank be 6 ft (1·8 m) deep, then area = 53 ft² (5 m²)
Make it then, say 4 ft wide × 13 ft 6 in long (1·2 × 4·1 m, or more sensibly, say, 1 × 5 m).

Example 2

Sewage digestion, i.e. biological treatment by digestion in septic tanks.

The capacity of septic tank required, taking 22 times average daily flow 22 × 2 000 = 44 000 gal (200 000 l).
Provide four tanks each 11 000 gal (50 000 l) = 1 760 ft³ (50 m³) of dimensions say 8 ft (2·5 m) deep, 7 ft (2 m) wide and 32 ft (10 m) long each.

Example 3

Imhoff tank, reckon required capacity of settling compartment to be one-third average flow, say 670 gal (3 000 l) and sludge compartment below slots to be two-thirds average flow, 1 330 gal (6 000 l) and above slots one-quarter or 500 gal (2 300 l).

Total is then 670 + 1 330 + 500 = 2 500 gal (11 300 l) corresponding to a detention period at 200 gal/h (0·25 l/s) (settling compartment) of 670/200 = $3\frac{1}{3}$ h, which is satisfactory. Therefore provide one tank designed accordingly.

Example 4

Trickling filtration of tank effluents. Assume pretreatment by Imhoff tank as for Example 3 and allow 2 persons/yd³ (2·5/m³).

Hence the volume of filter medium should be

80/2 = 40 yd³ = 1 080 ft³ (30 m³).

Taking a depth of 6 ft (1·8 m) the required area is

1 080/6 = 180 ft² (17 m²), say 15 ft (4·5 m) diameter.

Suppose the concrete floor is graded 1 in 20, filter medium 3 in (75 mm) nominal size, then use a rotary distributor fed by tipping trough, tipping every 4 min at peak hourly flow of 3·5 gal/min (0·25 l/s).

Hence capacity of troughs = 4 × 3·5 = 14 gal (64 l) when full.

Example 5

Biological treatment and disposal by subsurface irrigation. Assume pretreatment by septic tank treatment as in Example 1, an open sandy soil being available and groundwater table 7 ft (2·1 m) below surface. Percolation tests show fall of 1 in (25 mm) in respectively 4, 5 and $4\frac{1}{4}$ min; take time of fall as 5 min and read from Table 18.2 an allowance of 10 yd^2/hd (8 m^2/hd) and length of drain 12 ft/hd (4 m/hd). The total required area is then 80 × 10 = 800 yd^2 = 7 200 ft^2 (640 m^2) and length of drains = 80 × 12 = 960 ft = 320 yd (320 m). Divide the area into two sections, each 3 600 ft^2 (320 m^2), provide dosing chambers each of capacity 1 gal/yd (5 l/m) of drain, reckoned on one section, 320/2 = 160 gal (800 l) capacity each siphoning automatically into supply pipe. If this requires 3 min to empty, the capacity of the main supply pipe (distribution pipe) 160/3 = 53 gal/min or 0·14 ft^3/s (4 l/s).

BIBLIOGRAPHY

As Chapter 17, p. 367

REFERENCES

1. DARDEL, W. 'Disposal of Sewage from Small Communities and Institutions', *J. Proc. Inst. Sew. Purif.* (1952) Pt 3, 167
2. IMPEY, L. H. 'Sewage Treatment and Disposal for Small Communities and Institutions: The Development and Use of the Septic Tank', *J. Proc. Inst. Sew. Purif.* (1959) Pt 3, 311
3. U.S. Dept. of Health Education and Welfare, *Publ. Hlth Rept No. 526*, 'Manual of Septic Tank Practice' (1957) Washington, D.C.
4. KOLEGA, J. J., WHEELER, W. C. and HAWKINS, G. W., JR 'Current Septic Tank System Installation Practice in Connecticut', *J. Wat. Pollut. Control Fed.* **38** (1966) 1592
5. VIEHL, K. 'Über die Reinigung von häuslichem Abwasser durch Ausfaulen', *Gesundheitsingenieur* **65** (1942) 394
6. WILLIAMS, R. K. and WELLS, C. G. 'Some Notes on Aqua Privies', *J. Proc. Inst. Sew. Purif.* (1959) Pt 3, 308
7. McGAUHEY, P. H. and WINNEBERGER, J. H. 'Studies of the Failure of Septic Tank Percolation Systems', *J. Wat. Pollut. Control Fed.* **36** (1964) 593
8. See References Chapter 17

APPENDIX

CONVERSION FACTORS FOR CHANGING TO THE METRIC SYSTEM

(I) *Basic conversion factors*

British unit	Metric unit equivalent	Metric unit (symbol)	British unit equivalent
1 acre	= 0·4047 hectare	1 ha	= 2·471 acre
	= 4 047 square metres	1 ha = 10^4m^2	
1 bu (bushel)	= 36·37 litres	1 l	
1 Btu	= 1·055 kilojoules	1 kJ	= 0·9478 Btu
	= 0·2522 kilocalorie	1 kcal	
1 cwt	= 50·8 kilogrammes	1 kg	
1 ft	= 0·3048 metre	1 m	= 3·281 ft
1 ft^2	= 0·09290 square metre	1 m^2	= 10·76 ft^2
1 ft^3	= 0·02832 cubic metre	1 m^3	= 35·31 ft^3
	= 28·32 litres		
1 gal	= 4·546 litres	1 l	= 0·22 gal
1 US gal	= 3·785 litres	1 l	= 0·264 US gal
1 gr (grain)	= 0·0648 gramme	1 g	= 15·432 gr
1 hp	= 0·7457 kilowatt	1 kW	= 1·341 hp
1 in	= 25·4 millimetres	1 mm	= 0·03937 in
1 in^2	= 645·2 square millimetres	1 mm^2	= 1·55 × 10^{-3} in^2
1 in^3	= 16·39 × 10^3 cubic millimetres	1 mm^3	= 61 × 10^{-6} in^3
	= 16·39 millilitres	1 ml	
1 lb	= 0·4536 kilogramme	1 kg	= 2·205 lb
1 mile	= 1·609 kilometres	1 km	= 0·6214 mile
1 sq mile	= 2·590 square kilometres	1 km^2	= 0·387 sq mile
1 oz	= 28·349 grammes	1 g	= 0·0353 oz
1 p.s.i.	= 0·0703 atmosphere	1 atm	= 14·21 lbf/in^2
(lbf/in^2)	= 0·0703 kilogramme (force) per square centimetre	(1 kgf/cm^2)	
1 ton	= 1 016 kilogrammes	1 kg	= 0·9842 × 10^{-3} ton
	= 1·016 tonnes (metric ton)	1 t	
1 yd	= 0·9144 metre	1 m	= 1·094 yd
1 yd^2	= 0·8361 square metre	1 m^2	= 1·196 yd^2
1 yd^3	= 0·7646 cubic metre	1 m^3	= 1·308 yd^3

Temperature

British unit: Degree Fahrenheit = °F
Metric unit: Degree Celsius = °C
Ice point.. 32°F = 0°C
Boiling point of water under standard pressure conditions.................................. 212°F = 100°C

Conversion formulae:

$$°C = \frac{5}{9}(\text{degF} - 32)$$

$$°F = \frac{9}{5}\text{degC} + 32$$

(II) *Application of metric units*

Description	Unit	Symbol
Precipitation, run-off	millimetre (1 mm of rain = 1 l/m^2	mm
River flow	cubic metre per second	m^3/s
Flow in pipes	cubic metre per second	m^3/s
	or litre per second	l/s
Discharges	cubic metre per day	m^3/d
Usage of water	litre per person per day	l/person day
Density	kilogramme per cubic metre	kg/m^3
Concentration	milligramme per litre	mg/l
	or gramme per cubic metre	g/m^3
B.O.D. loading	kilogramme per cubic metre per day	kg/m^3d
Hydraulic load per unit area	cubic metre per square metre per day	m^3/m^2d
Hydraulic load per unit volume	cubic metre per cubic metre per day	m^3/m^3d
Velocity	metre per second	m/s

(III) *Conversion of common British units in use in sewage disposal to equivalent metric units* (Symbols as above)

British unit	Metric equivalent
1 p.p.m.	= 1 mg/l = 1 g/m^3
1 acre-foot	= 1 233·5 m^3
1 Btu/ft^3	= 37·26 kJ/m^3 = 8·9 kcal/m^3
1 Btu/ft^2h°F	= 4·9 kcal/m^2/h °C
1 ft^2/ft^3	= 3·28 m^2/m^3
1 ft^3/ft^2	= 0·3048 m^3/m^2
1 ft^3/gal	= 6·24 m^3/m^3
1 ft^3/lb	= 62·428 l/kg
1 c.f.s. (cusec)	= 0·02832 m^3/s (cumec)
1 mile gal/d	= 4 546 m^3/d = 0·0526 m^3/s
1 gal/acre	= 0·0112 m^3/ha
1 m.g.a.d.	= 11 200 m^3/ha
1 gal/ft	= 14·9 l/m
1 gal/ft^2	= 48·93 l/m^2
1 gal/yd^3	= 5·946 × 10^{-3} m^3/m^2
1 lb/acre	= 1·12 kg/ha
1 lb/1 000 gal	= 0·10 kg/m^3
1 lb/ft^2	= 4·88 kg/m^2
1 lb/1 000 ft^2	= 48·82 kg/ha
1 lb/ft^3	= 16·02 kg/m^3
1 lb/yd^2	= 0·543 kg/m^2
1 lb/yd^3	= 0·593 kg/m^3
1 yd^3/acre	= 1·89 m^3/ha
100 000 orifices to the in^2	= 155 orifices to the mm^2

REFERENCES

1. Ministry of Technology, National Physical Laboratory: *Changing to the Metric System*, 2nd edition. H.M.S.O., London (1967)
2. 'Metric Units, with Reference to Water, Sewage and Related Subjects, Report of Working Party', *Wat. Pollut. Control* **67** (1968) 475
3. 'Change-over to Metric System', *Wat. Pollut. Control* **69** (1970) 714

AUTHOR INDEX

SUBJECT INDEX

B

C

D

E

P

Q

R

S

T